Lecture Notes in Artificial Intelligence 9883

Subseries of Lecture Notes in Computer Science

More information about this series at http://www.springer.com/series/1244

Christo Dichev · Gennady Agre (Eds.)

Artificial Intelligence: Methodology, Systems, and Applications

17th International Conference, AIMSA 2016
Varna, Bulgaria, September 7–10, 2016
Proceedings

 Springer

Editors
Christo Dichev
Winston-Salem State University
Winston Salem, NC
USA

Gennady Agre
Institute of Information and Communication
  Technologies
Bulgarian Academy of Sciences
Sofia
Bulgaria

ISSN 0302-9743   ISSN 1611-3349 (electronic)
Lecture Notes in Artificial Intelligence
ISBN 978-3-319-44747-6   ISBN 978-3-319-44748-3 (eBook)
DOI 10.1007/978-3-319-44748-3

Library of Congress Control Number: 2016947780

LNCS Sublibrary: SL7 – Artificial Intelligence

Printed on acid-free paper

This Springer imprint is published by Springer Nature
The registered company is Springer International Publishing AG Switzerland

Preface

This volume contains the papers presented at the 17th International Conference on Artificial Intelligence: Methodology, Systems and Applications (AIMSA 2016). The conference was held in Varna, Bulgaria, during September 7–10, 2016 under the auspices of the Bulgarian Artificial Intelligence Association (BAIA). This long-established biannual international conference is a forum both for the presentation of research advances in artificial intelligence and for scientific interchange among researchers and practitioners in the field of artificial intelligence.

With the rapid growth of the Internet, social media, mobile devices, and low-cost sensors, the volume of data is increasing dramatically. The availability of such data sources has allowed artificial intelligence (AI) to take the next evolutionary step. AI has evolved to embrace Web-scale content and data and has demonstrated to be a fruitful research area whose results have found numerous real-life applications. The recent technological and scientific developments defining AI in a new light explain the theme of the 17th edition of AIMSA: "AI in the Data-Driven World."

We received 86 papers in total, and accepted 32 papers for oral and six for poster presentation. Every submitted paper went through a rigorous review process. Each paper received at least three reviews from the Program Committee. The papers included in this volume cover a wide range of topics in AI: from machine learning to natural language systems, from information extraction to text mining, from knowledge representation to soft computing, from theoretical issues to real-world applications. The conference theme is reflected in several of the accepted papers. There was also a workshop run as part of AIMSA 2016: Workshop on Deep Language Processing for Quality Machine Translation (DeepLP4QMT). The conference program featured three keynote presentations: one by Josef van Genabith, Scientific Director at DFKI, the German Research Centre for Artificial Intelligence, the second one from Benedict Du Boulay, University of Sussex, United Kingdom, and the third one by Barry O'Sullivan Director of the Insight Centre for Data Analytics in the Department of Computer Science at University College Cork.

As with all conferences, the success of AIMSA 2016 depended on its authors, reviewers, and organizers. We are very grateful to all the authors for their paper submissions, and to all the reviewers for their outstanding work in refereeing the papers within a very tight schedule. We would also like to thank the local organizers for their excellent work that made the conference run smoothly. AIMSA 2016 was organized by the Institute of Information and Communication Technologies Bulgarian Academy of Sciences, Sofia, Bulgaria, which provided generous financial and organizational support. A special thank you is extended to the providers of the EasyChair conference management system; the use of EasyChair for managing the reviewing process and for creating these proceedings eased our work tremendously.

July 2016

Christo Dichev
Gennady Agre

Organization

Program Committee

Gennady Agre	Institute of Information and Communication Technologies at Bulgarian Academy of Sciences, Bulgaria
Galia Angelova	Institute of Information and Communication Technologies at Bulgarian Academy of Sciences, Bulgaria
Grigoris Antoniou	University of Huddersfield, UK
Roman Bartak	Charles University in Prague, Czech Republic
Eric Bell	Pacific Northwest National Laboratory, USA
Tarek Richard Besold	Free University of Bozen-Bolzano, Italy
Maria Bielikova	Slovak University of Technology in Bratislava, Slovakia
Loris Bozzato	Fondazione Bruno Kessler, Italy
Justin F. Brunelle	Old Dominion University, USA
Ricardo Calix	Purdue University Calumet, USA
Diego Calvanese	Free University of Bozen-Bolzano, Italy
Soon Ae Chun	CUNY, USA
Sarah Jane Delany	Dublin Institute of Technology, Ireland
Christo Dichev	Winston-Salem State University, USA
Darina Dicheva	Winston-Salem State University, USA
Danail Dochev	Institute of Information and Communication Technologies at Bulgarian Academy of Sciences, Bulgaria
Benedict Du Boulay	University of Sussex, UK
Stefan Edelkamp	University of Bremen, Germany
Love Ekenberg	International Institute of Applied Systems Analysis, Austria
Floriana Esposito	University of Bari Aldo Moro, Italy
Albert Esterline	North Carolina A&T State University, USA
Michael Floyd	Knexus Research Corporation, USA
Susan Fox	Macalester College, USA
Geert-Jan Houben	TU Delft, The Netherlands
Dmitry Ignatov	National Research University, Higher School of Economics, Russia
Grigory Kabatyanskiy	Institute for Information Transmission Problems, Russia
Mehdi Kaytoue	INSA, France
Kristian Kersting	Technical University of Dortmund, Germany

Vladimir Khoroshevsky	Computer Center of Russian Academy of Science, Russia
Matthias Knorr	Universidade Nova de Lisboa, Portugal
Petia Koprinkova-Hristova	Institute of Information and Communication Technologies at Bulgarian Academy of Sciences, Bulgaria
Leila Kosseim	Concordia University, Montreal, Canada
Adila A. Krisnadhi	Wright State University, USA
Kai-Uwe Kuehnberger	University of Osnabrück, Germany
Sergei O. Kuznetsov	National Research University, Higher School of Economics, Russia
Evelina Lamma	University of Ferrara, Italy
Frederick Maier	Florida Institute for Human and Machine Cognition, USA
Riichiro Mizoguchi	Japan Advanced Institute of Science and Technology, Japan
Malek Mouhoub	University of Regina, Canada
Amedeo Napoli	LORIA, France
Michael O'Mahony	University College Dublin, Ireland
Sergei Obiedkov	National Research University, Higher School of Economics, Russia
Manuel Ojeda-Aciego	University of Malaga, Spain
Horia Pop	University Babes-Bolyai, Romania
Allan Ramsay	University of Manchester, UK
Chedy Raïssi	Inria, France
Ioannis Refanidis	University of Macedonia, Greece
Roberto Santana	University of the Basque Country, Spain
Ute Schmid	University of Bamberg, Germany
Sergey Sosnovsky	CeLTech, DFKI, Germany
Stefan Trausan-Matu	University Politehnica of Bucharest, Romania
Dan Tufis	Research Institute for Artificial Intelligence, Romanian Academy, Romania
Petko Valtchev	University of Montreal, Canada
Julita Vassileva	University of Saskatchewan, Canada
Tulay Yildirim	Yildiz Technical University, Turkey
David Young	University of Sussex, UK
Dominik Ślezak	University of Warsaw, Poland

Additional Reviewers

Boytcheva, Svetla	Loglisci, Corrado	Stoimenova, Eugenua
Cercel, Dumitru-Clementin	Rizzo, Giuseppe	Zese, Riccardo

Contents

Intelligent Agents and Planning

Posters

Machine Learning and Data Mining

Algorithm Selection Using Performance and Run Time Behavior

Tri Doan[⊠] and Jugal Kalita

University of Colorado Colorado Springs, 1420 Austin Bluffs Pkwy,
Colorado Springs, CO 80918, USA
{tdoan,jkalita}@uccs.edu

Abstract. In data mining, an important early decision for a user to make
is to choose an appropriate technique for analyzing the dataset at hand
so that generalizations can be learned. Intuitively, a trial-and-error approach becomes impractical when the number of data mining algorithms is
large while experts' advice to choose among them is not always available
and affordable. Our approach is based on meta-learning, a way to learn
from prior learning experience. We propose a new approach using regression to obtain a ranked list of algorithms based on data characteristics and
past performance of algorithms in classification tasks. We consider both
accuracy and time in generating the final ranked result for classification,
although our approach can be extended to regression problems.

Keywords: Algorithm selection · Meta-learning · Regression

1 Introduction

Different data mining algorithms seek out with different patterns hidden inside
a dataset. Choosing the right algorithm can be a decisive activity before a data
mining model is used to uncover hidden information in the data. Given a new
dataset, data mining practitioners often explore several algorithms they are used
to, to select the one to finally use. In reality, no algorithms can outperform all
others in all data mining tasks [22] because data mining algorithms are designed
with specific assumptions in mind to allow them to effectively work in particular
domains or situations.

Experimenting may become impractical due to the large number of machine
learning algorithms that are readily available these days. Our proposed solution uses a meta-learning framework to build a model to predict an algorithm's
behavior on unseen datasets. We convert the algorithm selection problem into a
problem of generating a ranked list of data mining algorithms so that regression
can be used to solve it.

The remainder of the paper is organized as follows. Section 2 presents related
work. Section 3 presents our proposed approach followed by our experiments
with discussion in Sect. 4. Finally, Sect. 5 summarizes the paper and provides
directions for future study.

© Springer International Publishing Switzerland 2016
C. Dichev and G. Agre (Eds.): AIMSA 2016, LNAI 9883, pp. 3–13, 2016.
DOI: 10.1007/978-3-319-44748-3_1

2   Related Work

Two common approaches to deal with algorithm selection are the learning curves approach and the dataset characteristics-based approach. While similarity between learning curves of two algorithms may indicate the likelihood that the two algorithm discover common patterns in similar datasets [15] in the learning curve approach, algorithm behavior is determined based on a dataset's specific characteristics [17] in the latter approach. Mapping dataset characteristics to algorithm behavior can be used in the meta-learning approach, which we find to be well suited for solving algorithm selection.

Some data mining practitioners may select an algorithm that achieves acceptable accuracy with a relatively short run time. For example, a classifier of choice for protein-protein interaction (PPI) should take run time into account as the computational cost is high when working with large PPI networks [23]. As a result, a combined measurement metric for algorithm performance (e.g., one that combines accuracy and execution time) should be defined as a monotonic measure of performance $mp(.)$ such that $mp(f_1) > mp(f_2)$ implies f_1 is better f_2 and vice versa. For example, the original ARR (Adjusted Ratio of Ratios) metric proposed by [3] uses the ratio between accuracy and execution time but does not guarantee monotonicity which has lead to others to propose a modified formula [1]. However, the use of single metric may not be desirable because it does not take into account the skew of data distribution and prior class distributions [4].

Model selection, on the other hand, focuses on hyper-parameter search to find the optimal parameter settings for an algorithm's best performance. For example, AUTO-WEKA [21] searches for parameter values that are optimal for a given algorithm for a given dataset. A variant version by [8] is a further improvement by taking past knowledge into account. The optimal model is selected by finding parameter settings for the same data mining algorithms and therefore model selection can be treated as a compliment to algorithm selection.

Our meta-learning approach can be distinguished from similar work in two ways: we use feature generation to obtain a fixed number of transformed features for each original dataset before generating meta-data, and we also use our proposed combined metric to integrate execution time with accuracy measurement.

3   Proposed Approach

The two main components of our proposed work are a regression model and a meta-data set (as training data). Regression has been used for predicting performance of data mining algorithms [2,11]. However, such works has either used a single metric such as accuracy or does not use data characteristics. A meta-data set is described in terms of features that may be used to characterize how a certain dataset performs with a certain algorithm. For example, statistical summaries have been used to generate new features for such meta-data. Due to varying number of features in real world datasets, the use of averages of statistical summaries as used in current studies may not suitable as a meta-data features. To overcome this problem, we transform each dataset into a fixed feature format in order to obtain the same number of statistical summaries to be used as features for a training meta-dataset.

We illustrate our proposed model in Fig. 1 with three main layers. A upper layer includes original datasets and corresponding transformed counterparts. The mid-layer describes the meta-data set(training data) where each instance (details in Table 4) includes three components. Two of these components are retrieved from the original dataset while the third component is a set of features generated from a transformed dataset in the upper layer. The label on each instance represents the performance of a particular algorithm in term of the proposed metric. The last layer is our regression model where we produced a predicted ranked list of algorithms sorted by performance for an unseen dataset.

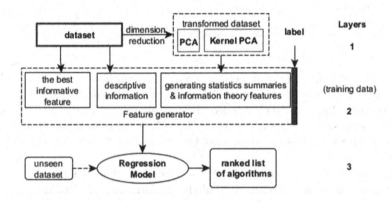

Fig. 1. Outline of the proposed approach

Using the knowledge of past experiments using m data mining algorithms on n known datasets, we can generate $m \times n$ training examples.

3.1   The Number of Features in Reduced Dataspace

The high dimensionality of datasets is a common problem in many data mining problems (e.g., in image processing, computational linguistics, and bioinformatics). In our work, dimensionality reduction produces a reduced data space with a fixed number of features to generate meta-features of training examples. Since different datasets have different number of features, our suggestion is to experimentally search for the number of features for a data mining application that does not cause a significant loss in performance. In our study, we want performance to be at least 80 % on the transformed dataset (although this number can be a parameter) compared to the full feature dataset.

Table 1 illustrates the performance when the size of feature space varies. We observe that the performances is the worst with one feature. With two features, there is an improvement in performance. However, using a low number of features

Table 1. Experiments of accuracy performance on reduced feature space

Dataset	acc	acc1	acc2	acc3	acc4	acc5	acc6	acc7	acc8	acc9	acc10	Features
leukemia	.72	.545	.590	.636	.636	.5	.5	.636	.590	.681	.595	7129
splice	.951	.5	.587	.766	.753	.747	.786	.705	.788	.773	.479	60
colon-cancer	.709	.631	.638	.631	.578	.631	.578	.684	.574	.631	.684	2000
digits	.643	.346	.595	.642	.602	.602	.602	.602	.602	.602	.602	64
usps	.829	.408	.615	.723	.726	.707	.708	.716	.7	.705	.711	256
ionosphere	.9	.266	.528	.528	.726	.566	.528	.584	.584	.518	.575	34
german-numer	.729	.49	.676	.633	.66	.613	.61	.633	.59	.673	.635	24
segment	.94	.342	.461	.810	.825	.822	.754	.772	.837	.834	.821	19
heart-statlog	.807	.230	.580	.469	.518	.456	.481	.530	.531	.456	.555	13
wbreast-cancer	.963	.573	.536	.541	.551	.551	.521	.560	.522	.551	.517	10
vowel	.688	.294	.474	.484	.521	.521	.547	.531	.536	.542	.518	10
glass	.635	0461	.507	.516	.523	.508	.485	.408	.477	.523		9
diabetes	.743	.666	.673	.703	.656	0.67	.633	.647	.656			8
live-disorder	.634	.567	.634	.557	.606	.586	.548					6
fourclass	.832	.595	.61									2

Note: acc, feature: refers accuracy, features with original dataset

may not be enough to perform well on a learning task in general. In addition, computing a relatively low number of features w.r.t the original dataset often requires higher computation time to avoid the non-convergence problem and gets lower performance [20]. Our experiments with datasets from biomedical, image processing, and text domains show that the use of four features in the reduced dataset works well.

The value of 4 is a good choice for the number of features as it satisfies our objective to secure at least 80 % of the performance of algorithms on the original datasets. We report the average of performances from all classification experiments as we change the number of features in Table 2 to evaluate how we choose the number dimension in our study. This choice is further evaluated with lowest run time in our assessment (accuracy and run time) compared to three or higher-than-4 dimensions.

Table 2. Average performance in different dimension space

No features	1	2	3	4	5	6	7	8	9	10
Avg accuracy	.5312	.581	.617	.6463	.6085	.5901	.6153	.6144	.6239	.6373

We perform classification to collect labels (accuracy measurement) for training examples. All available classification algorithms (refer to Table 3) are run on each of these reduced datasets.

Table 3. Algorithms used in our experiments

IBk	ZeroR	NaiveBayes
OneR	LWL	JRip
LibSVM	Bagging	SMO
Stacking	decisionStump	LogitBoost
RandomTree	Logistic	DecisionTable
MultiLayerPerceptron	J48	PART
RandomForests	RandomCommittee	AdaBoost
Vote	KStar	

3.2 Measurement Metric

One problem in data mining is the use of a variety of measurement metrics that leads to different comparisons. We will use the metric called SAR proposed by [4] instead of accuracy. SAR is defined as SAR = [Accuracy+AUC+(1−RMSE)]/3.

Inspired also by the A3R metric in [1], we develop a metric that we call the Adjusted Combined metric Ratio as a combined metric between SAR and execution time defined as follows $ACR = SAR/(\beta\sqrt{rt+1}+1))$ where $\beta \in [0,1]$, rt denotes run time.

The proposed ACR metric guarantees a monotonic decrease for execution time so that the longer run the time, the lower is the algorithm's performance. When time is ignored (when $\beta = 0$), the ACR formula becomes ACR = SAR which is a more robust evaluation metric than accuracy.

Figure 2 illustrates the monotonically decreasing function ACR as run time rt increases. When $\beta > 0$, our ACR formula reflects the idea of a penalization factor β if the run time is high. The performance of each algorithm computed using SAR and running time are recorded for each case as the performance feature. The remaining features are computed from a set of 4 features in the transformed dataset. This way, one instance of meta-data is generated from one transformed dataset. We record the result of running each algorithm on a dataset as a triple $< primary\ dataset, algorithm, performance >$ where the primary dataset is described in terms of its extracted characteristics or its meta-data, the algorithm represented simply by its name (in Weka), and computed performance on the dataset after feature reduction using the pre-determined β.

3.3 Meta-Features

We use a total of 30 meta-features for the training data, as given in Table 4. Each row of training data corresponds to meta-data obtained from a real dataset.

Let $LCoef^{ab}$ be a feature that measures the correlation coefficient between new transformed features a and b generated by a specific dimensionality reduction method (any 2 out of 3 features generated with PCA, similar to KPCA in each transformed dataset). We have 6 such features: $pcaLCoef^{12}$, $pcaLCoef^{13}$,

Table 4. Description of 30 meta-features

Feature	Description
ClassInt	Ratio of number of classes to instances
AttrClass	Ratio of number of features to number of classes
BestInfo	Most informative feature in original data
$pcaSTD^{1,2,3}$	Standard deviation for feature generated with PCA
$kpcaSTD^{1,2,3}$	Standard deviation for feature generated with KPCA
$pcaLCoef^{ab}$	Pearson linear coefficient
$pcaSkew^{1,2,3}$	Measure of asymmetry of the probability distribution
$pcaKurtosis^{1,2,3}$	Quantify shape of distribution
$kpcaLCoef^{ab}$	Pearson linear coefficient
$kpcaSkew^{1,2,3}$	Measure of asymmetry of the probability distribution
$kpcaKurtosis^{1,2,3}$	Quantify shape of distribution
nCEntropy	Normalized class entropy
entroClass	Class entropy for target attribute
TotalCorr	Amount of information shared among variables
Performance	ACR metric generated for each β setting

Note: the details of the above features explained below

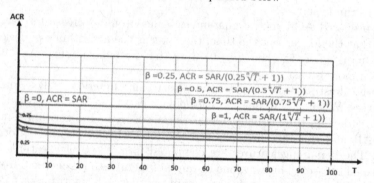

Fig. 2. Different β value plots for ACR measures

$pcaLCoef^{23}$, $kpcaLCoef^{12}$, $kpcaLCoef^{13}$, and $kpcaLCoef^{23}$. Each standard deviation value for the six new features is computed resulting in 6 standard deviation features (3 pcaSTD features and 3 kpcaSTD features). Similarly, 6 skewness and 6 kurtosis features are calculated for the 6 new transformed features.

4   Experimental Setup

We use datasets in Table 5 to generate transformed datasets and produce training data. We choose the best regression model among 6 candidates in Table 6 as our regression model. We introduce a parameter β to control the trade-off between the

Table 5. Datasets used

Arrhythmia	ionososphere	prnn-virus 3
bankrupt	japaneseVowels	RedWhiteWine
breastCancer	letter	segment
breastW	labor	sensorDiscrimination
cpu	liver disorder	solar flare
credit-a	lung cancer	sonar
cylinderBands	lymph	spambase
dermatology	sick	specfull
diabetes	molecularPromoter	splice
glass	monk-problem 1	spong
haberman	monk-problem 2	synthesis control
heart-cleveland	monk-problem 3	thyriod disease
heart-hungary	mushroom	tic-tac-toe
heart-stalog	page-blocks	vote
hepatitis	pen digits	vowels
horse-colic	post operation	wine
hypotheriod	primary tumor	

SAR metric (instead of Accuracy) and time. This metric is designed to measure the performance of a single algorithm on a particular dataset whereas A3R and ARR measure the performance of two algorithms on the same dataset. Figure 2 gives the ACR plots for 5 different values of β. For example, with $\beta = 0$, the user emphasizes the SAR metric (given as horizontal line) and accepts whatever the run time is. On the other hand, when $\beta = 1$, the user trades the SAR measurement for time. In this case, SAR is penalized more than half of its actual value.

Computed values of the proposed metric (ACR) are used as a performance (response) measurement corresponding to each training example. The generated meta-data are used as training examples in experiments with the 6 regression models 6 to select the best regression model to produce a ranked list of algorithms predicted performance.

To evaluate the performances among candidate regression algorithms in producing the final ranked list of algorithms, we use the RMSE metric and report the results in Table 6. This result is further assessed with Spearman's rank correlation test [13] to measure how close two ranks are, one predicted and the other based on actual performance. Our experiments (see Table 6) indicate that tree models, particularly CUBIST [18] obtain low RMSE compared to other non-linear regression models such as SVR [19], LARS [7] and MARS [10].

With the predicted performance of the ACR metric, our selected model (CUBIST) generates a ranked list of applicable algorithms. Note that we use $\beta = 0$ in the ACR formula to indicate the choice of performance based on SAR only.

Table 6. RMSEs by multiple regression models

Tree models	RMSE	Other models	RMSE
Model tree	0.9308	**SVR**	0.9714
Conditional D.T	0.9166	**LARS**	0.9668
Cubist	**0.9025**	**MARS**	0.9626

4.1 Experiment on Movie Dataset

In the section, we validate our proposed approach with a movie review dataset using algorithms in Table 3. Our goal is to show how a classifier algorithm can be picked by varying the importance of SAR (or Accuracy) vs. time. Our initial hunch is that Naive Bayes should be a highly recommended candidate because of both high accuracy and low run time [6], in particular for its ability to deal with non-numeric features. Naive Bayes was also used by [14] on the same dataset. Using top 5 algorithms from the ranked result list, we compute and compare results with those obtained by Naive Bayes classification. This collection of 50,000 movie reviews has at most 30 reviews for each movie [16] where a high score indicates a positive review. Each file is named $< counter >\_< score >$ .txt where the score is a value in the range (0..10). This "Large Movie Review Data Set" is present many challenges including high dimensionality with text features.

We perform the pre-processing steps including removal of punctuation, numbers and stop words before tokenization. Each token is considered a feature. There is a total of 117,473 features and 25,000 rows of reviews. The document term matrix ($117,473 \times 25000 = 2,936,825,000$) has only 2,493,414 non-zero entries or only $2,493,414/2,936,825,000 = 0.000849$ fraction of the entries is non-zero. After pre-processing the movie sentiment dataset, we obtain a meta-data instance for this dataset and apply the Cubist regression model. We use 3 different values of β to compute corresponding ACR to obtain three sets of labels for meta-data ($\beta = 0$, the higher SAR the better), a trade-off of SAR for time ($\beta = 0.5$ and $\beta = 0.75$), and in favor of time ($\beta = 1$, the shorter the better). We note that when $\beta > 0$, we take only run time into consideration (see Fig. 2) whereas using $\beta = 0$, we emphasize the SAR performance metric (as ACR equals SAR). We provide a short list of 5 top performers with 3 different values of β in Table 7.

4.2 Discussion

Our initial thought was that Naive Bayes was likely to be a good algorithm because of both high accuracy and low run time [6,14] particularly for its ability to deal with non-numeric features. We experiment with this algorithm of choice and compare with the Cubist method on the Movie Sentiment dataset [16] and report in Table 8.

As we see in Table 7, the list of top 5 algorithms changes significantly between $\beta = 0$ and $\beta = 1$ due to the trade-off between time and SAR. Two high performing

Table 7. Top 5 classifiers with different β

$\beta = 0$	$\beta = 0.5$	$\beta = 1$
Logistic	Random Tree	NaiveBayes
Random Forests	NaiveBayes	LWL
NaiveBayes	LWL	Logistic
SVM	Logistic	RandomCommittee
Bagging	Random Committee	RandomForest

classifiers, viz., Random Forests and SVM suffer from high computational time. When we prefer low run time, two other algorithms, LWL [9], and Random Committee [12] move into the top 5 ranked list. If we consider a compromise between SAR and time, we can use $\beta = 0.5$ which places Random Tree [5] and Naive Bayes on top of the ranked list and moves Random Forests and SVM down to 6th and 7th places in the final ranked list. It also shows that Naive Bayes is in 3rd place when we ignore run time. Given the fact that Naive Bayes is fast, if we are in favor of low execution time, we can increase β. The result shows that Naive Bayes classifier moves into 2nd place, or 1st place with $\beta = 0.5$, or $\beta = 1$, respectively.

Table 8 shows the performance using accuracy, AUC and RMSE with the SAR metric and run time for the top 5 algorithms. The lower AUC and higher RMSE of SVM compared to Naive Bayes explain the rank of SVM. Otherwise, SAR can be a good indicator for the corresponding accuracy.

We also note that our ranked results obtained with the combined SAR metric (with $\beta = 0$) are different from [3]. For instance, Brazdil et al. [3] rank Random Forests first but we rank it second due to lower AUC of this algorithm's performance on *abalone*. *SVM* in both methods ranks fourth but we are different in first place.

Table 8. Accuracy and SAR Performance on validation task

Algorithm	Accuracy	AUC	RMSE	SAR	Time
Logistic	0.856	0.93	0.325	0.820	171.88
Rand.Forest	0.838	0.916	0.393	0.787	237.7
NaiveBayes	0.816	0.9	0.388	0.779	10.03
SVM	0.848	0.848	0.389	0.769	529.16
Bagging	0.778	0.856	0.391	0.748	1170.3

5   Conclusion and Future Work

In this study, we demonstrate an alternative way to select suitable classification algorithms for a new dataset using the meta-learning approach. As the use of ensemble methods in real world applications becomes widespread, research on algorithm selection becomes more interesting but also challenging. We see the ensemble model as a good candidate to tackle the problem of big data when a single data mining algorithm may not be able to perform well because of limited computer resources. We want to expand this work to be able to provide performance values as well as estimated run time in the outcome.

References

1. Abdulrahman, M., Brazdil, P.: Measures for combining accuracy and time for meta-learning. In: ECAI, p. 49 (2014)
2. Bensusan, H., Kalousis, A.: Estimating the predictive accuracy of a classifier. In: Flach, P.A., De Raedt, L. (eds.) ECML 2001. LNCS (LNAI), vol. 2167, pp. 25–36. Springer, Heidelberg (2001)
3. Brazdil, P.B., Soares, C., Da Costa, J.P.: Ranking learning algorithms: using IBL and meta-learning on accuracy and time results. Mach. Learn. **50**(3), 251–277 (2003)
4. Caruana, R., Niculescu-Mizil, A.: Data mining in metric space: an empirical analysis of supervised learning performance criteria. In: Proceedings of the Tenth ACM SIGKDD. ACM (2004)
5. Cutler, A., Zhao, G.: Fast classification using perfect random trees. Utah State University (1999)
6. Dinu, L.P., Iuga, I.: The naive bayes classifier in opinion mining: in search of the best feature set. In: Gelbukh, A. (ed.) CICLing 2012, Part I. LNCS, vol. 7181, pp. 556–567. Springer, Heidelberg (2012)
7. Efron, B., Hastie, T., Johnstone, I., Tibshirani, R., et al.: Least angle regression. Ann. Stat. **32**(2), 407–499 (2004)
8. Feurer, M., Springenberg, J.T., Hutter, F.: Using meta-learning to initialize bayesian optimization of hyperparameters. In: ECAI Workshop (MetaSel) (2014)
9. Frank, E., Hall, M., Pfahringer, B.: Locally weighted naive bayes. In: Proceedings of the Nineteenth Conference on Uncertainty in Artificial Intelligence, pp. 249–256. Morgan Kaufmann Publishers Inc., Burlington (2002)
10. Friedman, J.: Multivariate adaptive regression splines. Ann. Stat. **19**(1), 1–141 (1991)
11. Gama, J., Brazdil, P.: Characterization of classification algorithms. In: Pinto-Ferreira, C., Mamede, N.J. (eds.) EPIA 1995. LNCS, vol. 990, pp. 189–200. Springer, Heidelberg (1995)
12. Hall, M., Frank, E.: The WEKA data mining software: an update. ACM SIGKDD Explor. Newslett. **11**(1), 10–18 (2009)
13. Kuhn, M., Johnson, K.: Applied Predictive Modeling. Springer, Berlin (2013)
14. Le, Q.V., Mikolov, T.: Distributed representations of sentences and documents. arXiv preprint (2014). arXiv:1405.4053
15. Leite, R., Brazdil, P., Vanschoren, J.: Selecting classification algorithms with active testing. In: Perner, P. (ed.) MLDM 2012. LNCS, vol. 7376, pp. 117–131. Springer, Heidelberg (2012)
16. Maas, A.L., Daly, R.E., Pham, P.T., Huang, D., Ng, A.Y., Potts, C.: Learning word vectors for sentiment analysis. In: 49th ACL, pp. 142–150 (2011)

17. Prudêncio, R.B.C., de Souto, M.C.P., Ludermir, T.B.: Selecting machine learning algorithms using the ranking meta-learning approach. In: Jankowski, N., Duch, W., Grąbczewski, K. (eds.) Meta-Learning in Computational Intelligence. SCI, vol. 358, pp. 225–243. Springer, Heidelberg (2011)
18. Quinlan, J.R.: Combining instance-based and model-based learning. In: Proceedings of the Tenth International Conference on Machine Learning (1993)
19. Smola, A.J., et al.: Regression estimation with support vector learning machines. Master's thesis, Technische Universit at München (1996)
20. Sorzano, C.O.S., Vargas, J., Montano, A.P.: A survey of dimensionality reduction techniques. arXiv preprint (2014). arXiv:1403.2877
21. Thornton, C., Hutter, F., Hoos, H.H., Leyton-Brown, K.: Auto-WEKA: combined selection and hyperparameter optimization of classification algorithms. In: 19th SIGKDD. ACM (2013)
22. Wolpert, D., Macready, W.: No free lunch theorems for optimization. IEEE Trans. Evol. Comput. 1(1), 67–82 (1997)
23. You, Z.H., Lei, Y.K., Zhu, L., Xia, J., Wang, B.: Prediction of protein-protein interactions from amino acid sequences with ensemble extreme learning machines and principal component analysis. BMC Bioinform. 14(8), 1 (2013)

A Weighted Feature Selection Method for Instance-Based Classification

Gennady Agre[1(✉)] and Anton Dzhondzhorov[2]

[1] Institute of Information and Communication Technologies,
Bulgarian Academy of Sciences, Sofia, Bulgaria
agre@iinf.bas.bg
[2] Sofia University "St. Kliment Ohrisdski", Sofia, Bulgaria
anton.dzhondzhorov@gmail.com

Abstract. The paper presents a new method for selecting features that is suited for the instance-based classification. The selection is based on the ReliefF estimation of the quality of features in the orthogonal feature space obtained after PCA transformation, as well as on the interpretation of these weights as values proportional to the amount of explained concept changes. The user sets a threshold defining what percent of the whole concept variability the selected features should explain and only the first "stronger" features, which combine weights together exceed this threshold, are selected. During the classification phase the selected features are used along with their weights. The experiment results on 12 benchmark databases have shown the advantages of the proposed method in comparison with traditional ReliefF.

Keywords: Feature selection · Feature weighting · k-NN classification

1 Introduction

Feature selection problem has been widely investigated by the machine learning and data mining community. The main goal is to select the smallest feature subset given a certain generalization error, or alternatively to find the best feature subset that yields the minimum generalization error [19]. Feature selection methods are usually classified in three main groups: wrapper, filter, and embedded methods. Wrappers use a concrete classifier as a black box for assessing feature subsets. Although these techniques may achieve a good generalization, the computational cost of training the classifier a combinatorial number of times becomes prohibitive for high-dimensional datasets. The filter methods select some features without involving any classifier relying only on general characteristics of the training data. Therefore, they do not inherit any bias of a classifier. In embedded methods the learning part and the feature selection part can not be separated - the structure of the class of functions under consideration plays a crucial role. Although usually less computationally expensive than wrappers, embedded methods are still much slower than filter approaches, and the selected features are dependent on the learning machine. One of the most popular filter methods is ReliefF [10], which is based on evaluating the quality of the features. The present paper describes an approach for improving ReliefF as a feature selection method by its combination with PCA algorithm.

© Springer International Publishing Switzerland 2016
C. Dichev and G. Agre (Eds.): AIMSA 2016, LNAI 9883, pp. 14–25, 2016.
DOI: 10.1007/978-3-319-44748-3_2

The structure of the paper is as follows: the next two sections briefly describe ReliefF and PCA algorithms. Section 4 presents the main idea of the proposed approach. The results of experiments testing the approach are shown in Sect. 5. Section 6 is devoted to discussion and related work and the final section is a conclusion.

2   Evaluation of Feature Quality by ReliefF

Relief [9] is considered as one of the most successful feature weighting algorithms. A key idea of it is to consider all features as independent ones and to estimate the relevance (quality) of a feature based on its ability to distinguish instances located near each other. In order to do this, the algorithm iteratively selects a random instance and then searches for its two nearest neighbours - a nearest hit (from the same class) and a nearest miss (from the different class). For each feature the estimation of its quality (weight) is calculated depending on the differences between the current instance and its nearest hit and miss along the corresponding attribute axis.

Sun and Li [21] have explained the effectiveness of Relief by showing that the algorithm is an online solution of a convex optimization problem, maximizing a margin-based objective function, where the margin is defined based on the nearest neighbour (1-NN) classifier. Therefore, compared with other filter methods, Relief usually performs better due to the performance feedback of a nonlinear classifier when searching for useful features. Compared with wrapper methods Relief avoids any exhaustive or heuristic combinatorial search by optimizing a convex problem and thus can be implemented very efficiently.

Igor Kononenko [10] proposed a more robust extension of Relief that was not limited to two class problems and could deal with incomplete and noisy data. Similarly to Relief, ReliefF randomly selects an instance $E = <\{e_1,...,e_p\}, c_E >$, but then searches for k of its hits $R^j = <\{r_1^j,...,r_p^j\}, c_E >$, $j = 1,...,k$ and misses $N^j(c) = <\{n_1^j,...,n_p^j\}, c >$, $c \neq c_E, j = 1,...,k$ for each class c using metrics $d(X,Y) = \left(\sum_{i=1}^{p} \delta^L(x_i,y_i)\right)^{\frac{1}{L}}$, $L = 1,2$.

The calculation of feature weight updates averages the contribution of all the hits and all the misses (m is the number of iterations):

$$w_i \leftarrow w_i - \sum_{j=1}^{k} \frac{\delta(e_i, r_i^j)}{k \cdot m} + \frac{1}{m \cdot k} \sum_{c \neq c_E} \frac{P(c)}{1 - P(c_E)} \sum_{j=1}^{k} \delta(e_i, n_i^j(c))$$

The weights w_i calculated by ReliefF are varied in interval $[-1, 1]$, as features with the weight equal to -1 are evaluated as absolutely irrelevant, and those with the weight of 1 – as absolutely relevant.

3 Selection of Features by PCA

Principal Component Analysis (PCA) is one of the most frequently used feature selection methods, which is based on extracting the basis axes (principle components) on which the data shows the highest variability [8]. The first principle component is in the direction of maximum variance of the given data. The remaining ones are mutually orthogonal and are ordered in a way of maximizing the remaining variance. PCA can be considered as a rotation of the original coordinate axes to a new set of axes that are aligned with the variability of the data - the total variability remains the same but the new features are now uncorrelated.

The dimensionality reduction (from n to k features, $k < n$) is done by selecting only a part of all principle components - those orthonormal axes that have the largest associated eigenvalues of the covariance matrix. The user sets the threshold h defining a part of the whole variability of data that should be preserved and first k new features, which eigenvalues in common exceed the threshold, are selected.

An optimization of the classical PCA was proposed by Turk and Pentland [18]. It is based on the fact that when the number of instances N is less than the dimension of the space n, that only $N-1$ rather than n meaningful eigenvectors will exist (the rest eigenvectors will have associated eigenvalues equal to zero). Having a normalized data matrix $\mathbf{D}$, the proposed optimization procedure allows calculating eigenvectors of original covariance matrix $\mathbf{S} = \mathbf{DD}^T$ based on eigenvectors of matrix $\mathbf{L} = \mathbf{D}^T\mathbf{D}$, which has the dimensions $N \times N$. This variant of PCA is very popular for solving tasks related to image representation, recognition and retrieval, when the number of images is significantly less than the number of features (pixels) describing an image.

4 Our Approach

Being very effective method for feature weighting, ReliefF has several drawbacks. One of them is that the algorithm assigns high relevance scores to all discriminative features, even if some of them are severely correlated [13]. As a consequence, redundant features might not be removed, when ReliefF is used for feature selection [21].

Although ReliefF has been initially developed as an algorithm for evaluating the quality of features, it has been commonly used as a feature subset selection method that is applied as a prepossessing step before the model is learnt [9]. In order to do that, a threshold h is introduced and only features, which weights are above this threshold, are selected. The selection of a proper value for the threshold is not an easy task and the dependence of ReleiefF from this parameter has been mentioned as one of its short-comings as a feature selection algorithm [6]. Several methods for selecting the threshold have been proposed - for example, Kira and Rendell [9] suggested the following bounds for this value: $0 < h \leq \frac{1}{\sqrt{\alpha m}}$, where α is the probability of accepting an irrelevant feature as relevant and m is the number of iterations used. However, as it is mentioned in [13], "the upper bound for h is very loose and in practice much smaller values can be used".

In its turn, PCA is traditionally used as a feature selection algorithm mainly in an unsupervised context – in most cases only first k uncorrelated attributes with total amount of eigenvalues above the user defined threshold h are selected. The main drawback of PCA, applied to the classification task, is that it does not take into account the class information of the available data. As it was mentioned in [4], the first few principal components would only be useful in those cases where the intra-class and inter-class variations have the same dominant directions or inter-class variations are clearly larger that intra-class variations. Otherwise, PCA will lead to a partial (or even complete) loss of discriminatory information. This statement was empirically confirmed for k-nearest neighbours (k-NN), decision trees (C4.5) and Naive Bayes classifiers tested on a set of benchmark databases [12].

Our approach is an attempt to explore the best features of the mentioned above methods for compensating their deficiencies. First, in order to improve the quality of ReliefF as *a feature weighting method* we apply it to a set of uncorrelated features found by PCA transformation decreasing in such a way the weights of redundant features and removing duplicating ones.

Second, in order to use ReliefF as *a feature selection method,* we apply a new method for selecting features, which is inspired by the interpretation of weights calculated by ReliefF as a portion of explained concept changes [13]. In such a way, the sum of the weights can be seen as an approximation to the value of concept variation - a measure of problem difficulty based on the nearest neighbor paradigm [13]. Similar to the approach used by PCA for selecting feature subsets in an unsupervised context, we select only the first k features (i.e. features with the biggest nonnegative ReliefF's weights) with *total amount of weights* above the user defined threshold h. Such an approach may be seen as a unified method for solving feature selection task based on evaluation of quality of features both in unsupervised and supervised context – in the first case we set a desired portion of explained total variability of the data, while in the second – the variability of the data is changed by the variability of concept to be learnt.

Lastly, the third aspect of our approach concerns the use of features selected by the proposed combination of algorithms in the context of the instance-based classification. As it has been shown in [20], ReliefF estimation of feature quality can be successfully used as weights in a distance metrics used for the instance-based classification:

$$d(X, Y) = \left(\sum_{i=1}^{n} w_i * \delta^L(x_i, y_i) \right)^{\frac{1}{L}}, w_i > 0$$

That is why, we use all features selected by our method *together with their weights,* when a classification task is going to be solved by an instance-based algorithm.

5 Experiments and Results

In order to test our ideas we have selected 12 databases described by numerical attributes (features) which number is varied from 4 to 16500 (see Table 1) – 9 benchmark bases are from UCI Machine Learning Repository (UCI)[1], 2 – from Kent Ridge Bio-medical Dataset (KR)[2] and one is our own database [15].

Table 1. Databases used in the experiments.

Database	Source	Atts	Examples	Classes	Missing attributes	Default accuracy
Diabetes (DB)	UCI	8	768	2	No	0.651
Breast Cancer Wisconsin (BCW)	UCI	10	699	2	Yes	0.655
Glass (GL)	UCI	9	214	6	No	0.327
Wine (WN)	UCI	13	178	3	No	0.399
Iris (IR)	UCI	4	150	3	No	0.333
Vehicle Silhouettes (VS)	UCI	18	846	4	No	0.258
Liver Disorders (LD)	UCI	6	345	2	No	0.560
Sonar (SN)	UCI	60	208	2	No	0.534
Banknote (BN)	UCI	4	1372	2	No	0.555
Faces (FC)	Custom	16500	102	2	No	0.559
Lung Cancer (LC)	KR	12533	181	2	No	0.829
Ovarian Cancer (OC)	KR	15154	253	2	No	0.640

All databases were used for evaluating a classifier by means of a hold-out cross-validation schema – for each database 70 % of randomly selected examples were used for training the classifier and the rest 30 % - for testing it. The experiments were repeated 70 times and the results were averaged. A traditional 5-NN algorithm with Euclidian distance was used as a basic classifier. The same algorithm was used for evaluating the quality of different feature weighting schemas but with the weighted Euclidian distance. The differences in classification accuracy of the 5-NN classifiers used different algorithms for calculating feature weights were evaluated by Student t-paired test with 95 % significance level.

5.1 Evaluation of Feature Weighing Schemas

The following algorithms for calculating feature weights were evaluated:

---

[1] https://archive.ics.uci.edu/ml/datasets.html.
[2] http://datam.i2r.a-star.edu.sg/datasets/krbd/.

- *PCA*: the transformation was applied to each training set and the calculated eigenvectors were used for transforming the corresponding testing set. Since PCA does not change the distance between examples, the classification accuracy of an instance-based classifier is not changed when it is applied to the PCA transformed data. That is why, in order to evaluate the quality of PCA as *a feature weighting algorithm*, we used the calculated eigenvalues as feature weights in classification.
- *ReliefF*: the feature weights were calculated by applying ReliefF algorithm to each training set. In our experiments we used 10 nearest neighbours and up to 200 iterations as the ReliefF parameters.
- *PCA+ReliefF*: each training set was initially transformed by application of PCA and the resulted dataset was then used for calculating feature weights by means of ReliefF.

The experiment results are shown in Table 2. Classification accuracy values are shown in bold (meaning statistically significant better performance than the basic (5-NN) classifier), in italics (for worse performance) or in regular fonts for cases with no statistically significant differences. The "Total" row summarises this information.

Table 2. Classification accuracy when the calculated weights are used during classification

Database	5-NN	PCA	ReliefF	PCA+ReliefF
DB	0.734 ± 0.025	*0.728 ± 0.024*	0.734 ± 0.026	0.733 ± 0.026
BCW	0.968 ± 0.013	*0.965 ± 0.011*	*0.966 ± 0.012*	*0.965 ± 0.012*
GL	0.626 ± 0.050	**0.645 ± 0.053**	**0.651 ± 0.055**	**0.646 ± 0.048**
WN	0.960 ± 0.024	**0.967 ± 0.023**	0.962 ± 0.024	**0.977 ± 0.018**
IS	0.961 ± 0.027	0.959 ± 0.029	0.959 ± 0.030	0.962 ± 0.028
VS	0.686 ± 0.022	*0.561 ± 0.028*	**0.693 ± 0.025**	**0.706 ± 0.025**
LD	0.607 ± 0.037	*0.593 ± 0.037*	0.613 ± 0.037	**0.644 ± 0.043**
SN	0.797 ± 0.053	0.786 ± 0.048	0.791 ± 0.054	0.799 ± 0.053
BN	0.998 ± 0.002	*0.994 ± 0.004*	**0.999 ± 0.001**	**0.999 ± 0.001**
FC	0.822 ± 0.055	*0.783 ± 0.074*	*0.811 ± 0.055*	0.821 ± 0.061
LC	0.925 ± 0.034	**0.978 ± 0.026**	**0.938 ± 0.034**	**0.963 ± 0.034**
OC	0.915 ± 0.034	*0.832 ± 0.042*	**0.959 ± 0.023**	**0.939 ± 0.028**
Total		*3+, 7−, 2=*	*5+, 2−, 5=*	*7+, 1−, 4=*

As it may be expected, PCA has been shown as a weak feature weighting algorithm in the classification context; however, it still lead to a significantly better accuracy on 3 databases and achieved a very high result on the Lung Cancer database.

The experiments have confirmed the commonly acknowledged statement that ReliefF is a rather strong feature weighting algorithm – in our case it has 5 statistically significant wins against 2 statistically significant loses and 5 statistically equal results in comparison with the unweighted variant of 5-NN algorithm.

The proposed combination of PCA and ReliefF has provided the best results – 7 statistically significant wins against only 1 statistically significant lose and 4 statistically equal results. The evaluation of differences in behaviour of pure ReliefF and ReliefF applied after PCA is shown in Table 3.

Table 3. Classification accuracy of PCA+ReliefF against ReliefF.

Database	ReliefF	PCA+ReliefF
DB	0.734 ± 0.026	0.733 ± 0.026
BCW	0.966 ± 0.012	0.965 ± 0.012
GL	0.651 ± 0.055	0.646 ± 0.048
WN	0.962 ± 0.024	**0.977 ± 0.018**
IS	0.959 ± 0.030	0.962 ± 0.028
VS	0.693 ± 0.025	**0.706 ± 0.025**
LD	0.613 ± 0.037	**0.644 ± 0.043**
SN	0.791 ± 0.054	0.799 ± 0.053
BN	0.999 ± 0.001	0.999 ± 0.001
FC	0.811 ± 0.055	0.821 ± 0.061
LC	0.938 ± 0.034	**0.963 ± 0.034**
OC	0.959 ± 0.023	*0.939 ± 0.028*
Total		*4+, 1−, 7=*

It can be seen, that in most cases the application of PCA before ReliefF has really improved the quality of ReliefF as a feature weighting algorithm.

5.2   Evaluation of Feature Selection Schemas

The results of the experiments for evaluating the ability of the mentioned above algorithms to select relevant features are presented in Table 4 (the first column presents the accuracy of 5-NN algorithm that uses the full set of features). *All selected features are used by 5-NN algorithm without taking into account their weights.*

Table 4. Classification accuracy when the calculated weights are used only for feature selection ($h = 0.95$)

Database	5-NN	PCA	ReliefF	PCA+ReliefF
DB	0.734 ± 0.025	**0.738 ± 0.026**	0.732 ± 0.026	0.734 ± 0.025
BCW	0.968 ± 0.013	0.969 ± 0.011	*0.966 ± 0.013*	0.968 ± 0.013
GL	0.626 ± 0.050	0.625 ± 0.051	**0.657 ± 0.052**	0.627 ± 0.048
WN	0.960 ± 0.024	**0.969 ± 0.021**	0.959 ± 0.026	**0.966 ± 0.023**
IS	0.961 ± 0.027	*0.950 ± 0.028*	0.961 ± 0.027	0.961 ± 0.029
VS	0.686 ± 0.022	*0.607 ± 0.025*	*0.683 ± 0.020*	0.688 ± 0.022
LD	0.607 ± 0.037	*0.575 ± 0.040*	0.609 ± 0.035	0.613 ± 0.041
SN	0.797 ± 0.053	0.800 ± 0.050	0.796 ± 0.049	0.801 ± 0.052
BN	0.998 ± 0.002	*0.973 ± 0.008*	0.999 ± 0.002	0.999 ± 0.002
FC	0.822 ± 0.055	0.818 ± 0.056	*0.811 ± 0.055*	0.814 ± 0.057
LC	0.925 ± 0.034	*0.898 ± 0.039*	*0.912 ± 0.045*	0.920 ± 0.048
OC	0.915 ± 0.034	*0.912 ± 0.034*	**0.931 ± 0.030**	0.914 ± 0.034
Total		*2+, 6−, 4=*	*2+, 4−, 6=*	*1+, 0−, 10=*

As it is expected, the behaviour of PCA as a feature selection algorithm in the supervised context still remains unsatisfactory, even though it is better than when it has been used for feature weighting.

A significant degradation can be observed in the behaviour of ReliefF. It should be mentioned that even when *all* features are used for weighted instance-based classification, in practice ReliefF *removes* highly irrelevant features, i.e. operates as (partially) feature selection algorithm. Such irrelevant features are those with the weights less or equal to zero. In our case, in addition to these irrelevant features ReliefF has been forced to remove slightly relevant or redundant attributes as well. Since the algorithm does not take into account the possible correlation between the features, it tends to underestimate less important (or redundant) features. As a result, the ordering of features by its importance created by ReliefF has occurred to be not very precise, which leads to removing some features that play important role for classification.

The mentioned above explanation is confirmed by the results shown in the last column of the table – the application of PCA eliminates the existing correlation between the features and allows ReliefF to evaluate the importance of the transformed features in a more correct way. As it can be seen, the accuracy of 5-NN algorithm running on the selected subset of features is statistically the same or even higher (for Wine database) than the accuracy of the same algorithm exploiting the whole set of features.

However, the comparison of results from Tables 2 and 4 has shown that as a whole, the classification accuracy of 5-NN algorithm used the combination of PCA and ReliefF for feature subset selection is *less* then the accuracy of the same algorithm that used the same combination for feature weighting. A possible explanation of this fact is that even after removing some irrelevant features the correct weighting of the rest features remains a very important factor for k-NN based classification. In order to prove this assumption we have conducted experiments in which feature weighting is combined with feature selection – during the classification phase all selected features are used with their weights calculated by the corresponding feature weighting algorithm at the pre-processing phase. The results of these experiments are shown in Table 5.

Table 5. Classification accuracy when the calculated weights are used both for feature selection and during classification ($h = 0.95$)

Database	5-NN	PCA	ReliefF	PCA+ReliefF
DB	0.734 ± 0.025	0.731 ± 0.022	0.733 ± 0.025	0.733 ± 0.025
BCW	0.968 ± 0.013	*0.965 ± 0.011*	*0.965 ± 0.012*	*0.965 ± 0.012*
GL	0.626 ± 0.050	**0.643 ± 0.052**	**0.656 ± 0.055**	**0.646 ± 0.048**
WN	0.960 ± 0.024	**0.968 ± 0.023**	0.961 ± 0.024	**0.978 ± 0.018**
IS	0.961 ± 0.027	0.956 ± 0.031	0.959 ± 0.029	0.961 ± 0.028
VS	0.686 ± 0.022	*0.549 ± 0.030*	0.690 ± 0.024	**0.704 ± 0.025**
LD	0.607 ± 0.037	*0.584 ± 0.042*	0.615 ± 0.038	**0.642 ± 0.041**
SN	0.797 ± 0.053	0.786 ± 0.048	0.793 ± 0.052	0.798 ± 0.051
BN	0.998 ± 0.002	*0.962 ± 0.010*	**0.999 ± 0.001**	**0.999 ± 0.001**
FC	0.822 ± 0.055	*0.784 ± 0.065*	0.811 ± 0.054	0.819 ± 0.062
LC	0.925 ± 0.034	**0.977 ± 0.026**	**0.939 ± 0.033**	**0.965 ± 0.033**
OC	0.915 ± 0.034	*0.832 ± 0.042*	**0.959 ± 0.024**	**0.940 ± 0.028**
Total		*3+, 6−, 3=*	*4+, 1−,7=*	*8+, 1−, 4=*

As one can see, in the context of the instance-based classification the best results have been achieved by the combination of ReliefF feature weighting method applied to PCA transformed databases with the proposed by us schema for feature selection.

5.3 Dimensionality Reduction

The last question that should be discussed is the dimensionality reduction that has been achieved by the proposed method. The number of removed features is shown in Table 6.

Table 6. Contribution of different algorithms to dimensionality reduction.

DB	All Atts	PCA Weighting	ReliefF Weighting	PCA +ReliefF Weighting	PCA Selection (h = 0.95)	ReliefF Selection (h = 0.95)	PCA+ReliefF Selection (h = 0.95)
DB	8	0	0	0	1	1	0
BCW	10	0	0	0	3	1	2
GL	9	0	0	0	3	1	1
WN	13	0	0	0	3	1	3
IS	4	0	0	0	2	0	1
VS	18	0	0	0	12	2	2
LD	6	0	1	1	1	1	1
SN	60	0	3	14	34	15	26
BN	4	0	0	0	1	0	0
FC	16500	16429	1799	16467	16460	5888	16475
LC	12533	12407	3130	12450	12433	6265	12469
OC	15154	14977	123	15075	15131	3277	15097

It should be mentioned that our implementation of PCA includes the optimization proposed in [18], which is very efficient in cases, when the number of features (n) is significantly greater than the number of examples (N). In such cases PCA behaves as a feature selection algorithm that preserves only N relevant (most important) attributes. The contribution of such implementation of PCA to the reduction of the final feature subset is shown in the table column named 'PCA Weighting'. The number of features evaluated by ReliefF as highly irrelevant (i.e. with non-positive values of feature weights) is shown in the column 'ReliefF Weighting'. The next column displays the number of features evaluated as highly irrelevant in cases when ReliefF has been applied after PCA transformation. The last three columns show the number of features that have been removed by the corresponding algorithm when the threshold h has been set.

The results show that when a database is described by a relatively small number of features (the first 9 databases in the table), our algorithm has succeeded to remove, in average, 10.8 % of them without compromising or even significantly raising in most cases (5+, 1−, 3=) the classification accuracy of 5-NN algorithm. The main contribution to this reduction belongs to ReliefF algorithm, which has evaluated (in average)

4.5 % of the features as highly irrelevant and 6.3 % of them – as weakly irrelevant or redundant.

In the last three databases, in which the number of attributes is significantly greater than the number examples, our algorithm has eliminated, in average, 99.7 % of the features evaluating 99.6 % of them as highly irrelevant and only 0.1 % - as redundant. The main contribution to this dimensionality reduction is due to the role of PCA implementation used. However, setting threshold h to 95 % of the total explained concept variability has forced ReliefF algorithm to evaluate 25.2 % of features remaining after PCA transformation as redundant. In the same time, the average classification accuracy of 5-NN algorithm using, in average, only 0.3 % of all features has significantly raised (2+, 0−, 1=).

All mentioned above have proved that the proposed method can be successfully used for dimensionality reduction without compromising the accuracy of instance-based classification.

6 Discussion and Related Work

The question that should be discussed is the scalability of the proposed approach. The first aspect of this problem is related to types of features that can be processed. ReliefF does not have any problems with processing nominal features by changing, for example, Euclidian distance with Manhattan distance [13]. Although the computation of principle components in PCA explores the apparatus of linear algebra, the algorithm can be easily adapted to work with nominal features by their binarization. [7, 17]. Since several binarization methods exist, we have not purposefully included any databases with nominal features into our experiments in order to exclude possible influence of such methods to the final quality of the proposed approach. However, the binarization allows applying our approach to databases with nominal features as well.

The other aspect is the dependence of the approach from the number of features and instances. For a database contained N instances described by n features, the complexity of Relief using m iterations is $O(mnN)$ – the same is valid for ReliefF as the most complex operation is finding k nearest neighbours for an instance. However, many different techniques for fast and efficient k-NN search have been developed [3], which can be applied in ReliefF implementation as well. Fast and scalable implementations exist also for PCA transformation (see e.g. [11]), so the proposed combination of PCA and ReliefF is also scalable.

Another question concerns the classification accuracy of k-NN algorithms that use the proposed approach for data pre-processing. In such a context our method could be considered as a wrapper feature selection method and the typical cross-validation approach can be applied as for selecting optimal k, as for selecting the proper value of threshold h optimizing the accuracy of the k-NN classifier.

The similar approach for feature selection based on ReliefF and PCA was proposed in [22] in the context of underwater sound classification. The authors also used PCA for removing correlation between the features and then ReliefF – for evaluating the feature quality. However, the selection of the features was done in the traditional manner (by comparison of each feature weight against a threshold), which led to unconvincing results. Moreover, the weights of the selected features were not used for classification. The approach was tested only on a single dataset of a small dimension (39 features) and was compared only with PCA without presenting any information about statistical significance of the results.

Considering our approach as a method for adapting PCA to the classification task, it can be related to [17] and works on class-dependent PCA methods (see e.g. [14]).

Considering the proposed method as an approach for improving ReliefF algorithm, it can be related to such works as [5, 21]. The first work proposes a so-called Orthogonal Relief, which is a combination of sequential forward selection procedure, the Gram-Schmidt orthogonalization procedure and Relief. The algorithm was tested only on 4 databases with 2 classes. The second work proposes a variant of Relief algorithm called WACSA, where correlations among features are taken into account to adjust the final feature subset. The algorithm was tested on five artificial well-known databases from the UCI repository.

Our future plans include more intensive testing of the proposed approach on a more diverse set of databases and comparing it with other state-of-the-art methods for feature selection.

7   Conclusion

The paper presents a new method for feature selection that is suited for the instance-based classification. The selection is based on the ReliefF estimation of the quality of attributes in the orthogonal attribute space obtained after PCA transformation, as well as on the interpretation of these weights as values proportional to the amount of explained concept changes. Only the first "strong" features, which combined ReliefF weights exceed the user defined threshold defining the desired percent of the whole concept variability the selected features should explain, are chosen. During the classification phase the selected features are used along with their weights calculated by ReliefF. The results of intensive experiments on 12 datasets have proved that the proposed method can be successfully used for dimensionality reduction without compromising or even raising the accuracy of instance-based classification.

References

1. Bins, J., Draper, B.: Feature selection from huge feature sets. In: Proceedings of the Eighth IEEE International Conference on Computer Vision, vol. 2, pp. 159–165 (2001)
2. Chang, C.-C.: Generalized iterative RELIEF for supervised distance metric learning. Pattern Recogn. **43**(8), 2971–2981 (2010)
3. Dhanabal, S., Chandramathi, S.: A review of various k-nearest neighbor query processing techniques. Intern. J. Comput. Appl. **31**(7), 14–22 (2011)
4. Diamataras, K.I., Kung, S.J.: Principal Component Neural Networks. Theory and Applications. Wiley, New York (1996)

5. Florez-lopez, R.: Reviewing RELIEF and its extensions: a new approach for estimating attributes considering high-correlated features. In: Proceedings of IEEE International Conference on Data Mining, Maebashi, Japan, pp. 605–608 (2002)
6. Freitag, D., Caruana, R.: Greedy attribute selection. In: Proceedings of Eleven International Conference on Machine Learning, pp. 28–36 (1994)
7. Hall, M.A.: Correlation-based feature selection of discrete and numeric class machine learning. In: Proceedings of International Conference on Machine Learning (ICML-2000), San Francisco, CA; pp. 359–366. Morgan Kaufmann, San Francisco (2000)
8. Jolliffe, I.T.: Principal Component Analysis. Springer, New York (1986)
9. Kira, K., Rendell, L.A.: The feature selection problem: traditional methods and a new algorithm. In: Proceedings of AAAI 1992, San Jose, USA, pp. 129–134 (1992)
10. Kononenko, I.: Estimating attributes: analysis and extensions of RELIEF. In: Proceedings of European Conference on Machine Learning, Catania, Italy, vol. 182, pp. 171–182 (1994)
11. Ordonez, C., Mohanam, N., Garcia-Alvarado, C.: PCA for large data sets with parallel data summarization. Distrib. Parallel Databases 32(3), 377–403 (2014)
12. Pechenizkiy, M.: The impact of feature extraction on the performance of a classifier: kNN, Naïve Bayes and C4.5. In: Kégl, B., Lee, H.-H. (eds.) Canadian AI 2005. LNCS (LNAI), vol. 3501, pp. 268–279. Springer, Heidelberg (2005)
13. Robnik-Sikonja, M., Kononenko, I.: Theoretical and empirical analysis of ReliefF and RReliefF. Mach. Learn. J. 53, 23–69 (2003)
14. Sharma, A., Paliwala, K., Onwubolu, G.: Class-dependent PCA, MDC and LDA: a combined classifier for pattern classification. Pattern Recogn. 39, 1215–1229 (2006)
15. Strandjev, B., Agre, G.: On impact of PCA for solving classification tasks defined on facial images. Intern. J. Reason. Based Intell. Syst. 6(3/4), 85–92 (2014)
16. Sun, Y., Li, J.: Iterative RELIEF for feature weighting: algorithms, theories, and applications. IEEE Trans. Pattern Anal. Mach. Intell. 29(6), 1035–1051 (2007)
17. Tsymbal, A., Puuronen, S., Pechenizkiy, M., Baumgarten, M., Patterson, D.W.: Eigenvector-based feature extraction for classification. In: Proceedings of FLAIRS Conference, pp. 354–358 (2002)
18. Turk, M., Pentland, A.: Eigenfaces for recognition. J. Cogn. Neurosci. 3(1), 71–86 (1991)
19. Vergara, J., Estevez, P.: A review of feature selection methods based on mutual information. Neural Comput. Appl. 24, 175–186 (2014)
20. Wettschereck, D., Aha, D.W., Mohri, T.: A review and empirical evaluation of feature weighting methods for a class of lazy learning algorithms. Artif. Intell. Rev. 11, 273–314 (1997)
21. Yang, J., Li, Y.-P.: Orthogonal relief algorithm for feature selection. In: Huang, D.-S., Li, K., Irwin, G.W. (eds.) ICIC 2006. LNCS, vol. 4113, pp. 227–234. Springer, Heidelberg (2006)
22. Zeng, X., Wang, Q., Zhang, C., Cai, H.: Feature selection based on ReliefF and PCA for underwater sound classification. In: Proceedings of the 3rd International Conference on Computer Science and Network Technology (ICCSNT), Dalian, pp. 442–445 (2013)

Handling Uncertain Attribute Values in Decision Tree Classifier Using the Belief Function Theory

Asma Trabelsi[1(✉)], Zied Elouedi[1], and Eric Lefevre[2]

[1] Université de Tunis, Institut Supérieur de Gestion de Tunis,
LARODEC, Tunis, Tunisia
trabelsyasma@gmail.com, zied.elouedi@gmx.fr
[2] Univ. Artois, EA 3926, Laboratoire de Génie Informatique et d'Automatique
de l'Artois (LGI2A), 62400 Béthune, France
eric.lefevre@univ-artois.fr

Abstract. Decision trees are regarded as convenient machine learning techniques for solving complex classification problems. However, the major shortcoming of the standard decision tree algorithms is their unability to deal with uncertain environment. In view of this, belief decision trees have been introduced to cope with the case of uncertainty present in class' value and represented within the belief function framework. Since in various real data applications, uncertainty may also appear in attribute values, we propose to develop in this paper another version of decision trees in a belief function context to handle the case of uncertainty present only in attribute values for both construction and classification phases.

Keywords: Decision trees · Uncertain attribute values · Belief function theory · Classification

1 Introduction

Decision trees are one of the well known supervised learning techniques applied in a variety of fields, particulary in artificial intelligence. Indeed, decision trees have the ability to deal with complex classification problems by producing understandable representations easily interpreted not only by experts but also by ordinary users and providing logical classification rules for the inference task. Numerous decision tree building algorithms have been introduced over the years [2,9,10]. Such algorithms take as inputs a training set composed with objects described by a set of attribute values as well as their assigned classes and output a decision tree that enables the classification of new objects. A significant shortcoming of the classical decision trees is their inability to handle data within an environment characterized by uncertain or incomplete data. In the case of missing values, several kinds of solutions are usually considered. One of the most popular solutions is dataset preprocessing strategy which aims at removing the missing values. Other solutions are exploited by some systems implementing decision tree

C. Dichev and G. Agre (Eds.): AIMSA 2016, LNAI 9883, pp. 26–35, 2016.
DOI: 10.1007/978-3-319-44748-3_3

learning algorithms. Missing values may also be considered as a particular case of uncertainty and can be modeled by several uncertainty theories. In the literature, various decision trees have been proposed to deal with uncertain and incomplete data such as fuzzy decision trees [15], probabilistic decision trees [8], possibilistic decision trees [5–7] and belief decision trees [4,16,17]. The main advantage that makes the belief function theory very appealing over the other uncertainty theories, is its ability to express in a flexible way all kinds of information availability from full information to partial ignorance to total ignorance and also it allows to specify the degree of ignorance in a such situation. In this work, we focus our attention only on the belief decision trees approach developed·by authors in [4] as an extension of the classical decision tree to cope with the uncertainty of the objects' classes and also allows to classify new objects described by uncertain attribute values [3]. In such a case, the uncertainty about the class' value is represented within the Transferable Belief Model (TBM), one interpretation of the belief function theory for dealing with partial or even total ignorance [14]. However, in several real data applications, uncertainty may appear in the attribute values [11]. For instance, in medicine, symptoms of patients may be partially uncertain. In this paper, we get inspired from the belief decision tree paradigm to handle data described by uncertain attribute values. Particulary, we tackle the case where the uncertainty occurs in both construction and classification phases. The reminder of this paper is organized as follows: Sect. 2 highlights the fundamental concepts of the belief function theory as interpreted by the TBM framework. In Sect. 3, we detail the building and the classification procedures of our new decision tree version. Section 4 is devoted to carrying out experiments on several real world databases. Finally, we draw our conclusion and our main future work directions in Sect. 5.

2   Belief Function Theory

In this Section, we briefly recall the fundamental concepts underlying the belief function theory as interpreted by the TBM [13].

Let us denote by Θ the frame of discernment including a finite non empty set of elementary events related to a given problem. The power set of Θ, denoted by 2^{Θ} is composed of all subsets of Θ.

The basic belief assignment (bba) expressing beliefs on the different subsets of Θ is a function $m : 2^{\Theta} \rightarrow [0,1]$ such that:

$$\sum_{A \subseteq \Theta} m(A) = 1. \tag{1}$$

The quantity $m(A)$, also called basic belief mass (bbm), states the part of belief committed exactly to the event A. All subsets A in Θ such that $m(A) > 0$ are called focal elements.

Decision making within the TBM framework consists of selecting the most probable hypothesis for a given problem by transforming beliefs into probability measure called the pignistic probability and denoted by $BetP$. It is defined as follows:

$$BetP(A) = \sum_{B \subseteq \Theta} \frac{|A \cap B|}{|B|} \frac{m(B)}{1 - m(\emptyset)} \ \forall \ A \ \in \ \Theta \tag{2}$$

Let m_1 and m_2 be two bba's provided by fully reliable distinct information sources [12] and defined in the same frame of discernment Θ. The resulting bba using the conjunctive rule is defined by:

$$(m_1 \bigcirc m_2)(A) = \sum_{B,C \subseteq \Theta : B \cap C = A} m_1(B).m_2(C) \tag{3}$$

It is important to note that some cases require the combination of bba's defined on different frames of discernment. Let Θ_1 and Θ_2 be two frames of discernment, the vacuous extension of belief functions consists of extending Θ_1 and Θ_2 to a joint frame of discernment Θ defined as:

$$\Theta = \Theta_1 \times \Theta_2 \tag{4}$$

The extended mass function of m_1 which is defined on Θ_1 and whose focal elements are the cylinder sets of the focal elements of m_1 is computed as follows:

$$m^{\Theta_1 \uparrow \Theta}(A) = m_1(B) \ where \ A = B \times \Theta_2, B \subseteq \Theta_1 \tag{5}$$
$$m^{\Theta_1 \uparrow \Theta}(A) = 0 \ otherwise$$

3   Decision Tree Classifier for Partially Uncertain Data

Authors in [4], have proposed what is called belief decision trees to handle real data applications described by known attribute values and uncertain class's value, particulary where the uncertainty is represented by belief functions within the TBM framework. However, for many real world applications, uncertainty may appear either in attribute values or in class value or in both attribute and class values. In this paper, we propose a novel decision tree version to tackle the case of uncertainty present only in attribute values for both construction and classification phases. Throughout this paper, we use the following notations:

- T: a given training set composed by J objects I_j, $j = \{1, \ldots, J\}$.
- $\mathcal{L}$: a given testing of L objects O_l, $l = \{1, \ldots, L\}$.
- S: a subset of objects belonging to the training set T.
- $C = \{C_1, \ldots, C_q\}$: represents the q possible classes of the classification problem.
- $A = \{A_1, \ldots, A_n\}$: the set of n attributes.
- Θ^{A_k}: represents the all possible values of an attribute $A_k \in A$, $k = \{1, \ldots, n\}$.
- $m^{\Theta^{A_k}}\{I_j\}(v)$: expresses the bbm assigned to the hypothesis that the actual attribute value of object I_j belongs to $v \subseteq \Theta^{A_k}$.

3.1   Decision Tree Parameters for Handling Uncertain Attribute Values

Four main parameters conducted to the construction of our proposed decision trees approach:

- **The attribute selection measure:** The attribute selection measure is relied on the entropy calculated from the average probability obtained from the set of objects in the node. To choose the most appropriate attribute, we propose the following steps:

  1. Compute the average probability relative to each class, denoted by $Pr\{S\}(C_i)$, by taking into account the set of objects S. This function is obtained as follows:

  $$Pr\{S\}(C_i) = \frac{1}{\sum_{I_j \in S} P_j^S} \sum_{I_j \in S} P_j^S \gamma_{ij} \qquad (6)$$

  where γ_{ij} equals 1 if the object I_j belongs to the class C_i, 0 otherwise and P_j^S corresponds to the probability of the object I_j to belong to the subset S. Assuming that the attributes are independent, the probability P_j^S will be equal to the product of the different pignistic probabilities induced from the attribute bba's corresponding to the object I_j and enabling I_j to belong to the node S.

  2. Compute the entropy $Info(S)$ of the average probabilities in S which is set to:

  $$Info(S) = -\sum_{i=1}^{q} Pr\{S\}(C_i) log_2 Pr\{S\}(C_i) \qquad (7)$$

  3. Select an attribute A_k. For each value $v \in \Theta^{A_k}$, define the subset $S_v^{A_k}$ composed with objects having v as a value. As the A_k values may be uncertain, $S_v^{A_k}$ will contain objects I_j such that their pignistic probability corresponding to the value v is as follows:

  $$BetP^{\Theta^{A_k}}\{I_j\}(v) \neq 0 \qquad (8)$$

  4. Compute the average probability, denoted by $Pr\{S_v^{A_k}\}$, for objects in subset $S_v^{A_k}$, where $v \in \Theta^{A_k}$ and $A_k \in A$. It will be set as:

  $$Pr\{S_v^{A_k}\}(C_i) = \frac{1}{\sum_{I_j \in S_v^{A_k}} P_j^{S_v^{A_k}}} \sum_{I_j \in S_v^{A_k}} P_j^{S_v^{A_k}} \gamma_{ij} \qquad (9)$$

  where $P_j^{S_v^{A_k}}$ is the probability of the object I_j to belong to the subset $S_v^{A_k}$ having v as a value of the attribute A_k (its computation is done in the same manner as the computation of P_j^S).

5. Compute $Info_{A_k}(S)$ as discussed by Quinlan [9], but using the probability distribution instead of the proportions. We get:

$$Info_{A_k}(S) = \sum_{v \in \Theta^{A_k}} \frac{|S_v^{A_k}|}{|S|} Info(S_v^{A_k}) \qquad (10)$$

where $Info(S_v^{A_k})$ is calculated from Eq. 7 using $Pr\{S_v^{A_k}\}$ and we define $|S| = \sum_{I_j \in S} P_j^S$ and $|S_v^{A_k}| = \sum_{I_j \in S_v^{A_k}} P_j^{S_v^{A_k}}$.

6. Compute the information gain yielded by the attribute A_k over the set of objects S such that:

$$Gain(S, A_k) = Info(S) - Info_{A_k}(S) \qquad (11)$$

7. Compute the *Gain Ratio* relative to the attribute A_k by the use of the *SplitInfo*

$$GainRatio(S, A_k) = \frac{Gain(S, A_k)}{SplitInfo(S, A_k)} \qquad (12)$$

where the *SplitInfo* value is defined as follows:

$$SplitInfo(S, A_k) = - \sum_{v \in \Theta^{A_k}} \frac{|S_v^{A_k}|}{|S|} log_2 \frac{|S_v^{A_k}|}{|S|} \qquad (13)$$

8. Repeat the same process for each attribute $A_k \in A$ (from step 3 to step 7) and then select the one that has the maximum *GainRatio*.

- **Partitioning Strategy:** The partitioning strategy, also called the splitting strategy, consists of splitting the training set according to the attribute values. As we only deal with categorical attributes, we create an edge for each attribute value chosen as a decision node. Due to the uncertainty in the attribute values, after the partitioning step each training instance may belong to more than one subset with a probability of belonging calculated according to the pignistic probability of its attribute values.

- **Stopping criteria:** Four key strategies are suggested as stopping criteria:
  1. The treated node contains only one instance.
  2. The treated node contains instances belonging to the same class.
  3. There is no further attribute to test.
  4. The gain ratio of the remaining attributes are equal or less than zero.

- **Structure of leaves:** Leaves, in our proposed decision tree classifier, will be represented by a probability distribution over the set of classed computed from the probability of instances belonging to these leaves. This is justified by the fact that leaves may contain objects with different class values called heterogeneous leaves. Therefore, the probability of the leaf L relative to each class $C_i \in C$ is defined as follows:

$$Pr\{L\}(C_i) = \frac{1}{\sum_{I_j \in L} P_j^L} \sum_{I_j \in L} P_j^L \gamma_{ij} \qquad (14)$$

where P_j^L is the probability of the instance I_j to belong to the leaf L.

3.2 Decision Tree Procedures to Deal with Uncertain Attribute Values

By analogy to the classical decision tree, our new decision tree version will be composed mainly of two procedures: the construction of the tree from data present uncertain attributes and the classification of new instances described by uncertain attribute values.

A. Construction procedure

Suppose that $\mathcal{T}$ is our training set composed by J objects characterized by n uncertain attributes $A = \{A_1, \ldots, A_n\}$ represented within the TBM framework. Objects of $\mathcal{L}$ may belong to the set of classes $C = \{C_1, \ldots, C_q\}$. The different steps of our building decision tree algorithm are described as follows:

1. Create the root node of the decision tree that contain all the training set objects.
2. Check if the node verify the stopping criteria presented previously.
   - If yes, declare it as a leaf node and compute its probability distribution.
   - If not, the attribute that has the highest *GainRatio* will be designed as the root of the decision tree related to the whole training set.
3. Perform the partitioning strategy by creating an edge for each attribute value chosen as a root. This partition leads to several training subsets.
4. Create a root node for each training subset.
5. Repeat the same process for each training subset from the step 2.
6. Stop when all nodes of the latter level are leaves.

B. Classification procedure

Once our decision tree classifier is constructed, it is possible to classify new objects of the testing set $\mathcal{L}$ described by uncertain attribute values [3]. As previously mentioned, the uncertainty about a such attribute values A_k relative to a new object to classify can be defined by a bba $m^{\Theta^{A_k}}$ expressing the part of beliefs committed exactly to the different values of this attribute. The bba $m^{\Theta^{A_k}}$ will be defined on the frame of discernment Θ^{A_k} including all the possible values of the attribute A_k. Let us denote by Θ^A the global frame of discernment relative to all the attributes. It is equal to the cross product of the different Θ^{A_k}:

$$\Theta^A = \underset{k=1,\ldots,n}{\times} \Theta^{A_k}. \tag{15}$$

Since an object is characterized by a set of combination of values where each one corresponds to an attribute, we have firstly to look for the joint bba representing beliefs on the different attribute values relative to the new object to be classified. To perform this goal just have to apply the following steps:

- Extend the different bba's $m^{\Theta^{A_k}}$ to the global frame of attributes Θ^A. Thus, we get the different bba's $m^{\Theta^{A_k} \uparrow \Theta^A}$.

– Combine the different extended bba's through the conjunctive rule of combination:

$$m^{\Theta^A} = \bigodot_{k=1,\ldots,n} m^{\Theta^{A_k} \uparrow \Theta^A} \qquad (16)$$

Once we have obtained the joint bba denoted by m^{Θ^A}, we consider individually the focal elements of this latter. Let x be a such focal element. The next step in our classification task consists of computing the probability distribution $Pr[x](C_i)(i = 1, \ldots, q)$. It is important to note that the computation of this latter depends on the subset x and more exactly on the focal elements of the bba m^{Θ^A}:

- If the treated focal element x is a singleton, then $Pr[x](C_i)$ is equal to the probability of the class C_i corresponding to the leaf to which the focal element is attached.
- If the focal element is not a singleton (some attributes have more than one value), then we have to explore all possible paths relative to this combination of values. Two possible cases may arise:
  * If all paths lead to the same leaf, then $Pr[x](C_i)$ is equal to the probability of the class C_i relative to this leaf.
  * If these paths lead to distinct leaves, then $Pr[x](C_i)$ is equal to the average probability of the class C_i relative to the different leaves.
- Finally, each test object's probability distribution over the set of classes will be computed as follows:

$$Pr_l(C_i) = \sum_{x \subseteq \Theta^A} m^{\Theta^A}(x) Pr[x](C_i) \ \forall \ C \in \{C_1, \ldots, C_q\} \text{ and } l = \{1, \ldots, L\}$$

$$(17)$$

The most probable class of the object O_l is the one having the highest probability Pr_l.

4   Experimentations

To evaluate the feasibility of our novel decision trees approach, we have carried our experiments on real categorical databases obtained from the UCI repository [1]. Due to the computational cost of our proposed approach, we have performed our experiments on several small databases. Table 1 provides a brief description of these data sets where #Instances, #Attributes and #Classes denote respectively the total number of instances, the total number of attributes and the total number of classes. It is important to note that our approach can also be applied in the case of numerical databases when applying some kinds of data preprocessing such as discretization, etc.

Let us remind that our purpose is to construct our decision tree classifier from datasets characterized by uncertain attribute values. Thus, we propose to

Table 1. Description of databases

Databases	#Instances	#Attributes	#Classes
Tic-Tac-Toe	958	9	2
Parkinsons	195	23	2
Balloons	16	4	2
Hayes-Roth	160	5	3
Balance	625	4	3
Lenses	24	4	3

include uncertainty in attribute values by tacking into consideration the original data sets and a degree of uncertainty P such that:

$$m^{\Theta^{A_k}}\{I_j\}(v) = 1 - P \tag{18}$$
$$m^{\Theta^{A_k}}\{I_j\}(\Theta^{A_k}) = P$$

The degree of uncertainty P takes value in the interval $[0, 1]$:

- Certain Case: $P = 0$
- Low Uncertainty: $0 \leq P < 0.4$
- Middle Uncertainty: $0.4 \leq P < 0.7$
- High Uncertainty: $0.7 \leq P \leq 1$

The performance of our novel decision tree paradigm when classifying new objects can be measured through several classification accuracies. In this work, we relied on:

- The PCC criterion that represents the percent of correct classification of objects belonging to the test set. It is computed as follows:

$$PCC = \frac{\text{Number of well classified instances}}{\text{Number of classified instanced}} \tag{19}$$

The number of well classified instances corresponds to the number of test instances for which the most probable classes obtained through our proposed decision tree classifier are the same as the real ones.
- The distance criterion: the main idea underling this criterion is to perform a comparison between a test instance's probability distribution over the set of classes and its real class. It is set as follows:

$$DistanceCriterion_j = Distance(Pr_j(C_i), C(I_j)) \tag{20}$$
$$= \sum_{i=1}^{q}(Pr_j(C_i) - \gamma_{ij})^2$$

where $C(I_j)$ corresponds to the real class of the test instance I_j and γ_{ij} equals 1 when $C(I_j) = C_i$ and 0 otherwise.

Note that this distance satisfies the following property:

$$0 \leq DistanceCriterion_j \leq 2 \qquad (21)$$

Besides, we just have to compute the average distance yielded from all test instances to get a total distance.

We run our proposed classifier using the 10-folds cross validation technique that randomly split the original data set into 10 equal sized subsets. Of the 10 subsets, a single subset is used as a test data and the remaining subsets are used as training data. The cross-validation process is then repeated 10 times where each subset is used exactly once as a test set. Our experimental results in terms of classification accuracy and distance are depicted in Figs. 1 and 2 for the different mentioned databases.

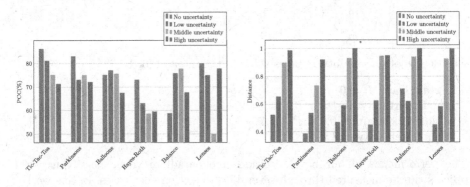

Fig. 1. PCC results   **Fig. 2.** Distance results

From the results given in Fig. 1, we can remark that our proposed decision tree classifier has yielded good classification accuracy for the different uncertainty levels for the different databases. For instance, for Balloons database, we have 75.2 %, 77.1 %, 75.7 % and 67.5 % as PCCs relative respectively to no, low, middle and high uncertainties. Concerning the distance criterion, from Fig. 2, we deduce that our classifier has given interesting results in term of distance criterion. In fact, all distance values belong to the closed interval [0.386, 1]. Mostly, the distance increases with the increasing of the uncertainty degree. For example, the distance results relative to Balloons database are 0.46, 0.59, 0.93 and 1 for respectively no, low, middle and high uncertainties. This interpretation is available for the major remaining databases.

5   Conclusion

Tackling classification problem in the case of uncertainty present in attribute values remains a challenging task but currently very under-studied. Thus, in this paper, we have proposed a new decision tree classifier to handle uncertainty present in the attribute values. Since we have obtained promising results, time complexity is still a critical problem, especially for large or even medium sized databases. So, as a future work we look forward reducing time complexity. We intend also to apply a pruning technique to reduce the dimensionality space and improve the classification accuracy.

References

1. Lichman, M.: UCI machine learning repository (2013). University of California, Irvine, School of Information and Computer Sciences. http://archive.ics.uci.edu/ml
2. Breiman, L., Friedman, J., Olshen, R., Stone, C.: Classification and Regression Trees. Wadsworth and Brooks, Monterey (1984)
3. Elouedi, Z., Mellouli, K., Smets, P.: Classification with belief decision trees. In: Cerri, S.A., Dochev, D. (eds.) AIMSA 2000. LNCS (LNAI), vol. 1904, pp. 80–90. Springer, Heidelberg (2000)
4. Elouedi, Z., Mellouli, K., Smets, P.: Belief decision trees: theoretical foundations. Int. J. Approximate Reasoning **28**(2), 91–124 (2001)
5. Hüllermeier, E.: Possibilistic induction in decision-tree learning. In: Elomaa, T., Mannila, H., Toivonen, H. (eds.) ECML 2002. LNCS (LNAI), vol. 2430, pp. 173–184. Springer, Heidelberg (2002)
6. Jenhani, I., Amor, N.B., Elouedi, Z.: Decision trees as possibilistic classifiers. Int. J. Approximate Reasoning **48**(3), 784–807 (2008)
7. Jenhani, I., Elouedi, Z., Ben Amor, N., Mellouli, K.: Qualitative inference in possibilistic option decision trees. In: Godo, L. (ed.) ECSQARU 2005. LNCS (LNAI), vol. 3571, pp. 944–955. Springer, Heidelberg (2005)
8. Quinlan, J.R.: Decision trees as probabilistic classifiers. In: 4th International Machine Learning, pp. 31–37 (1897)
9. Quinlan, J.R.: Induction of decision trees. Mach. Learn. **1**(1), 81–106 (1986)
10. Quinlan, J.R.: C4. 5: Programs for Machine Learning. Elsevier, Amsterdam (2014)
11. Samet, A., Lefèvre, E., Yahia, S.B.: Evidential data mining: precise support and confidence. J. Intell. Inf. Syst. **47**(1), 135–163 (2016). Springer
12. Smets, P.: Application of the transferable belief model to diagnostic problems. Int. J. Intell. Syst. **13**(2–3), 127–157 (1998)
13. Smets, P.: The transferable belief model for quantified belief representation. In: Smets, P. (ed.) Quantified Representation of Uncertainty and Imprecision, pp. 267–301. Springer, Heidelberg (1998)
14. Smets, P., Kennes, R.: The transferable belief model. Artif. Intell. **66**(2), 191–234 (1994)
15. Umano, M., Okamoto, H., Hatono, I., Tamura, H., Kawachi, F., Umedzu, S., Kinoshita, J.: Fuzzy decision trees by fuzzy ID3 algorithm and its application to diagnosis systems. In: 3rd IEEE Conference on Fuzzy Systems, pp. 2113–2118. IEEE (1994)
16. Vannoorenberghe, P.: On aggregating belief decision trees. Inf. Fusion **5**(3), 179–188 (2004)
17. Vannoorenberghe, P., Denoeux, T.: Handling uncertain labels in multiclass problems using belief decision trees. In: IPMU 2002, vol. 3, pp. 1919–1926 (2002)

Using Machine Learning to Generate Predictions Based on the Information Extracted from Automobile Ads

Stere Caciandone[1] and Costin-Gabriel Chiru[2(✉)]

[1] University Politehnica Bucharest,
Splaiul Independenței 313, Bucharest, Romania
stere.caciandone@gmail.com
[2] ADAPTIVEBEE S.R.L., 64-66 Dionisie Lupu Str., Bucharest, Romania
costin.chiru@iminent.com

Abstract. In this paper we address the issue of predicting the reselling price of cars based on ads extracted from popular websites for reselling cars. To obtain the most accurate predictions, we have used two machine learning algorithms (multiple linear regression and random forest) to build multiple models to reflect the importance of different combinations of features in the final price of the cars. The predictions are generated based on the models trained on the ads extracted from such sites. The developed system provides the user with an interface that allows navigation through ads to assess the fairness of prices compared to the predicted ones.

Keywords: Multiple linear regression · Random forest · Feature selection · Car reselling price prediction

1 Introduction

The idea of using machine learning to classify objects or make different decisions under certain circumstances is in fact popular today. There are many applications of predictive algorithms which achieve high accuracy in predicting values in different domains: from applications in the economic field, where they may be used to determine the future value of goods [1], to the medical area, where they are used for example in Epidemiology [2], to meteorology, to forecast the weather in time, following the analysis of data from several decades, or to predict the outcome of different sport events [3].

The aim of this research was to predict the resale value of cars, particularly considering the fact that the Romanian market is dominated by previously owned cars: for example, between January–February 2014, there were registered 30,600 previously owned cars, compared to only 8,770 new cars [4]. In each of the previous 5 years, the number of second-hand cars that were registered exceeded the number of new ones, in some years the difference between them being substantial (for example, in 2013 the number of registered second-hand cars is almost 4 times higher than that of new cars).

According to this data, we can say that in Romania the number of people who purchase second-hand cars is considerably higher than the one of new cars buyers.

C. Dichev and G. Agre (Eds.): AIMSA 2016, LNAI 9883, pp. 36–45, 2016.
DOI: 10.1007/978-3-319-44748-3_4

Thus, potential customers need more information to determine whether the asked price for the car they are interested in reflects its value or is unreasonably high.

Regarding the car brands that are registered, the most popular brands of registered cars made abroad are German brands, the main representative being Volkswagen with a share of 25.8 %, followed by Opel with 17.7 %, Ford with 11.7 % and BMW with 6 % [4]. The national brand, Dacia, is leading the new cars market with a 30 % share and it ranks on the third position in the second-hand cars market with a 10 % share.

Considering the above aspects, the current work aims to analyze the car sale ads to determine the links between the price of a car and its features using two different machine learning algorithms: multiple linear regression and random forest. In order to do that, we first analyzed the individual relationship of each feature to the car price to see if it was relevant or not. Following this step, we assigned different weights to the relevant features, obtaining a rule that provided better predictions than when only individual features were used.

For data collection we used one of the largest online add platform for cars resale in Romania (autovit.ro). This platform allowed us to mine the data needed to build a data model with all the characteristics that we required.

The paper continues with the presentation of the similar approaches. Then, we present the application architecture and the data that was used in our experiments. Afterwards, we describe a case study and the obtained results. The paper ends with our final conclusions.

2 Similar Approaches

As already mentioned, similar analysis were made in different domains, such as trying to predict the stock price, house prices, gas prices or the level of wages/salaries. An approach similar to ours had the purpose to predict if the gold price for the next day will be higher or lower than the one from the current day [5]. To achieve this, the author used logical regression [6]. This approach used 25 features which were considered relevant for the gold price (US Stock Indices, World Stock Indices, COMEX Futures, Forex Rates, Bond rates, Dollar Index, etc.) and built a model that resulted in a 54 % accuracy rate. Observing that the precision was not very good, the author inspected the 25 features and discarded the noisy ones, remaining with only 10 features. Using these features and the linear regression model (LRM), the precision increased to 69.30 %. Applying support vector machines (SVM), a different learning algorithm, with the same features, lead to 69.08 % precision.

In the domain of car prices prediction there are similar approaches that are using automated learning algorithms. One approach investigated the market from the United States with the goal of predicting the selling price of a certain model (Corolla) from a certain brand (Toyota) [7]. Initially, the author made a preliminary set of observations, to see which features are important to predict the price and which are irrelevant. The analysis was made on a set of 1000 cars, and after applying linear regression a prediction with an error of 9.2 % was obtained, concluding that this is a good approach to predict the price.

Another study had the goal to evaluate the vehicles from the Mauritius market using automated learning techniques [8]. This paper used multiple automated learning techniques such as: Multiple Linear Regression, K-Nearest Neighbors, Decision Trees and Naïve Bayes in the attempt to make a good prediction. The author discovered certain limits in applying some of the algorithms caused by the data shortage; for example, for many announces the number of kilometers of the car was missing, which was an important feature to apply the multiple linear regression algorithm. The conclusion was that the small size of data and its poor quality limited the analysis on this subject and that a bigger database was needed in order to apply more complex algorithms, such as neural networks, genetic algorithms or fuzzy logic.

A comparison between the automated learning algorithms is made in [9], where the authors observe that non-linear learning algorithms are superior to the linear ones because they are more general. Thus, independently to the brand and model of the vehicle, the non-linear algorithms generate a single rule, while the linear ones need specific approaches to generate predictions for each brand and model in part. However, the linear algorithms obtain superior results if they are used for each brand and model in part. The conclusion of the study was that it is better to use the random forest learning algorithm for prediction, as this algorithm generates a single formula for all the models of all brands, making it easier to build the model, monitor and maintain it.

3   Application Architecture

The developed application has several steps. In the first step, the data needed for the prediction algorithms is collected and saved in a database.

Afterwards, two different machine learning algorithms are used for evaluating car resale prices. The first algorithm is multiple linear regression, one of the most popular algorithms for predicting car resale prices, according to what we could find in the analyzed literature. The second one is random forest, as suggested by the research done in [9]. Since both these algorithms require lots of data in order to be accurate, we must be sure that we have the needed data available, in order to avoid problems like overfitting or being unable to learn a good model, which translates in low accuracy.

The next step is to present the results to the user. This is done using a web interface that allows the user to see the price estimated by the machine learning algorithms and notifies them if the price is fair, underrated or too high. At the same time, it allows the user to enquire the system about the price for a particular model of car, with applicability for the case (s)he is interested in a possible assessment of her/his own car. Based on the data from the database and the models learnt by the machine learning algorithms, the system will return the value of the input car. Moreover, the user will be able to send to her/his e-mail account links with ads that have been identified as having fairly priced/undervalued cars if (s)he is interested in them.

4   Used Data

As already mentioned, quantitative algorithms, such is the case of linear regression or random forest, require large sets of training data in order to achieve the best possible prediction. In order to meet this need, the first step of our research was to acquire a large enough database of car resale announcements. Thus, we mined the content of the autovit.ro website, and we extracted from it over 16,000 ads. To retrieve the data from the site we used a web crawler (scrappy) implemented in Python.

The next step was sorting and cleaning the data, thus removing the ads that did not contain all the necessary data (for example, they didn't specify the number of kilometers/horsepower) and the outliers - announcements about selling damaged cars for parts or ones containing different mistypes.

After these steps, the database consisted of 15,500 ads that met the basic criteria: all the features were set and we consider the ad not to be an outlier. The entries from the database contain the following information about the car: price, year of manufacture, mileage, horsepower, engine capacity, fuel type, make, model, transmission norm euro, color, number of doors and two other items that are not needed for the algorithm, but are required for the results interface: the link to the page containing the image of the car and the one used for the sale announcement.

5   Case Study

In order to achieve a predictive model as accurately as possible we must consider all the features that we have available. However, it is possible that not all the features will be relevant in determining the price of a car. Thus, in a first step, we need to identify on which features the price is linearly dependent. For this we will chose a model that has a large enough number of ads in the database and we will plot its features to see if they are in a linear relationship with the price or not.

Since the linear regression algorithm is a very specific algorithm, it must be used for a specific brand and model. During the training, the algorithm is building a model that will be afterwards used for prediction. Thus, for this case study we had to choose a model from the database with a large enough number of ads. We decided to select the Volkswagen brand, which is one of the most popular car on the market, while the chosen model was the Passat, which had 820 ads in the database.

5.1   The Features Analysis

The first feature that was analyzed is the car mileage and, as it can be seen in the graph in Fig. 1a, there is a linear relationship between the mileage of a car and its price (the greater the number of kilometers the lower the price drops). After applying simple linear regression to obtain a coefficient for the mileage feature, as expected, a negative coefficient of -0.069 is obtained. The resulting predicting formula would then be given by (1).

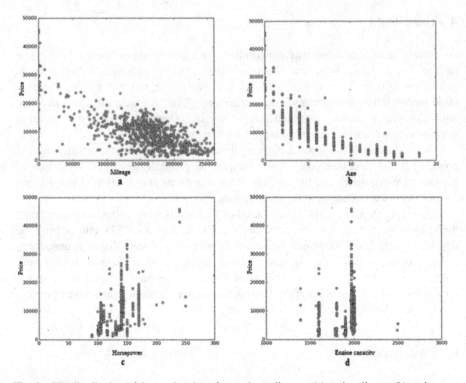

Fig. 1. The distribution of the resale price of a car depending on: (a) car's mileage; (b) car's age; (c) car's horsepower and (d) car's engine capacity.

$$Price \; = \; Base \; price \; + \; No. \; km. \; * \; (-0.069) \qquad (1)$$

The coefficient of determination (R^2) of (1) is 0.45, which means that the mileage is an important feature that we will use further in our model. R^2 has a value between 0 and 1; the closer it is to 1, the better is the model's accuracy.

The next analyzed feature was the car's horsepower. In Fig. 1b can be seen that the car becomes more expensive as it has more horsepower. Applying linear regression, we obtain a coefficient of 114.47 for this feature. Consequently, the dependence of price on the horsepower is given by (2).

$$Price \; = \; Base \; price \; + \; Horsepower \; * \; 114.4 \qquad (2)$$

The R^2 score for this dependence is 0.2, indicating that there is a weak correlation between them. This feature will also be used in the final model.

The age of the car is the next feature we inspected. We can see that the graph from Fig. 1c shows a direct relationship between age and price. As the car is older, its price decreases. Applying simple linear regression a coefficient of −1118.89 was obtained for this criterion and a R^2 score of 0.61 for the prediction, which indicates a very good correlation between these two features. Based on this criterion, the price should be computed as in (3):

$$Price = Base\ price + Age * (-1118.89) \qquad (3)$$

The next feature to be evaluated was the cylinder capacity of the vehicle. Apparently, the engine capacity does not have a linear relationship with the price, as we can see from Fig. 1d. When applying simple linear regression for it, we obtain a value of 5.58 for the coefficient, and an R^2 prediction score of 0.027, indicating that there is no correlation between the car's cylinder capacity and its price. Therefore, this criterion should not be included in the final model. However we observe that the cars with a greater cilinder capacity tend to be more expensive. It is possible that since a substantial part of the car models that were considered have a cilindrical capacity of 2,000 cm^3, this criterion becomes irrelevant in the analysis. We will still investigate how its presence or absence will influence the accuracy of the final model.

The other features that we investigated were the colour of the vehicle, the amount of gas emissions, the type of fuel the car uses and the transmission type. When applying simple linear regression for each of them, we obtained the follwing R^2 scores: 0.003 for colour, 0.011 for the amount of gas emissions, 0.049 for the type of fuel the car uses, and 0.13 for the transmission type. Even though the first 3 features have a too low correlation to be considered in the final model, except for the colour, we will use them in order to analyze their impact on the overall prediction. The fourth feature will also be used in the final model.

Next, we will analyze the results obtained using the selected features for determining the selling price of a Volkswagen Passat car and see how the features used in the prediction model influence the system's accuracy.

5.2   Building the Model with the Finest Accuracy

We used the following features for the first prediction model: the age of the car, mileage, horsepower, engine capacity, fuel type (petrol or diesel) and the type of transmission (automatic, manual) and obtained the coefficients corresponding to these features shown in (4). It should be noted that the starting price was evaluated at 16,240 euro and that the transmission and fuel type had binary values (0 or 1), while the other features had natural values.

$$Price = 16240 + Age * (-974.86) + No.\ km. * (-0.04) + Horse\ power * 54.29 +$$
$$Diesel\ Fuel * 887.53 + Petrol\ Fuel * (-887.53) + Manual\ transm. * (-646.88) +$$
$$Automated\ transm. * 424.31$$

$$(4)$$

The prediction model was tested using 10-folds cross-validation method (repeatedly trained on 90 % of the ads and tested on the remaining ones, until all the data was used as test data one time). The obtained average accuracy was 70.9 %, which gives a decent prediction model. Next, we will try to obtain the best accuracy that we can get using different subsets of these features.

The first subset consisted of the two most relevant features: mileage and age. Using multiple linear regression, we obtained the coefficients from (5).

$$\text{Price} = 23055 + \text{Age} * (-1108.55) + \text{No. km.} * (-0.04) \qquad (5)$$

After cross-validation, we obtained an average of 67.8 % accuracy, so this model is less accurate than the original one, the two features failling to predict the price as well as the complete set of features. We continued our analysis by adding, besides these two features, the horsepower. This time, the prediction accuracy was 71.2 %, which was slightly better than the original model. By also adding to this model the the transmission type, the accuracy continued to improve to 71.4 %. The coefficients of the new model are the ones from (6).

$$\text{Price} = 15040 + \text{Age} * (-1047.05) + \text{No. km.} * (-0.038) + \text{Horsepower} * 49.3$$
$$+ \text{Automated transm.} * 974.07 + \text{Manual transm.} * 19.01$$

$$(6)$$

When introducing the type of fuel in the prediction model, the average accuracy sligthly dropped to 71.3 %. Therefore, we supposed that this feature does not help in improving the model and we removed it from the model, while continuing to add other features to it. The next feature added to the model, the engine capacity feature, also lead to a drop in accuracy to 71.1 %, so it was also removed.

From (6) it can be seen that the best prediction model so far is containing the features that have achieved the best scores in the individual simple linear regression analysis between them and the price (age, mileage, horsepower and type of transmission). This model's accuracy is 71.4 % using cross-validation.

However, this is not the best model that can be built with this dataset. An additional improvement can be made by changing the values given for the age feature: as it can be seen from Fig. 1b, there is a logarithmic relationship between age and price. Thus, we applied again multiple linear regression on the most succesful data model obtained so far to see if by using the logarithm of the age feature instead of the actual values improves the accuracy or not. An average accuracy of 80.6 % was obtained, which means that we brought a significant improvement in the prediction model (by more than 9 %).

Encouraged by this improvement, we applied the same principle on the feature describing the mileage of the car starting from the observation that apart from a few outliers, there is also a logarithmic relationship between this feature and the cars prices. The obtained average accuracy for the cross-validation testing on this new model was 82.3 %, which means an additional improvement of its predecessor by almost 2 %. This was the best result that we obtained using the considered features.

In order to validate the correctness of the improved model, we applied the obtained models on some different brands/models of cars. We will further use the term "initial prediction model" to denote the model using the age, mileage, horsepower and type of transmission and the term of "improved prediction model" for the same model, but on which we applied logarithm on the features age and mileage.

For the validation, we have chosen the car models that have a significant number of ads in order to provide a consistent basis for training and testing.

The first car model that we validated the models against was Volkswagen Golf, for which we had 945 ads in the database. The average accuracy obtained using the initial prediction model was 77 %, while for the improved one it was 86.6 %, which is almost 10 % better than the original. The validation continued with Audi A4, which had 509 ads. After running cross-validation we obtained 78.2 % accuracy for the initial prediction model and 87 % for the improved one. Again, we notice a difference in favor of the improved model of almost 9 %. The next car model for validation was Skoda Octavia, for which we had a training base of 398 ads. For the initial prediction model an accuracy of 65.8 % was obtained, and a 75.1 % accuracy for the improved model. Again this highlights the advantage of the improved model that makes a prediction by almost 10 % more accurate than the original model. In the case of Opel Astra, for which we had 632 ads, the obtained average accuracy was 75.1 % for the initial model and 81.1 % for the improved prediction model. Same tendency was seen for Ford Focus (474 ads), even though the type of transmission was excluded from the prediction model because there were no models with automatic transmission. The average accuracy was 69.8 % for the first prediction model and 82.7 % for the improved model.

As it can be seen, the improved prediction model gets good accuracy rates of over 80 % for some models. Furthermore, we intend to analyze how much the data set size influence the prediction model performance.

To test the power of the prediction model, we needed a test set of at least 50 + 8 * m data, where m is the number of predictors used, while for a single feature we need at least 104 + m data samples, according to the recommendations from [10]. Thus, in our case the minimum number of ads that we could fully test the model against was 82, obtained when m = 4 (it should be noted that manual/automatic features represent in fact a single predictor - transmission type - having binary values).

Therefore, for testing the accuracy of our multiple linear regression prediction model we will analyze only the vehicle models for which we have at least 82 ads in our database.

Similar tests were done for the random forest regression algorithm and we noticed that the model that obtains the best score is the one that uses all the features in the prediction. Besides these features, the available implementation in scikit takes the brand and vehicle model as separate features and thus there is no longer needed to run the algorithm for each vehicle model separately.

6 Obtained Results

The results obtained for the two algorithms are shown in Tables 1 and 2. For multiple linear regression algorithm, we used the improved model consisting of log(age), log (mileage), horsepower and type of transmission and applied it only for the car models having at least 82 ads.

Table 1. Results obtained using the improved multiple linear regression algorithm.

Brand, model	Average	Brand, model	Average	Brand, model	Average
Audi A3	88.5 %	BMW X5	78.4 %	Skoda Fabia	90,4 %
Audi A4	87 %	BMW X6	72.1 %	Opel Astra	81,1 %
Audi A5	77.3 %	Ford Focus	82,7 %	Opel Corsa	89,6 %
Audi A6	84.2 %	Ford Fiesta	83,1 %	Mercedes C class	85.6 %
BMW Seria 3	76.3 %	Ford Mondeo	78,7	Mercedes S class	82.1 %
BMW Seria 5	74.2 %	Skoda Octavia	77,2 %	Renault Megane	76.2 %
VW Golf	86,6 %	VW Touareg	83,9 %	Dacia Logan	72.1 %
VW Passat	82,3 %	VW Polo	87.9 %	Dacia Sandero	68.1 %

Table 2. Results obtained using the random forest regression algorithm.

Brand	Audi	Bmw	Ford	Dacia	Mercedes	Volkswagen	Skoda	Renault
Avg.	93.3 %	92.2 %	90.8 %	89.3 %	92.3 %	93.3 %	90.9 %	91.2 %

7   Conclusion

We obtained good results by applying multiple linear regression on car models having a large number of ads (over 200). We found that the prediction accuracy is good (somewhere between 80–90 %) and that the car models with a little over the minimum number of data samples required (as mentioned in the case study) had a prediction average accuracy of at least 70 %. For the rest of car models, that did not have sufficient number of ads, the algorithm did not achieved satisfactory results, thus highlighting the need for a large data set for training and testing and confirming the formula suggested in [10].

Predictions achieved by the random forest regression were much better - around 90 %. The advantage of this algorithm being the ease with which it can be applied to all car models of the analyzed brands and how it elaborates the prediction based on the average predictions obtained from all its trees.

In conclusion we can say that these algorithms are useful in determining the selling price of a car, the most important part being in the case of multiple linear regression the choice of the features with which the prediction model is built and observing the type of relationship between the independent and dependent features (logarithmic, linear, exponential, polynomial, etc.) as this relation type proved to greatly improve the obtained resutls. Applying algorithms of this type can be extended to all cases in which it appears that there is a correlation between different features.

References

1. Baumeister, C., Kilian, L., Lee, T.K.: Inside the crystal ball: new approaches to predicting the gasoline price at the pump. CFS Working Paper No. 500 (2015). doi:10.2139/ssrn. 2550731
2. Dasgupta, A., Sun, Y.V., König, I.R., Bailey-Wilson, J.E., Malley, J.D.: Brief review of regression-based and machine learning methods in genetic epidemiology: the Genetic Analysis Workshop 17 experience. Genet. Epidemiol. 35(S1), S5–S11 (2011)
3. Aditya, S.T., Aditya, P., Vikesh, K.: Game ON! predicting english premier league match outcomes. Stanford University, Project for course CS229 (2013)
4. Information Portal of the Directorate for Driving Licenses and Vehicle Registrations. http://www.drpciv.ro/info-portal/displayStatistics.do
5. Potoski, M.: Predicting gold prices. Stanford University. Project for course CS229 (2013)
6. Cox, D.R.: The regression analysis of binary sequences. J. R. Stat. Soc. Ser. B 820, 215–242 (1958)
7. Chen, P.: Predicting Car Prices Part 1: Linear Regression. www.datasciencecentral.com/profiles/blogs/predicting-car-prices-part-1-linear-regression
8. Pudaruth, S.: Predicting the price of used cars using machine learning techniques. Int. J. Inf. Comput. Technol. 4(7), 753–764 (2014)
9. Voß, S., Lessmann, S.: Resale price prediction in the used car market. Tristan Symposium VIII (2013)
10. Tabachnick, B.G., Fidell, L.S.: Using Multivariate Statistics, 6th edn. Pearson Education Limited, Boston (2013)

Estimating the Accuracy of Spectral Learning for HMMs

Farhana Ferdousi Liza$^{(\boxtimes)}$ and Marek Grześ$^{(\boxtimes)}$

School of Computing, University of Kent, Canterbury CT2 7NZ, UK
{fl207,mgrzes}@kent.ac.uk

Abstract. Hidden Markov models (HMMs) are usually learned using the expectation maximisation algorithm which is, unfortunately, subject to local optima. Spectral learning for HMMs provides a unique, optimal solution subject to availability of a sufficient amount of data. However, with access to limited data, there is no means of estimating the accuracy of the solution of a given model. In this paper, a new spectral evaluation method has been proposed which can be used to assess whether the algorithm is converging to a stable solution on a given dataset. The proposed method is designed for real-life datasets where the true model is not available. A number of empirical experiments on synthetic as well as real datasets indicate that our criterion is an accurate proxy to measure quality of models learned using spectral learning.

Keywords: Spectral learning · HMM · SVD · Evaluation technique

1 Introduction

Learning parameters of dynamical systems and latent variable models using spectral learning algorithms is fascinating because of their capability of globally optimal parameter estimation. The inherent problem of local optima in many existing local search methods, such as Expectation Maximisation (EM), Gradient Descent, Gibbs Sampling, or Metropolis Hastings, led to the development of the spectral learning algorithms which are based on the method of moments (MoM). The goal of the MoM is to estimate the parameters, θ, of a probabilistic model, $p(x|\theta)$, from training data, $X = \{x_n\}$ where $n = 1 \ldots N$. The basic idea is to compute several sample moments (empirical moments) of the model on the training data, $\phi_i(X) = \frac{1}{N} \sum_{n=1}^{N} f_i(x_n)$, and then to alter the parameters so that the expected moments under the model, $\langle f_i(x) \rangle_{p(x|\theta)} = \int f_i(x)p(x|\theta)dx$, are identical with the empirical moments, that is $\phi_i(X) = \langle f_i(x) \rangle_{p(x|\theta)}$. In short, the method of moments involves equating empirical moments with theoretical moments.

Hsu et al. [9] proposed an efficient and accurate MoM-based algorithm for discrete Hidden Markov Models (HMMs) that provides a theoretical guarantee for a unique and globally optimal parameter estimation. However, the algorithms that are based on MoM require large amounts of data to equate empirical moments with the theoretical moments [7,8].

© Springer International Publishing Switzerland 2016
C. Dichev and G. Agre (Eds.): AIMSA 2016, LNAI 9883, pp. 46–56, 2016.
DOI: 10.1007/978-3-319-44748-3_5

In real-life experiments, the practitioners would like to know what are the minimal data requirements that can guarantee that a particular model is learned near-optimally. When spectral learning does not have access to a sufficient amount of data, the estimates may be far from the global optima, and in some cases, the parameter estimates may lie outside the domain of the parameter space [6,15]. The practitioners are not able to judge how far their parameters can be from the optimal solution of a given model when the true model is not available, which is a normal scenario in practice. As a consequence, when the empirical model learned using spectral learning does not perform well, the practitioner does not know whether the solution is sub-optimal and the model is still correct, or whether the model is simply wrong. In this paper, we design a new method that can approximate the convergence of spectral learning.

The contribution of this work is that we provide a way to verify whether a particular dataset is sufficient to train a HMM using spectral learning [9]. In the current big data era, the proposed criterion can also be deployed in a system where there is possibility of more incoming data over time. In particular, the proposed measure can comfortably be incorporated into the online spectral learning algorithms, such as [2]. A number of authors have been dealing with the other features of spectral leaning, such as low rank, scalability and insufficient statistics; however, to the best of our knowledge, there is no work where the basis vector rotation-based measure would be used as a proxy method to estimate the convergence of spectral learning algorithms.

2   Background

A hidden Markov model (HMM) is a probabilistic system that models Markov processes with unobserved (hidden) states. A discrete HMM can be defined as follows: let $(x_1, x_2, \dots)$ be a sequence of discrete observations from a HMM where $x_t \in \{1, \dots, n\}$ is the observation, and $(h_1, h_2, \dots)$ be a sequence of hidden states where $h_t \in \{1, \dots, m\}$ is the hidden state at time step t. The parameters of an HMM are $\langle \pi, T, O \rangle$ where $\pi \in \mathbb{R}^m$ is the initial state distribution, $T \in \mathbb{R}^{m \times m}$ is the transition matrix, and $O \in \mathbb{R}^{n \times m}$ is the observation matrix. $T(i, j) = P(i|j)$ is the probability of going to state i if the current state is j, $\pi(j)$ is the probability of starting in state j, and $O(q, j) = P(q|j)$ is the probability of emitting symbol q if the current state is j.

In general, the term 'spectral' refers to the use of eigenvalues, eigenvectors, singular values and singular vectors. Singular values and singular vectors can be obtained from the singular value decomposition (SVD) [1]. The SVD of a matrix A is a factorisation of A into a product of three matrices $A = UDV^\top$ where the columns of U and V are orthonormal and the matrix D is diagonal with positive real entries known as singular values. U and V are called left and right singular vectors. The k dimensional subspace that best fits the data in A can be specified by the top k left singular vectors in matrix U. The spectral algorithm for HMMs uses the SVD to retrieve a tractable subspace where the hidden state dynamics are preserved [9].

A convenient way to calculate the probability of a HMM sequence $(x_1, x_2, \ldots, x_t)$ using the matrix operator [5,10], $A_x = T \text{diag}(O_{x,1}, \ldots, O_{x,m})$ for $x = 1, \ldots, n$, is as follows:

$$Pr(x_1, \ldots, x_t) = 1_m^\top A_{x_t} \ldots A_{x_1} \pi. \tag{2.1}$$

The spectral learning algorithm for HMMs learns a representation that is based on this observable operator view of HMMs; this is an observable view because every A_x represents state transitions for a given observation x. However, in this case, the set of 'characteristic events' revealing a relationship between the hidden states and observations have to be known or estimated from data that requires the knowledge of the T and O matrices. For a real dataset, we don't have exact T and O matrices. To relax this requirement, Hsu et al. [9] has used a transformed set of operators (in a tractable subspace) based on low order empirical moments of the data. In practice, the following empirical moment matrices have to be estimated from the data:

$$P_1 \in \mathbb{R}^n \quad [P_1]_i = Pr(x_1 = i) \tag{2.2}$$

$$P_{2,1} \in \mathbb{R}^{n \times n} \quad [P_{2,1}]_{ij} = Pr(x_2 = i, x_1 = j) \tag{2.3}$$

$$P_{3,x,1} \in \mathbb{R}^{n \times n} \quad [P_{3,x,1}]_{ij} = Pr(x_3 = i, x_2 = x, x_1 = j). \tag{2.4}$$

The resulting transformed operators for the HMM are then computed as follows:

$$\hat{b}_1 = \hat{U}^\top \hat{P}_1; \ \hat{b}_\infty = (\hat{P}_{2,1}^\top \hat{U})^+ \hat{P}_1; \ \hat{B}_x = \hat{U}^\top \hat{P}_{3,x,1}(\hat{U}^\top \hat{P}_{2,1})^+ \ \forall x \in [n] \tag{2.5}$$

where $\hat{U}$ is computed by performing SVD on the correlation matrix $P_{2,1}$. If T and O in the underlying HMM are of rank m and $\hat{\pi} > 0$, then it can be shown that

$$\hat{Pr}(x_1, \ldots, x_t) = b_\infty^\top B_{x_t} \ldots B_{x_1} b_1 \tag{2.6}$$

which means that using the parameters learned by the spectral learning algorithm, a full joint probability of a sequence can be computed without knowing the exact T and O matrices. In a real situation, one does not have the optimal empirical moments and has to approximate the moment matrices, $\hat{P}_1$, $\hat{P}_{2,1}$ and $\hat{P}_{3,x,1}$, from a finite amount of data and consequently to approximate the operators $\hat{b}_1, \hat{b}_\infty$ and $\hat{B}_x$. Hsu et al. [9] has proved that the joint probability estimates of a HMM sequence are consistent (i.e. they converge to the correct model) when the number (N) of sampled observations tends to infinity:

$$\lim_{N \to \infty} \sum_{x_1, \ldots, x_t} |Pr(x_1, \ldots, x_t) - \hat{Pr}(x_1, \ldots, x_t)| = 0.$$

3 Main Method

Based on our empirical observations, in this paper, we have proposed a convergence criterion for a spectral learning algorithm for HMMs [9] with finite data.

Since the true convergence cannot be determined with certainty when the true model is unavailable, our method is a proxy measure that approximates the difference from the true model. Technically speaking, here convergence means that the minimal training data requirement is satisfied to yield empirical moments that are sufficiently close to the real moments. Note that when the essential amount of data is not available, the parameter estimates based on empirical moments may not even be in the domain of the parameter space (e.g. probabilities can be negative [6,15]). Our observations lead to a straightforward methodology that can approximate whether the algorithm has access to a sufficient amount of data. Specifically, we apply the spectral learning algorithm on a number of sub-sets of the training data where the size of the sub-sets is increased in subsequent iterations and each subset contains data of the previous sub-set to observe the effect of the increasing dataset. The spectral learning solution uses one of the orthonormal matrices, $\hat{U}$, as described in the previous section, to define a solution to the overall learning problem. Our main method is based on an observation that the bases contained in $\hat{U}$ rotate when the algorithm is executed on different sub-sets of the training data. By rotation, we mean the angle change between any two subsequent basis vectors contained in corresponding $\hat{U}$. Note that every sub-set defines one basis. The key point is that the magnitude of those rotations diminishes when the size of the sub-set grows. In Sect. 5, experimental results in (Fig. 2) confirm this claim, where the angle change differences become smaller as the size of the training sub-sets becomes larger.

Therefore, our hypothesis is that the magnitude of those rotations (measured as an angle change difference between two successive basis) can be a good proxy to determine the convergence (or equivalently data sufficiency) of the spectral learning algorithm. In order to justify our hypothesis, we show empirically on synthetic data that the learned model is an accurate approximation of the true model, when the angle that quantifies the magnitude of the rotations of the successive bases is sufficiently small. In short, we show empirically that when the rotation is small, the error is small as well. So, we can treat the rotation as a proxy to quantify the error.

In our approach, the original one-shot spectral learning algorithm for HMMs has been converted to a multi-step procedure for multiple training sub-sets described above. When the basis rotation between two successive corresponding basis vectors is less than a required value (e.g. 10^{-5}), our empirical experiments show that the spectral learning solution converges to a stable solution in the parameter space. However, if the required rotation (angle change difference) between two corresponding bases is not achievable with the training data at hand, then the spectral learning solution cannot be considered as reliable, and in that case another suitable parameter estimation technique should be used for the task. On convergence, the original spectral algorithm should be applied to the whole dataset to compute the final parameters. The rotation (angle change difference) is calculated using the maximum value of the dot product between each successive corresponding basis vectors in $\hat{U}$.

The next sections will show empirical evidence that smaller basis rotations are correlated with the real error on data when the true model is known, and therefore the magnitude of the basis rotation can be considered as a proxy that can assess the quality of the learned parameters for a particular model.

4   Experimental Methodology

4.1   Evaluation

In order to seek empirical evidence to support our hypothesis, we need a notion of an error function that can measure the quality of the HMM model learned from data, and we want to show empirically that our proxy measure is correlated with that error function. Then, we will conclude that our proxy method is a good indicator of the quality of the learned model on real data where the true error cannot be computed because the ground truth is not known.

Zhao and Poupart [15] used a normalized L_1 error that uses the sum of tth roots of absolute errors, where t is the length of test sequences and τ is the set of all test sequences. This approach relies on the probability of seeing a certain sequence of outputs, $Pr(x_1, \ldots, x_t)$.

$$L_1 = \sum_{(x_1, \ldots, x_t) \in \tau} |Pr(x_1, \ldots, x_t) - \hat{Pr}(x_1, \ldots, x_t)|^{\frac{1}{t}}. \tag{4.1}$$

The error bounds for spectral learning in HMMs were derived in Hsu et al. [9, Sect. 4.2.3] for a similar measure. Moreover, unlike other approaches, such as Kullback-Leibler divergence [14], this method does not use a logarithm in its computation, and is robust against negative probabilities.

Spectral learning for HMMs [9] uses the transformed operators and Eq. 2.6 to calculate joint probabilities. Certainly, when one knows the exact model (which is true in the case of synthetic HMMs), the exact probability $Pr(x_1, ..., x_t)$ can be calculated using Eq. 2.1, which either involves multiplication of the exact transition, T, and emission matrices, O, or combined matrices, A_x. Calculating the error by comparing such exact measures reveals how close the estimated model is to the exact model. The normalized L_1 error serves this purpose without using the exact T and O because, as shown in [15], it uses probabilities of sequences and handles negative values. As a result, this leads to a measure that can indicate whether a model is well-fitted or not. We use this error to show that the angle change difference can indicate that the model is well-fitted. For a model to be well-fitted, in the spectral algorithms, the empirical moments have to be sufficient. In that sense, the angle change difference can also be a valid indication of the sufficiency of empirical moments.

4.2   Experimental Settings

The performance of HMM learning algorithms, in general, depends on the linear independence of the rows of $(T$ and $O)$ for a maximum discrimination of the state and observation symbols. However, as T and O are inextricably linked to the model execution, [3] defined Inverse Condition Number (ICN) for indicating the linear independence of T and O. ICN was calculated as a ratio between the smallest and largest singular value of a row augmented matrix of T and O. A row augmented matrix is a matrix obtained by appending the columns of two given matrices. Such ICN was also demonstrated in [4] with Local Search method (LM) based parameter estimation techniques for HMMs. If ICN is close to 1 then the HMM is well-conditioned; if the ICN is close to 0 then HMM is ill-conditioned. While experimenting with the synthetic HMM systems, ICN was used to verify how our proposed convergence measure works with ill-conditioned HMMs.

Table 1. Description of Benchmark and some random HMMs (here Ex. or ex. are abbreviation of the word Example. Ex. are used in our analysis and plots)

HMM	Reference	ICN	HMM	Reference	ICN
Ex.1 ($m = 2$, $n = 3$)	[11, p. 26 ex. 1]	0.5731	Ex.2 ($m = 2$, $n = 6$)	[11, p. 40 ex. 2]	0.7338
Ex. 3 ($m = 2$, $n = 2$)	[12, p. 79 ex. 1]	0.5881	Ex. 4 ($m = 2$, $n = 10$)	Random	0.6756
Ex. 5 ($m = 3$, $n = 8$)	[11, p. 26 ex. 3]	0.6158	Ex. 6 ($m = 3$, $n = 3$)	Random	0.6101
Ex. 7 ($m = 3$, $n = 10$)	[11, p. 26 ex. 4]	0.6219	Ex. 8 ($m = 3$, $n = 3$)	[12, p. 80 ex. 4]	0.2070
Ex. 9 ($m = 3$, $n = 3$)	[12, p. 80 ex. 5]	0.3305	Ex. 10 ($m = 3$, $n = 3$)	[12, p. 81 ex. 6]	0.4612
Ex. 11 ($m = 3$, $n = 3$)	[12, p. 81 ex. 7]	0.3678	Ex. 12 ($m = 3$, $n = 3$)	[12, p. 79 ex. 2]	0.3715
Ex. 13 ($m = 3$, $n = 10$)	Random	0.6355	Ex. 14 ($m = 3$, $n = 12$)	Random	0.6528
Ex. 15 ($m = 3$, $n = 20$)	Random	0.5475	Ex. 16 ($m = 2$, $n = 2$)	Random	0.9800
Ex. 17 ($m = 2$, $n = 2$)	Random	0.1800	Ex. 18 ($m = 2$, $n = 2$)	Random	0.0200

The proposed convergence measure was tested on both synthetic datasets and on one real dataset. The normalised L_1 error can be computed on synthetic data only since the true model is required. The synthetic datasets are the HMM benchmarks used in [11–13]. All benchmark HMMs are summarised in the Table 1 where HMM column has the number of states (m) and the number of possible observations (n) used in the experiment, the Reference column has the source of O and T matrices, and the ICN column has the value of the ICN for the corresponding HMM. In our analysis, we have also used additional random HMM systems generated by rejection sampling method to obtain different ICN values.

The observation triples for training [9] were generated by sampling from the corresponding HMMs. The real dataset was arranged into triples using a sliding window approach. This dataset is based on web-navigation data from msnbc.com[1] and consists of 989818 time-series sequences and 17 observable symbols. The testing data to compute the L_1 error for every synthetic HMM consists of 20000 observation sequences of length 50.

To show that the basis vector angle change difference can be a good indication of the sufficiency of the empirical moments, each HMM was trained incrementally (with subsets of training data) for different required maximum angle change difference (θ), and for each case the normalised L_1 error was calculated.

---

[1] https://archive.ics.uci.edu/ml/datasets/MSNBC.com+Anonymous+Web+Data.

In all our experiments, we added 20 additional training examples in each subsequent sub-set. Our goal is to show that, when maximum requirement for angle change difference is smaller, L_1 is also smaller on well-conditioned HMMs. The chosen θ were 30, 10, 5, 2, 1, 0.5, 0.1, 0.01, 0.001, 0.0001, and 0.00001 degrees. To visualise and validate the convergence on smaller θs, the T and O matrices were recovered from spectral parameters using Appendix C of [9]. The next section will show the quality of those matrices as a function the ICN and the required maximum angle change difference.

5   Experimental Results

The L_1 error was calculated for different θ values for each benchmark HMM system as described in the previous section. When θ is large (30, 10, 5, 2, 1, 0.5, 0.1, 0.01), the spectral learning solution of joint probability generates many negative probabilities for test sequences and the error pattern is thus inconclusive. As a negative probability for a sequence makes L_1 larger, for large θs, small spikes can be seen in Fig. 1b. However, for well-conditioned HMMs, the error reduces when the θ becomes smaller (0.001, 0.0001, and 0.00001 in Fig. 1b). This indicates that a smaller θ implies a well-fitted model. In a practical application where the exact model is not known, our angle change measure can inform a practitioner about the quality of the current parameters for her model. For the same experiment, the basis vector angle change difference for different θ were plotted in (Fig. 2). To achieve a larger θ, a small training dataset is sufficient. However, to achieve a small θ, comparatively larger training data is required. Thus, the angle change difference is correlated with training data requirements as well as the L_1 measure of correctness. A similar result of the angle change difference for different θ was found on all other synthetic HMMs systems and on a real dataset.

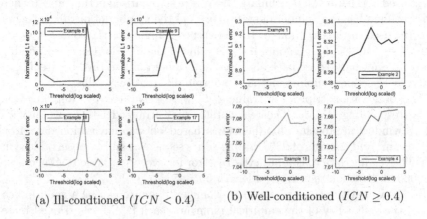

(a) Ill-condtioned ($ICN < 0.4$)   (b) Well-conditioned ($ICN \geq 0.4$)

Fig. 1. L_1 error with chosen θs for different example HMMs

Table 2. Recovered T and O matrices for a well-conditioned HMM (ICN ≥ 0.4) with different θs. (See that smaller θ retrieves T and O closer to the true parameters)

$\theta = 5$	$\theta = 2$	$\theta = 0.01$
$T = \begin{bmatrix} 1.6486 & -0.5497 \\ 0.3941 & 0.5716 \end{bmatrix}$	$T = \begin{bmatrix} 0.9695 + 0.4851i & 1.5889 - 0.0733i \\ 0.2789 - 0.8469i & -0.7950 - 0.5653i \end{bmatrix}$	$T = \begin{bmatrix} 0.7074 & 0.8103 \\ 0.1287 & -0.0615 \end{bmatrix}$
$O = \begin{bmatrix} 0.5590 & 0.1565 \\ 0.1537 & 0.3937 \\ 0.2541 & 0.4830 \end{bmatrix}$	$O = \begin{bmatrix} 0.3430 + 0.1563i & 0.3430 - 0.1563i \\ 0.5470 + 0.0000i & 0.2667 + 0.0000i \\ 0.2502 + 0.0860i & 0.2502 - 0.0860i \end{bmatrix}$	$O = \begin{bmatrix} 0.5332 & 0.1293 \\ 0.5322 & 0.1239 \\ 0.2178 & 0.4636 \end{bmatrix}$
$\theta = 0.0001$	$\theta = 0.00001$	**True parameter**
$T = \begin{bmatrix} 0.9336 & 0.3400 \\ 0.0628 & 0.6582 \end{bmatrix}$	$T = \begin{bmatrix} 0.8874 & 0.3503 \\ 0.1106 & 0.6330 \end{bmatrix}$	$T = \begin{bmatrix} 0.9000 & 0.3000 \\ 0.1000 & 0.7000 \end{bmatrix}$
$O = \begin{bmatrix} 0.2873 & 0.9300 \\ 0.4739 & 0.0198 \\ 0.2391 & 0.0499 \end{bmatrix}$	$O = \begin{bmatrix} 0.2435 & 0.8298 \\ 0.5080 & 0.1066 \\ 0.2476 & 0.0646 \end{bmatrix}$	$O = \begin{bmatrix} 0.2500 & 0.8000 \\ 0.5000 & 0.1000 \\ 0.2500 & 0.1000 \end{bmatrix}$

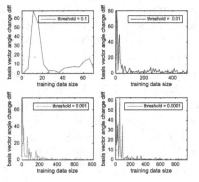

Fig. 2. Synthetic data (Ex. 1)   **Fig. 3.** Real data ($\theta = 0.001$)

The actual number of hidden states is usually not known for a real dataset. The angle change difference for different numbers of hidden states (m) (Fig. 3) shows that more hidden states leads to a higher model complexity (i.e. higher m) and as a result more training data is required to achieve the same θ, 0.001. For instance, to achieve $\theta = 0.001$ with the number of hidden states $m = 8$, $20 \times 4 \times 10^4$ training examples are required, whereas to achieve the same θ with $m = 4$, 20×10^4 training examples are sufficient. Here, 20 is the number of training examples (observations) added to each subsequent training sub-set. Thus, the angle change difference is also correlated with model complexity.

L_1 is not possible to calculate for the real dataset because of the absence of the ground truth as the exact model for real dataset is not known. This is true for all real datasets in general. However, the angle change based criterion can be used with ease to determine the sufficiency of the training data (consequently, empirical moments), and therefore the fitness of the model for real data. Therefore, by using angle change difference as a measure of convergence, it is possible to determine the required training data for sufficient empirical moments based on model complexity, m. From the empirical evidence, we observed that the θ value of 10^{-5} gives satisfactory result in most cases. However, by taking smaller θ, we would be more confident about the solution.

In our experiments, the T and O matrices were extracted to visualize the convergence. It was also observed that when the angle change is small, the recovered T and O matrices are in the parameter space and they are close to the exact model (Table 2). This is another confirmation that leads to a conclusion that the model fitness or empirical moment sufficiency for a well-conditioned HMM can be determined using a certain small angle change difference as a convergence criterion.

On well-conditioned HMMs, when the empirical moments are sufficient, the L_1 error is reduced monotonously (e.g. for a particular HMM, L_1 gets smaller when the θ becomes smaller (Fig. 1b). This is not the case on ill-conditioned HMMs (Fig. 1a) because of the ICN uncertainty existing in the observable space. If a HMM is ill-conditioned (i.e. if ICN is close to 0 or $ICN < 0.4$), L_1 is not correlated in the same way with angle change difference, and it does not lead to conclusive results. For example, in Fig. 4a, for a HMM with $ICN = 0.2$, the L_1 error is lower for $\theta = 0.1$ whereas the error is higher for $\theta = 0.0001$.

This is a feature of a particular HMM, and not a problem with the parameter estimation techniques, because ill-conditioned HMMs are almost impossible to learn even with substantial amount of data at hand (Table 3). Therefore, for ill-conditioned HMMs, the angle change difference is not a proxy for model fitness. This is due to the ambiguity and the uncertainty feature of HMMs [3]. However, on well-conditioned HMMs with $ICN \geq 0.4$, the L_1 is monotonously consistent (Fig. 4a), and shows that the model is fitted well. In Fig. 4b, the non-monotonous L_1 corresponds to a model that was not trained enough, and the angle change difference is large. That means a smaller θ will result in convergence when the error will follow the monotonous trend. If the error is not monotonous even with small θ then its means the HMM is ill-conditioned. In that case, HMM will not be able to model the data. Our empirical experiments confirm that the angle change difference is a useful proxy for sufficient empirical moments and consequently model fitness for well-conditioned HMMs.

Table 3. Recovered T and O matrices for an ill-conditioned HMM (ICN ≤ 0.2) with the required maximum angle change difference $\theta = 0.00001$

True system	Estimated system	True system	Estimated system
$T = \begin{bmatrix} 0.51 & 0.49 \\ 0.49 & 0.51 \end{bmatrix}$	$T = \begin{bmatrix} 1.0372 & 1.3084 \\ -0.0372 & -0.3084 \end{bmatrix}$	$T = \begin{bmatrix} 0.50 & 0.10 & 0.20 \\ 0.20 & 0.60 & 0.40 \\ 0.30 & 0.30 & 0.40 \end{bmatrix}$	$T = \begin{bmatrix} 0.9793 & -0.5696 & 3.5463 \\ -0.3589 & 2.2337 & -4.6428 \\ 0.2551 & -0.4065 & 1.3400 \end{bmatrix}$
$O = \begin{bmatrix} 0.49 & 0.51 \\ 0.51 & 0.49 \end{bmatrix}$	$O = \begin{bmatrix} 0.4899 & 0.0979 \\ 0.5101 & 0.9121 \end{bmatrix}$	$O = \begin{bmatrix} 0.20 & 0.40 & 0.70 \\ 0.70 & 0.40 & 0.10 \\ 0.10 & 0.20 & 0.20 \end{bmatrix}$	$O = \begin{bmatrix} 0.2686 & 0.5274 & 1.0800 \\ 0.5980 & 0.2478 & 0.0552 \\ 0.2275 & 0.1330 & -0.1376 \end{bmatrix}$

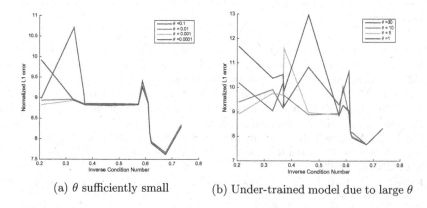

(a) θ sufficiently small   (b) Under-trained model due to large θ

Fig. 4. Angle change difference and model error

6   Conclusion

In this paper, we have proposed a basis vector angle change based convergence
criterion for spectral learning that is known to require large amounts of data.
These algorithms usually work in one-shot and on real data there is no indication
for the practitioner about the convergence. As a result, it is likely that a practitioner will end up using an under-trained model. We showed how a one-shot
algorithm can be trained several times for several sub-sets of the available training data of a growing size. The advantage is that in this way one can approximate
the convergence of the algorithm to check whether the dataset provides sufficient
empirical moments that allow the algorithm to produce parameters that lead to
a small error. We demonstrated our method on spectral learning for HMMs and
showed empirically that our claims are justified in that case. We have tested
our proposed method on synthetic and real data and showed empirically that
our method can indicate sufficiency of the empirical moments. As a result, the
practitioners can check whether they need more data or whether they need a
different model if the predictive power of their solution is not satisfactory.

For the sake of computational efficiency, one could apply our method in conjunction with an online algorithm [2] which relies on incremental SVD. However,
in this study, our goal was to investigate the convergence of spectral learning
and, thus, using standard, non-incremental SVD had better methodological jus-
tification.

We know that, in theory [9], spectral learning for HMMs will converge to an
optimal solution given a sufficient amount of data. In the face of our results,
it would be interesting to compare spectral learning with local search methods,
such as EM, for different magnitudes of the angle change.

References

1. Baker, K.: Singular Value Decomposition Tutorial. Ohio State University (2005)
2. Boots, B., Gordon, G.: An online spectral learning algorithm for partially observable nonlinear dynamical systems. In: Proceedings of AAAI (2011)
3. Caelli, T., McCane, B.: Components analysis of hidden Markov models in computer vision. In: Proceedings of 12th International Conference on Image Analysis and Processing, pp. 510–515, September 2003
4. Davis, R.I.A., Lovell, B.C.: Comparing and evaluating HMM ensemble training algorithms using train and test and condition number criteria. Pattern Anal. Appl. **6**(4), 327–336 (2003)
5. Even-Dar, E., Kakade, S.M., Mansour, Y.: The value of observation for monitoring dynamic systems. In: IJCAI, pp. 2474–2479 (2007)
6. Glaude, H., Enderli, C., Pietquin, O.: Spectral learning with proper probabilities for finite state automaton. In: Proceedings of ASRU. IEEE (2015)
7. Hall, A.R., et al.: Generalized Method of Moments. Oxford University Press, Oxford (2005)
8. Hansen, L.P.: Large sample properties of generalized method of moments estimators. Econometrica: J. Econometric Soc. **50**, 1029–1054 (1982)
9. Hsu, D., Kakade, S.M., Zhang, T.: A spectral algorithm for learning hidden Markov models. J. Comput. Syst. Sci. **78**(5), 1460–1480 (2012)
10. Jaeger, H.: Observable operator models for discrete stochastic time series. Neural Comput. **12**(6), 1371–1398 (2000)
11. Mattfeld, C.: Implementing spectral methods for hidden Markov models with real-valued emissions. CoRR abs/1404.7472 (2014). http://arxiv.org/abs/1404.7472
12. Mattila, R.: On identification of hidden Markov models using spectral and non-negative matrix factorization methods. Master's thesis, KTH Royal Institute of Technology (2015)
13. Mattila, R., Rojas, C.R., Wahlberg, B.: Evaluation of Spectral Learning for the Identification of Hidden Markov Models, July 2015. http://arxiv.org/abs/1507.06346
14. Vanluyten, B., Willems, J.C., Moor, B.D.: Structured nonnegative matrix factorization with applications to hidden Markov realization and clustering. Linear Algebra Appl. **429**(7), 1409–1424 (2008)
15. Zhao, H., Poupart, P.: A sober look at spectral learning. CoRR abs/1406.4631 (2014). http://arxiv.org/abs/1406.4631

Combining Structured and Free Textual Data of Diabetic Patients' Smoking Status

Ivelina Nikolova[1]([✉]), Svetla Boytcheva[1], Galia Angelova[1], and Zhivko Angelov[2]

[1] Institute of Information and Communication Technologies,
Bulgarian Academy of Sciences, 25A Acad. G. Bonchev Str., 1113 Sofia, Bulgaria
{iva,galia}@lml.bas.bg, svetla.boytcheva@gmail.com
[2] ADISS Ltd., 4 Hristo Botev Blvd., 1463 Sofia, Bulgaria
angelov@adisss-bg.com

Abstract. The main goal of this research is to identify and extract risk factors for Diabetes Mellitus. The data source for our experiments are 8 mln outpatient records from the Bulgarian Diabetes Registry submitted to the Bulgarian Health Insurance Fund by general practitioners and all kinds of professionals during 2014. In this paper we report our work on automatic identification of the patients' smoking status. The experiments are performed on free text sections of a randomly extracted subset of the registry outpatient records. Although no rich semantic resources for Bulgarian exist, we were able to enrich our model with semantic features based on categorical vocabularies. In addition to the automatically labeled records we use the records form the Diabetes register that contain diagnoses related to tobacco usage. Finally, a combined result from structured information (ICD-10 codes) and extracted data about the smoking status is associated with each patient. The reported accuracy of the best model is comparable to the highest results reported at the i2b2 Challenge 2006. These method is ready to be validated on big data after minor improvements.

Keywords: Biomedical language processing · Machine learning · Diabetes risk factors · Preventive healthcare

1 Introduction

Chronic diseases have become epidemiology in last decades and main cause for increasing mortality risks and the rapid increase of health care costs. Recently at the national level was started initiative for development of new technologies and data repositories for retrospective analyses in order to support the health management. In 2015 in Bulgaria a Diabetes Registry (DR) was created automatically [2] with the help of natural language processing techniques applied on the outpatient records (ORs). The ORs are submitted from all kinds of professionals to the Bulgarian National Health Insurance Fund (NHIF), for the period 2012–2014. Diabetes Mellitus is a major cause of cardiovascular diseases,

© Springer International Publishing Switzerland 2016
C. Dichev and G. Agre (Eds.): AIMSA 2016, LNAI 9883, pp. 57–67, 2016.
DOI: 10.1007/978-3-319-44748-3_6

and leading cause of adult blindness, kidney failure, and non traumatic lower-extremity amputations [14], its treatment is costly. Thus the prevention and early diagnostics have crucial importance. Therefore in this project we are focusing on analysis of the risk factors for Diabetes Mellitus and its complications as they are priority tasks of preventive health care.

The ORs in the DR are partially structured, all diagnoses and the data for drugs, only in case they are reimbursed by NHIF are available as XML fields. However the risk factors are mostly encoded in the plain text fields of the document. Here we present results of our work on automatic smoker status identification based on natural language processing (NLP) techniques combined with structured data.

Information Extraction (IE) has proven to be effective technology which provides access to important facts about the patient health and disease development in large volumes of plain text patient records. IE is a matured technology and now widely applied in industrial applications however its application to biomedical data is often in narrow domain only, it is tied to specific languages and medical practices. These particularities hamper the easy transfer of technologies for biomedical text processing between different languages and tasks. Machine learning and rule-based approaches integrated in various hybrid systems are common and with the development of new resources in the field these methods become more and more robust [6]. Most developed are the methods for English medical text processing boosted by the US initiatives for secondary use of medical health records.

The contents of the article is structured as follows: Sect. 2 describes the related studies on the topic, Sect. 3 outlines the materials the study was performed on, in Sect. 4 we brief our methods, the results are presented in Sect. 5 and in Sect. 6 some conclusions are drawn.

2   Related Work

Different aspects of electronic health records analyses have been explored last decade for Bulgarian language. Most comprehensive work was done on hospital discharge letters of patients with endocrinology disorders where some high performance extractors for symptoms, lab test values, diagnoses, and medication [2] are developed. In the recent years these analyses have been extended towards ORs from various practitioners who submit their records to the NHIF. Medications are being extracted and normalized to ATC codes with comparatively high accuracy. Analyses on chronicle diseases comorbidity have also been done [3]. One of the most significant works in this direction is the automatic development of the Diabetes Registry from the outpatient records available in the NHIF, again with the help of natural language processing techniques [2]. Persons with potential health hazards related to family history of Diabetes Mellitus are studied in [12]. The current work is part of a larger project for exploration of the DR, personal history and certain conditions influencing health status.

The most considerable work on automatic smoker status identification from discharge letters was done at the First i2b2 De-identification and Smoking Challenge 2006 [16]. Similarly we limited the scope of our study only to understanding of the explicitly stated smoking information. The smoker categories were defined for the challenge as follows: (i) past smoker - somebody who quit smoking more than a year ago; (ii) current smoker - somebody who is currently smoking or has quit smoking less than a year ago; (iii) smoker - it is clear that the patient is a smoker but there is no sufficient information to be classified as (i) or (ii); (iv) non-smoker - somebody who never smoked; (v) unknown - no mentions of smoking status in the discharge letter.

Most of the teams which participated in the challenge apply a two step strategy: (*i*) identifying sentences in the records discussing smoker status and (*ii*) classifying only these sentences into the predefined categories. Most often the first step was performed based on trigger terms. The authors report that excluding the irrelevant sentences increased the performance of their algorithms significantly. Similarly in this study we classify only samples which contain trigger terms. The system which achieved highest results on the test set is presented in [4]. They annotated additional data thus increased their training data sample and used linguistic and engine specific features. The latter ones had major contribution to the system performance. They include semantic features such as semantic types of some medical entities - medication, diagnoses, negation and anti-smoking medication. Similarly we introduce in our system semantic features by assigning category to the terms available in our categorical dictionaries. These will be explained in detail in Sect. 4. Aramaki et al. [1] at the second step apply comparison of each sentence with sentences from the training set. The sum of the similarity measures between each extracted sentence and the most similar sentences in the training set is used to determine the smoking status of the extracted sentence. Another systems incorporating rule-based and machine learning approaches also achieved good results [5]. The authors perform an intermediate filtering of records which are not meaningful to the task. Smoker status identification is an important task in automated structuring of patient records and it is still under development for various languages [9].

3   Materials

The DR contains outpatient records in Bulgarian language provided by the Bulgarian NHIF in XML format. The available records for 2014 are nearly 8 mln. for about 462,000 patients. Although the major part of the information necessary for the health management is available as structured fields, some of the important factors for the patient status and the disease development are only available in the free-text sections like anamnesis, status, clinical examination, therapy. All texts are in Bulgarian but contain variety of terms in Latin (in Latin alphabet) or Latin terms transliterated in Cyrillic alphabet. We process raw data that contain many spelling and punctuation errors. Due to the limited number of language resources for Bulgarian and the telegraphic style of the message in the ORs some of the traditional methods for text analysis are not applicable e.g. sentence splitting, dependency parsing etc. Only very focused narrow context information extraction techniques can be helpful in these settings.

Similarly to the i2b2 challenge, in our study we define 4 smoker categories:

- **smoker** - the text explicitly states that the patient has recently smoked (*yes*);
- **past smoker** - the text has evidences about the successful smoking cessation (*ex*);
- **non-smoker** - the text explicitly states that the patient has never smoked (*no*);
- **unknown** - there is no explicit statement in the text regarding the patient's smoking status (*unkn*).

Following the good practices from the i2b2 challenge initially we extract from all ORs only 256 characters concordances around the trigger words: "пуш" (*push*, root of smoke), "цигар" (*cigar*, root of cigarette) and "тютюн" (*tyutyun*, tabacco). This task is performed by BITool [2] over the DR records from 2014. This context is necessary for the human to judge and annotate the data. However when we train our model we strip out only a narrow context of 7 tokens to the left and to the right of the trigger. The OR sections in which these strings occur and are taken in consideration are: anamnesis, patient status, diagnosis, clinical examinations, treatment recommendations. Then we annotated manually some randomly selected 3,092 concordances (Set 1 in Table 1) and additionally add to Set 1 about 200 concordances (Set 2 in Table 1) mainly for more complicated cases of past smokers, that contain rich temporal information about the smoking status progress (Fig. 1). The first example has class "smoker" and the second one - "past smoker". The annotation is performed per record level with the classes explained above. We annotated with current, past or non-smoker only explicit statements about the smoking status. Expressions like "отказва пушенето" (*quits smoking*) we consider *unknown* since they do not state clearly the smoking status at the moment.

Table 1. Class distribution in the annotated data set.

Class	ex	no	yes	unkn
Set 1 concordances	56	2,059	941	37
Set 2 concordances	220	2,066	966	40

In the ORs the smoker status is expressed with various expressions like:

- пушач (*pushach*, smoker) - class **smoker**
- тютюнопушене цигари/ден: 5 (*tyutyunopushene cigari/den: 5*, tabacco smoking cigarettes/day: 5) - class **smoker**
- тютюнопушене цигари/ден: 0 (*tyutyunopushene cigari/den: 0*, tabacco smoking cigarettes/day: 0) - class **non-smoker**
- тютюнопушене(-) (*tyutyunopushene*, tabacco smoking) - class **non-smoker**
- бивш пушач (*bivsh pushach*, past smoker) - class **past smoker**

Fig. 1. Examples for rich temporal information about the smoking status progress.

- пушач до преди 3 мес. (*pushach do predi 3 mes.*, smoker until 3 months ago) - class **past smoker**
- цигарите! (*tsigarite!*, the cigarettes!) - class **unknown**

The distribution of the classes in the annotated data is shown on Table 1. The classes of current smokers (*yes*) and non-smokers (*no*) are considerably bigger than the past smokers (*ex*) and the unknown cases (*unkn*). The imbalance of the data presupposes that the smaller classes will be more difficult to predict. Among them the ex-smokers are of our interest.

4   Method

The workflow of this study is shown on Fig. 2.

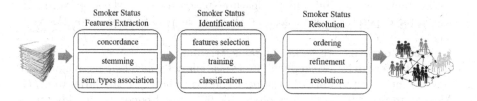

Fig. 2. Workflow.

We perform three stages pipeline. The first stage is responsible for preprocessing of the input data - extracting concordances for the trigger words related to smoking status, stemming these concordances and association of the words with semantic types with the help of 12 vocabularies. In the next stage we perform feature selection and supervised training using manually annotated data. Later this model is refined with additional features extracted from the DR records and the smoker status is being determined.

As explained earlier we focus our work on classifying only ORs containing trigger words signaling smoking. We extract phrases from the free text sections of the OR in the near context of a trigger word. We annotated manually 3,292 of these so called "concordances" and we train a supervised model from the labeled data. The development and training corpus is 66 % of our records and the remaining is test data.

Fig. 3. The vocabulary coverage

We use the following types of features for this task:

- **Linguistic features** - we use the stemmed form [11] of the tokens. Each token stem is an attribute in our feature space except for the stop words. In the latter experiments we also add the verb tense information for the verb smoke.
- **Context features** - these are bigrams, trigrams.
- **Semantic features** - we apply a set of vocabularies which help us to figure out the semantics of the words in the near context. The 12 vocabularies are: *(1)*. Markup terms; *(2)*. Vocabulary of the 100,000 most frequent Bulgarian terms [13]; *(3)*. Generic medical terms in Bulgarian; *(4)*. Anatomical terms in Latin; *(5)*. Generic names of drugs for Diabetes Mellitus Treatment; *(6)*. Laboratory tests; *(7)*. Diseases; *(8)*. Treatment; *(9)*. Symptoms; *(10)*. Abbreviations; *(11)*. Stop words; *(12)*. Negation terms. These are applied in the specified order and the annotations of the latter ones override the previous ones. The categories matched within the concordance are used as features as well as is the number of occurrence of each category. For each concordance is generated single binary vector with bits signaling whether the given attribute is present in the current concordance or not.

The vocabulary coverage is shown on Fig. 3 and Table 2. In the columns are shown the size of each vocabulary (Size), the number of tokens matched in the text by this vocabulary (Tokens), the percentage of tokens in the text matched by this vocabulary (Tokens %), the number of vocabulary entries - types which were matched in the text (Type). The largest coverage has the vocabulary of stop words, then diagnoses, next is the vocabulary of most frequent Bulgarian words followed by the markup words.

Table 2. Lexical profile statistics.

Category	Size	Tokens	Tokens %	Type
1. tags	99	20,684	7.87	29
2. btb	102,730	41,582	15.83	1,051
3. bg med	3,624	1,545	0.59	91
4. term anat	4,382	3,792	1.44	8
5. drugs	154	12	0.01	5
6. lab test	202	18	0.01	5
7. diagnoses	8,444	54,431	20.72	941
8. treatment	339	4,170	1.59	57
9. symptoms	414	4,180	1.59	173
10. abbrev	477	14,404	5.48	83
11. stop words	805	67,153	25.56	166
unknown		50,744	19.32	3,757
TOTAL	121,670	262,715		6,366

The vocabularies lookup and some statistics which helped us for better understanding of the data in means of collocations and terminology are done with AntWordProfiler [10].

5   Results and Discussion

The results shown below are achieved after experiments with various features and instance data size. We narrowed our feature space iteratively starting from a very large space of over 20,000 features. When we restricted the token features only to the ones which appear in 7-token window from the focal term, the attribute space decreased significantly. Then we applied a few rules for filtering out attributes which are not related to smoking and we arrived to about 7,000 attributes in our first experiments. In order to reduce them even more, we applied automatic attribute selection by subset evaluation with default parameters as provided in Weka [7] however the results of the classification in the reduced space were less satisfactory.

Among the algorithms we applied are JRip, LibLINEAR, SMO and SVM with RBF kernel through their Weka implementations or wrappers. In our initial experiments SMO outperformed the other algorithms with 2 to 9 points in F1 for most of the classes therefore the feature engineering phase and final experiments were done with it. SMO is Weka's implementation of John Platt's sequential minimal optimization algorithm for training a support vector classifier. The results reported here are obtained with it only.

We trained our model with 67 % of the data and tested it on the other 33 %. Experiment SMO-1, Table 3 was done on Set1 of the corpus and achieved quite

Table 3. Classification evaluation. SMO-1 - 7,334 attr., SMO-2 - 8,205 attr., SMO-3 - 8,368 attr., SMO-4 - 8,427 attr.

	Precision	Recall	F1	Class
SMO-1	0.92	0.50	0.65	ex
	0.93	0.98	0.96	no
	0.92	0.84	0.88	yes
	0.44	0.36	0.40	unkn
	0.92	0.92	0.92	w. avg
SMO-2	0.89	0.68	0.77	ex
	0.93	0.99	0.96	no
	0.88	0.84	0.86	yes
	0.83	0.33	0.48	unkn
	0.91	0.91	0.91	w. avg
SMO-3	0.85	0.70	0.77	ex
	0.93	0.99	0.96	no
	0.89	0.84	0.86	yes
	0.83	0.33	0.48	unkn
	0.91	0.92	0.91	w. avg
SMO-4	0.88	0.75	0.81	ex
	0.93	0.99	0.96	no
	0.89	0.83	0.86	yes
	0.83	0.33	0.48	unkn
	0.92	0.92	0.92	w. avg

high accuracy for the big classes, however the small classes like *ex* and *unkn* remained hard for guessing. We searched for the reasons not only in the features trained on our development set but also by exploring the data in the DR. Since *ex* is of major importance we analyzed new examples of this class and added them to the corpus. Our expectations were that additional data will lead to improvement of the recognition rate for this class. However the explanations in the ORs of type *ex* are often quite complex and contain a chain of several events related to smoking as shown on Fig. 1. As result the recall indeed improved but the precision has dropped (SMO-2, Table 3). In the next experiments the goal was to improve the precision for *ex* while preserving the achieved accuracy for the big classes. Often past smokers are confused with current ones and less often with non-smokers. Thus some temporality features to distinguish between current and past event have been added. The prepositions which clarify the event smoking were removed from the stop list and added as features to enable bigrams like *smoked until* to enter in the feature set. The results are shown on SMO-3, Table 3. In SMO-4 we added the tense of the verb *smoke* to the feature set. We must mention that the verb *smoke* is used in past tense mostly in records for *ex-smokers* but also in records for *smoker* such as "was smoking 2 packs a day, now smokes only 10 cigarettes". It appears also in records of *non-smokers* such as "never smoked". Still, introducing this feature lead to higher accuracy for both classes *ex* and *yes*. In these 4 steps we improved the recognition of *ex* with 16 points in F1 while preserving the scores for the majority class *no* and with a minor compromise of 2 points in F1 for class yes.

Table 4. ICD-10 diagnoses for tobacco abuse and NLP. ORs - outpatient records; Ps - patients; non cl. - not classified records; *yes, no, ex, unkn* - manually annotated records with the respective class. Z72.0, F17, Z81.2 - ICD-10 codes.

	ICD-10 only			ICD-10 + NLP			NLP only					
	Z72.0	F17	Z81.2	Z72.0	F17	Z81.2	non cl.	yes	no	ex	unkn	Total
ORs	1,007	23	1	1,113	17	122	820,360	942	2,065	220	39	**825,909**
Ps	609	11	1	968	14	121	457,032	851	1,973	175	36	**461,791**

The instances of class "unknown" are underrepresented in the data set and that is why they are extracted with lower recall. However the precision of the extraction module is comparatively good which means that the features describe well the observed examples. And when dealing with medical data, high precision is a must. Oversampling often helps to increase precision and for real world application it could also be applied. The results we present here are comparable to the ones reported on the i2b2 challenge for smoker status identification from discharge letters in English.

Additional improvement of classification results is possible by taking into account contextualization information. For instance, the concordances extracted from Treatment section refer either to past smoker in case some medication name contains searched key string, or current smoker - in case the searched key string was found in explanations for diet, nutrition and life style recommendation.

In addition to the free text sections of the OR, we analyze also the diagnoses sections. It is not strange that the diagnoses may also contain the triggers we used for extracting the concordances because there are ICD-10 diagnoses [8] like Z71.6 "Tobacco abuse counseling", Z81.2 "Family history of tobacco abuse", P04.2 "Fetus and newborn affected by maternal use of tobacco", T65.2 "Tobacco and nicotine", Z58.7 "Exposure to tobacco smoke", Z72.0 "Tobacco use", Z86.4 "Personal history of psychoactive substance abuse" and F17 "Mental and behavioral disorders due to use of tobacco". Unfortunately these diagnoses are rarely used by professionals, because in the Bulgarian standard for ORs the number of coded diagnoses is at most 5. Another reason for non presence of these diagnoses in ORs is that not all professional encode them explicitly, for instance Ophthalmology. In the DRs for 2014 there are singular ORs that contain codes: T65.2, Z71.6 and Z86.4. Diagnoses P04.2 and Z58.7 are presented in none of the ORs. The majority of markers for smoking status are presented only as free text in the ORs (Table 4). For all patients from the DR we have at least one OR containing information about their smoking status in 2014. Our ultimate goal is to enrich the patients record in the DR with risk factors information and as of this study - with his/her smoking status. We combine information extracted by NLP techniques and ICD-10 codes (if any). For those patients for who only ICD-10 codes are available - we can resolve the current smoker status as: "smoker" - for T65.2, Z71.6 and Z72.0; "past smoker" - for Z86.4 and "unknown" - for the rest. We can add also status "passive smoker" for Z81.2. And vice versa - for patients without ICD-10 codes for Tobacco use in their ORs we can add the following diagnoses: Z72.0 for "smoker", and Z86.4 for "past smoker". In case both ICD-10 and ORs text contain information about the patient's smoking status - the ICD-10 code can be used for classifier validation. Further investigation of how smoking is influencing health status can be performed on the basis of other diagnosis in the patient's OR and analysis of the temporal information. Similar research was presented in [17], but for ICD-9 codes that include also procedures, however in this study only two classes are considered - ever-smoker and never-smoker.

6   Conclusion

We built a highly accurate model for smoker status identification in Bulgarian outpatient records. Although no rich semantic resources for Bulgarian exist, we were able to enrich our model with semantic features based on categorical vocabularies. The results from this study are comparable to the highest results reported at the i2b2 Challenge 2006. We succeed to improve our model by identifying specific features of the underrepresented classes while preserving the extraction accuracy of the bigger classes. Our next challenge is to apply this model to big data.

There are several risk factors for Diabetes Mellitus that are in the focus of researchers [15] and we plan to continue this work by investigating other potential health hazards like alcohol, drugs, lifestyle and etc. These could be approached with similar means, because ICD-10 provides also diagnoses codes for problems related to lifestyle.

Acknowledgements. This study is partially financed by the grant DFNP-100/04.05.2016 "Automatic analysis of clinical text in Bulgarian for discovery of correlations in the Diabetes Registry" with the Bulgarian Academy of Sciences.

References

1. Aramaki, E., Imai, T., Miyo, K., Ohe, K.: Patient status classification by using rule based sentence extraction and BM25 kNN-based classifier. In: i2b2 Workshop on Challenges in Natural Language Processing for Clinical Data (2006)
2. Boytcheva, S., Angelova, G., Angelov, Z., Tcharaktchiev, D.: Text mining and big data analytics for retrospective analysis of clinical texts from outpatient care. Cybern. Inf. Technol. 15(4), 58–77 (2015)

3. Boytcheva, S., Angelova, G., Angelov, Z., Tcharaktchiev, D.: Mining clinical events to reveal patterns and sequences. In: Margenov, S., Angelova, G., Agre, G. (eds.) Innovative Approaches and Solutions in Advanced Intelligent Systems. Studies in Computational Intelligence, vol. 648, pp. 95–111. Springer, Heidelberg (2016)
4. Clark, C., Good, K., Jezierny, L., Macpherson, M., Wilson, B., Chajewska, U.: Identifying smokers with a medical extraction system. J. Am. Med. Inform. Assoc. **15**, 36–39 (2008)
5. Cohen, A.M.: Five-way smoking status classification using text hot-spot identification and error-correcting output codes. J. Am. Med. Inform. Assoc. **15**, 32–35 (2008)
6. Cohen, K.B., Demner-Fushman, D.: Biomedical Natural Language Processing, vol. 11. John Benjamins Publishing Company, Amsterdam (2014)
7. Hall, M., Frank, E., Holmes, G., Pfahringer, B., Reutemann, P., Witten, I.: The WEKA data mining software: an update. SIGKDD Explor. Newsl. **11**(1), 10–18 (2009)
8. International Classification of Diseases and Related Health Problems 10th Revision. http://apps.who.int/classifications/icd10/browse/2015/en
9. Jonnagaddala, J., Dai, H.-J., Ray, P., Liaw, S.-T.: A preliminary study on automatic identification of patient smoking status in unstructured electronic health records. In: ACL-IJCNLP 2015, pp. 147–151 (2015)
10. Laurence, A.: AntWordProfiler (Version 1.4.0w) (Computer software). Waseda University, Tokyo, Japan (2014). http://www.laurenceanthony.net/
11. Nakov, P.: BulStem : Design and evaluation of inflectional stemmer for Bulgarian. In: Proceedings of Workshop on Balkan Language Resources and Tools (1st Balkan Conference in Informatics) (2003)
12. Nikolova, I., Tcharaktchiev, D., Boytcheva, S., Angelov, Z., Angelova, G.: Applying language technologies on healthcare patient records for better treatment of Bulgarian diabetic patients. In: Agre, G., Hitzler, P., Krisnadhi, A.A., Kuznetsov, S.O. (eds.) AIMSA 2014. LNCS, vol. 8722, pp. 92–103. Springer, Heidelberg (2014)
13. Osenova, P., Simov, K.: Using the linguistic knowledge in BulTreeBank for the selection of the correct parses. In: Proceedings of The Ninth International Workshop on Treebanks and Linguistic Theories, Tartu, Estonia, pp. 163–174 (2010)
14. Rice, D., Kocurek, B., Snead, C.A.: Chronic disease management for diabetes: Baylor Health Care System's coordinated efforts and the opening of the Diabetes Health and Wellness Institute. Proc. (Bayl. Univ. Med. Cent.) **23**, 230–234 (2010)
15. Stubbs, A., Uzuner, Ö.: Annotating risk factors for heart disease in clinical narratives for diabetic patients. J. Biomed. Inform. **58**, S78–S91 (2015)
16. Uzuner, Ö., Goldstein, I., Luo, Y., Kohane, I.: Identifying patient smoking status from medical discharge records. J. Am. Med. Inform. Assoc.: JAMIA **15**(1), 14–24 (2008)
17. Wiley, L.K., Shah, A., Xu, H., Bush, W.S.: ICD-9 tobacco use codes are effective identifiers of smoking status. J. Am. Med. Inform. Assoc. **20**(4), 652–658 (2013)

Deep Learning Architecture for Part-of-Speech Tagging with Word and Suffix Embeddings

Alexander Popov[✉]

Institute of Information and Communication Technologies, BAS,
Akad. G. Bonchev. 25A, 1113 Sofia, Bulgaria
alex.popov@bultreebank.org

Abstract. This paper presents a recurrent neural network (RNN) for part-of-speech (POS) tagging. The variation of RNN used is a Bidirectional Long Short-Term Memory architecture, which solves two crucial problems: the vanishing gradients phenomenon, which is architecture-specific, and the dependence of POS labels on sequential information both preceding *and* subsequent to them, which is task-specific.

The approach is attractive compared to other machine learning approaches in that it does not require hand-crafted features or purpose-built resources such as a morphological dictionary. The study presents preliminary results on the BulTreeBank corpus, with a tagset of 153 labels. One of its main contributions is the training of distributed word representations (word embeddings) against a large corpus of Bulgarian text. Another is complementing the word embedding input vectors with distributed morphological representations (suffix embeddings), which are shown to significantly improve the accuracy of the system.

Keywords: Deep learning · Recurrent neural networks · Long Short-Term Memory · Part-of-speech tagging · Word embeddings · Suffix embeddings

1 Introduction

Recurrent neural networks (RNNs) have been used successfully at solving tasks where the input is presented sequentially in time (for an overview see [1]). More specifically, in relation to NLP they have been used for text generation ([2,3]), translation ([4]), language understanding ([5]), etc. Part-of-speech tagging, another NLP task involving input and output sequences, is well defined for the application of a similar solution to it: an input list of words must be tagged with corresponding POS labels, which can also express morphological features.

This paper presents an RNN architecture adapted to the problem of POS tagging. To overcome the inherent difficulty in training RNNs, and more specifically the *vanishing gradients phenomenon*, the hidden layer of the network is constructed with *Long Short-Term Memory* (LSTM) cells. To account for the fact that POS categories depend on the whole sentence and not just on the previously seen words (e.g., consider the sentence "I made her *duck*... to avoid the incoming ball."), a specific type of RNN is used – a *bidirectional RNN*.

© Springer International Publishing Switzerland 2016
C. Dichev and G. Agre (Eds.): AIMSA 2016, LNAI 9883, pp. 68–77, 2016.
DOI: 10.1007/978-3-319-44748-3_7

The architecture can in principle take in any kind of numerical representation of the individual words in the input sentences, such as one-hot vector
encoding of the words themselves, or of features associated with them. However,
following numerous demonstrations of the effectiveness of distributed representations of words (more popular in NLP as *word embeddings*), such real-valued vec-
tored encoding has been chosen for the inputs. The paper describes the process
of obtaining such representations for Bulgarian word forms, as well as that of
analogously obtaining distributed representations of morphological suffixes. The
latter are shown to significantly improve POS tagging.

2   Related Work

LSTM networks were first introduced in [6]. For a good introduction to both
LSTM and bidirectional RNNs, see [1]. A brief description of the principles
underlying these constructs is included in the next section.

One of the most popular methods for obtaining distributed representations of
words in a continuous space is described in [7]. It introduces two efficient methods
for calculating word embeddings: the *Continuous Bag-of-Words Model* (CBOW)
and the *Continuous Skip-gram Model* (Skip-gram). Both employ shallow neural
networks to predict textual context on the basis of textual input. While the
CBOW model tries to predict one word looking at a left- and right-bounded
window of words around the target, the Skip-gram model does the reverse – it
predicts the context of words surrounding an input word, one context and word
at a time. These two methods are much faster than previously used techniques,
without sacrificing the accuracy of the representations, and consequently make
it possible to train networks on huge amounts of text in workable time frames.

One popular neural network solution to POS tagging is [8]. That study uses
a convolutional network to carry out several NLP tasks (such as POS tagging,
chunking, named entity recognition, semantic role labeling) in a unified manner. The POS tagging task is not solved sequentially, since convolutional networks rather examine global input snapshots and learn to discover patterns in
them; this in fact helps the network in solving a number of NLP tasks in parallel – because shared information about different linguistic aspects helps across
the tasks.

Another study on neural networks that approaches POS tagging as a
sequence-to-sequence tagging problem is [9]. In it a bidirectional LSTM network
is shown to perform competitively with state-of-the-art solutions. It presents several ways to encode the input data. For instance, one-hot vector representations
of the input words combined with three-dimensional binary vectors indicating
case are reported to yield a good accuracy score. The authors also describe
a novel way of training word embeddings – rather than predicting particular unknown words, they generate artificial contexts where certain words are
replaced with others, then the network tries to predict the artificially swapped
("incorrect") ones. Adding the word embeddings further boosts accuracy, while
an additional morphological feature (bigram suffix encoded as a one-hot vector) helps the network achieve state-of-the-art results. Another study by the

same authors [10] shows that the same architecture can be successfully applied to other sequential tagging NLP tasks: chunking and named entity recognition. [11] combines a bidirectional LSTM network with a CRF layer for POS tagging, chunking and NER.

[12] describes a hybrid approach to POS tagging in Bulgarian: a simple RNN combined with a rule-based approach. This system achieves "95.17 % accuracy for POS disambiguation and 92.87 % for all grammatical features". [13] reports on the accuracy of several POS taggers against the BulTreeBank data: 95.91 % (BLL Tagger), 94.92 % (Mate morphology tagger), and 93.12 % (TreeTagger). A direct comparison with the current system is not possible, however, due to the difference in the tagsets used (the older work uses the full tagset of 581 labels).

3   Bidirectional LSTM Architecture for POS Tagging

3.1   Recurrent Neural Networks

An RNN is one whose hidden layer neurons have cyclic connections to themselves (Fig. 1). This allows the network to keep a kind of dynamic memory, by maintaining an internal state that depends not just on the input data, but also on the previous states of the neurons. For a sequence of inputs $x_1, x_2, ..., x_n$, the network computes the corresponding output vector y_t via the equation:

$$\mathbf{y_t} = W_{hy}h_t + b_y \qquad (1)$$

where W is a matrix holding the connection weights between two layers (in this case, between the hidden and the output layer) and b_y is the bias for the output layer. This is standard for any neural network with hidden layers. The difference lies in the way h_t, the output of the hidden layer, is obtained:

$$\mathbf{h_t} = F_{act}(W_{xh}x_t + W_{hh}h_{t-1} + b_h) \qquad (2)$$

where F_{act} is the activation function in the hidden layer, the first term in the summation is the input multiplied by the relevant connection weight, the second one is the vector of the previous hidden state multiplied by the relevant hidden-to-hidden connection weight and the last one is the hidden layer bias.

Theoretically, such architectures should be able to remember information from many time steps ago and integrate it with incoming data. In practice, RNNs are extremely difficult to train, one of the main reasons being that when the error gradients are backpropagated via the chain rule, they tend to grow either very large or very small, effectively precluding the network from training. This is known as the *exploding/vanishing gradients problem*. There is a simple way to solve the explosion issue – the backpropagated gradients are capped at some values, thus allowing updates within reasonable upper bounds.

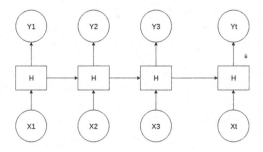

Fig. 1. An unrolled recurrent neural network

3.2 Long Short-Term Memory

The vanishing gradients problem can be solved through the use of a special type of RNN architecture – Long Short-Term Memory. The module in the plain RNN that is looped to itself has a single hidden layer inside; an *LSTM cell* is internally more complex. Its cell state is modified by the outputs of several gates: *forget gate, input gate* and *output gate*. The LSTM cell has four hidden layers inside, which learn the appropriate weights needed to correctly forget and update (parts of) the cell state. Their outputs are calculated via the following equations ([14]):

$$\mathbf{i_t} = \sigma(W_{xi}x_t + W_{hi}h_{t\text{-}1} + W_{ci}c_{t\text{-}1} + b_i) \tag{3}$$

$$\mathbf{f_t} = \sigma(W_{xf}x_t + W_{hf}h_{t\text{-}1} + W_{cf}c_{t\text{-}1} + b_f) \tag{4}$$

$$\mathbf{c_t} = f_t c_{t-1} + i_t tanh(W_{xc}x_t + W_{hc}h_{t-1} + b_c) \tag{5}$$

$$\mathbf{o_t} = \sigma(W_{xo}x_t + W_{ho}h_{t\text{-}1} + W_{co}c_t + b_o) \tag{6}$$

$$\mathbf{h_t} = o_t tanh(c_t) \tag{7}$$

where i, f, c, o correspond to the input gate, forget gate, cell state and output gate activations, σ denotes the sigmoid activation function and *tanh* – the hyperbolic tangent. The memory learns to selectively decide which pieces of information to keep and which to forget, thus making it possible to remember input from many steps ago and to discard input that is not relevant to the current state.

3.3 Bidirectional LSTM

To address the specifics of the task at hand, the LSTM-RNN is converted to a *bidirectional RNN* (Fig. 2). This means simply that instead of one LSTM cell in the layer between input and output, there are two. One is fed with the sequential inputs as they appear normally in time; the other receives the inputs in reverse – from end to beginning. This is done so that the network can look forward *as well as* backward. The two LSTM cells perform their calculations independently, but their output vectors are concatenated and then fed to the next layer.

Finally, a softmax layer takes the output of the Bi-LSTM layer and converts it to a probability vector, where each position corresponds to a tagset label.

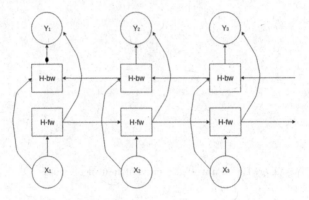

Fig. 2. A bidirectional RNN

4   Input Data

4.1   Word Embeddings

Neural network approaches are very appealing because they circumvent the need for crafting a lot of features to train on. It has been demonstrated that distributed representation vectors capture a lot of semantic and syntactic information about words. To obtain such representations, however, large corpora are needed (typically good results, as measured on word analogy and similarity tasks, are obtained by training on billions of words).

This study is a first step in exploring word embeddings for processing Bulgarian. As a basis for these preliminary experiments, a corpus of Bulgarian texts was compiled, consisting of roughly 220 million words (including most of the Bulgarian Wikipedia, news articles and literary texts). It contains about 457000 unique frequent words (the total number of unique words is about 1.6 million; words are considered frequent if they occur at least 5 times in the corpus). Word embedding vectors were calculated with the Skip-gram Word2Vec model distributed with TensorFlow [15]. Due to the relatively small size of the corpus, the dimensionality of the vectors was set to 200 (in fact, some cursory experiments with a dimensionality of 600 suggested that such an increase results in no gains when trained with this corpus, but this should be investigated separately).

4.2   Suffix Embeddings

In this subsection is presented an attempt to capture morphological information in a distributed manner, analogously to the way word embeddings capture syntactico-semantic information, the hypothesis being that this could complement word representations in tasks sensitive to the shape of the words.

A corpus of 10 million words was extracted from the Bulgarian Wikipedia data and used to calculate the suffix embeddings. The same algorithm (Skip-gram) was used with no changes made to it, only the input data was processed so as the suffixes should appear as individual words in the corpus, i.e. the words in the running text were exchanged with parts of them, here termed *suffixes*.

The suffixes are normalizations of the words which were carried out in the following way. All the letters in the words are converted to lowercase, but information about the original case is kept for the next steps. If the word token is longer than 3 characters, only the trailing three symbols are left and the beginning is deleted and replaced by an underscore ("_"). If the token is 3 or fewer characters long, it is kept as it is. If the original token starts with a capital letter (and continues in lowercase), the symbol "@" is concatenated to the beginning. If all characters in the original token are capital letters and the token contains more than one character, "$" is added to the beginning. If the original token contains lower- *and* uppercase characters, "#" is added to the beginning. Here are come examples:

По = @по
по = по
държава = _ава
Държава = @_ава
АП = $ап
БГНЕС = $_нес
вик = вик
Вик = @вик
ВиК = #вик

Bulgarian morphology is relatively rich and most of the important markers occur at the end of words. This normalization captures the morphological information coded in the last three symbols of each word form (if it has as many) together with information about letter case, which is much needed to handle the initial words in sentences, proper names, etc. There are about 20000 unique frequent suffixes in the training corpus, for which vector representations have been calculated (the total number of unique suffixes is about 37000). The dimensionality of the suffix embeddings is set to 50, since they are far fewer than the unique frequent words and are trained on a smaller corpus.

4.3   Bulgarian Gold Corpus

The BulTreeBank POS tagged corpus [16] was used for training and testing purposes. It contains approximately 38000 sentences with POS tagged words. 3500 of those were used for validation purposes and another 3500 were used for testing. The rest (~30000) were used as training data. The tagset used for training and testing has 153 labels; it is more coarse-grained than the full tagset, which numbers 680 labels, but significantly more fine-grained than tagsets used for morphologically poorer languages, such as the Penn Treebank tagset for English, which has only 36 labels.

5 Experiments

The network described was implemented for the present study with the use of the open source library TensorFlow [15], which provides ready-made implementations of LSTMs and bidirectional RNNs. For the experiments reported here, only a vanilla version of the implementation was used. That is, the network has only one hidden layer (which contains the two LSTM cells, each of them in turn containing four hidden layers; however, more such bidirectional LSTM layers can be added in future investigations), it does not make use of popular optimization tricks such as adding regularization, momentum, dropout, etc.

Neural networks may not require hand-crafted features to train on, but they do tend to be very complicated in the sense that the architectures are defined by a lot of hyperparameters that need to be set according to heuristics and a lot of empirical data. It is not surprising, due to these reasons given above, that the results presented here do not achieve the state-of-the-art for the task – a lot of further work on parameter and architecture optimization is necessary. However, the purpose of this preliminary study is to demonstrate the viability of the approach and the potential gains of experimenting with different encodings of the input data.

5.1 Size of the Word Embeddings

In order to justify the decision to set the dimensionality of the word embedding vectors to 200, an experiment was conducted to compare the performance of the network with the two sets of representations. In the first case, words from the input data are converted to their corresponding word embeddings of size 200 and these vectors are passed into the hidden layer. In the second case the dimensionality of the vectors obtained via the Skip-gram model is 600. Both experiments were run for 10000 training iterations of the network, with a learning rate of 0.3 and 100 neurons per hidden layer in the LSTM cells. Table 1 shows the results.

Table 1. POS tagging accuracy depending on the dimensionality of the input word embeddings (after 10000 training iterations)

Word embedding size	Accuracy
200	77.67 %
600	71.89 %

These numbers do not constitute a conclusive proof that the 200-position vectors are superior than the 600-position ones when trained on the currently available corpus, since it is possible that some other configuration of the hyperparameters might result in a different comparison. For instance, it is possible that the larger vectors would benefit more from a larger hidden layer. Initial experimenting does not support this hypothesis, but more extensive tests are needed nevertheless. Also, the model with the larger vectors might need more training iterations to reach its full capacity. Therefore further experimentation with different hyperparameter settings and longer training is due before a verdict can be passed. However, it is reasonable to suppose that 600 positions are simply too many with regards to the size of the training corpus ($\sim$220 million words) and for that reason an encoding in 200 position is more sensible. It is possible that a number somewhere in-between might be a better choice when training with such an amount of text.

5.2 Suffix Embedding

An additional small experiment is presented here that demonstrates the viability of the approach involving suffix embeddings. The neural network is fed only with vectors from the distributed representations obtained for the suffixes, not making any use at all of the word embeddings. Table 2 shows that for a smaller number of training iterations the suffix embeddings model is very competitive compared with the word embeddings one. Perhaps this is so because it provides more condensed and abstract, grammatical information, in contrast to the more lexically-infused information encoded in the word embeddings. Two different sets of vectors are used: one with vectors of size 50 (the set used in the main experiment reported here) and of size 200. The hidden layers are again of size 100, the learning rate is set to 0.3 and the experiments are run for 10000 training iterations. The accuracy numbers are relatively close, which seems to confirm the intuition that suffixes, being a lot fewer and trained on a smaller corpus, need not be represented in a very large space.

Table 2. POS tagging accuracy depending on the dimensionality of the input suffix embeddings (after 10000 training iterations)

Suffix embedding size	Accuracy
50	77.11 %
200	78.16 %

5.3 Word Embeddings + Suffix Embeddings

The central experiment in this article is a comparison between a model trained on word embeddings only and one trained on a combination of word and suffix embeddings. In the first case the input vectors are again of size 200. In the second case the 200-position vectors of the word representations are concatenated with the suffix embedding vectors of size 50. The learning rate is set to 0.3 for both setups; for the first one the hidden layers in the LSTMs are initialized with 100 neurons, while for the second one the initialization is with 125 neurons to reflect

Table 3. POS tagging accuracy when using word embeddings only, and when complementing them with suffix embeddings (after 100000 training iterations)

Model	Accuracy
Word embeddings (100 neurons)	91.45 %
Word embeddings (125 neurons)	91.13 %
Word + suffix embeddings (125 neurons)	94.47 %

the increased size of the input vectors. A third experiment was conducted with word embeddings only, but with hidden layers of 125 units, so as to demonstrate that the difference in size is not putting the word-embeddings-only model at a disadvantage. The models were tested for their accuracy after 100000 training iterations. Table 3 shows the results.

The difference is significant enough to suggest that the morphological information from the suffix embeddings complements that which is present in the word embeddings. It is possible that a different configuration of the network hyperparameters might yield the same results without using the suffix embeddings, but additional experimentation and optimization is needed to check that. It is also important to determine how much of the usefulness of the suffix embeddings comes from encoding the actual word endings and how much from encoding information about the case of the words. An interesting observation is that the model that uses both types of vector representations initially learns much faster (at an earlier stage of the training the accuracy of the model is about 8 % higher than that of the word-embeddings-only model). This is another indirect confirmation of the intuition that morphological information is much easier to learn than lexical information.

6   Conclusion and Future Work

This paper has presented a deep learning architecture with bidirectional LSTM cells for POS tagging. It successfully learns to translate word sequences to POS sequences. The study reports on the training of Bulgarian word embeddings. It also reports on the training of distributed representations of morphological information for Bulgarian. The two sets of distributed representations are shown to complement each other.

There are numerous directions that need to be explored in order to improve the models and to understand them better. In terms of the architecture itself, momentum, regularization and dropout could be added; more hidden LSTM modules could also improve accuracy, as well as additional hidden layers before and after the LSTM module (furnished with appropriate activation functions). Hyperparameter tuning (with respect to the learning rate, batch size, hidden layer size, etc.) should also be explored in much greater depth.

In terms of preprocessing, there is a lot of work that can be done as well. The corpus currently used to train the word embeddings is too small in comparison to billion-word corpora used for other languages. Increasing the corpus should allow for a meaningful increase in the dimensionality of the vectors (i.e. encoding of more relevant features). Additional experimentation and analysis of the suffix embedding method is also needed – to improve their quality (which includes implicitly devising a good evaluation method apart from POS tagging) and to obtain a good understanding how exactly they complement word embeddings.

Another important task is to train the system on the full tagset of 581 POS labels attested in the BulTreeBank corpus.

References

1. Graves, A.: Supervised Sequence Labelling. Springer, Heidelberg (2012)
2. Sutskever, I., Martens, J., Hinton, G.E.: Generating text with recurrent neural networks. In: Proceedings of the 28th International Conference on Machine Learning (ICML-11) (2011)
3. Graves, A.: Generating sequences with recurrent neural networks. arXiv preprint (2013). arXiv:1308.0850
4. Bahdanau, D., Kyunghyun, C., Bengio, Y.: Neural machine translation by jointly learning to align and translate. arXiv preprint (2014). arXiv:1409.0473
5. Yao, K., Zweig, G., Hwang, M., Shi, Y., Yu, D.: Recurrent neural networks for language understanding. In: INTERSPEECH, pp. 2524–2528 (2013)
6. Hochreiter, S., Schmidhuber, J.: Long short-term memory. Neural Comput. 9(8), 1735–1780 (1997)
7. Mikolov, T., Chen, K., Corrado, G., Dean, J.: Efficient estimation of word representations in vector space. arXiv preprint (2013). arXiv:1301.3781
8. Collobert, R., Weston, J.: A unified architecture for natural language processing: deep neural networks with multitask learning. In: Proceedings of the 25th International Conference on Machine Learning, pp. 160–167. ACM (2008)
9. Wang, P., Yao, Q., Soong, F.K., He, L., Zhao, H.: Part-of-speech tagging with bidirectional long short-term memory recurrent neural network. arXiv preprint (2015). arXiv:1510.06168
10. Wang, P., Qian, Y., Soong, F.K., He, L., Zhao, H.: A unified tagging solution: bidirectional LSTM recurrent neural network with word embedding. arXiv preprint (2015). arXiv:1511.00215
11. Huang, Z., Xu, W., Yu, K.: Bidirectional LSTM-CRF models for sequence tagging. arXiv preprint (2015). arXiv:1508.01991
12. Simov, K., Osenova, P.: A hybrid system for morphosyntactic disambiguation in Bulgarian. In: Proceedings of the RANLP 2001 Conference, Tzigov Chark, pp. 288–290 (2001)
13. Simova, I., Vasilev, D., Popov, A., Simov, K., Osenova, P.: Joint ensemble model for POS tagging and dependency parsing. In: SPMRL-SANCL 2014 (2014)
14. Graves, A., Mohamed, A., Hinton, G.: Speech recognition with deep recurrent neural networks. In: 2013 IEEE International Conference on Acoustics, Speech and Signal Processing (ICASSP), pp. 6645–6649. IEEE (2013)
15. Abadi, M., Agarwal, A., Barham, P., Brevdo, E., Chen, Z., Citro, C., Corrado, G., Davis, A., Dean, J., Devin, M.: TensorFlow: large-scale machine learning on heterogeneous distributed systems. arXiv preprint (2016). arXiv:1603.04467
16. Simov, K., Osenova, P., Slavcheva, M., Kolkovska, S., Balabanova, E., Doikoff, D., Ivanova, K., Simov, A., Kouylekov, M.: Building a linguistically interpreted corpus of Bulgarian: the BulTreeBank. In: LREC (2002)

Response Time Analysis of Text-Based CAPTCHA by Association Rules

Darko Brodić[1](✉), Alessia Amelio[2], and Ivo R. Draganov[3]

[1] University of Belgrade, Technical Faculty in Bor, V.J. 12, 19210 Bor, Serbia
dbrodic@tf.bor.ac.rs
[2] DIMES, University of Calabria, Via Pietro Bucci Cube 44, 87036 Rende (CS), Italy
aamelio@dimes.unical.it
[3] Technical University of Sofia, Blvd. Kliment Ohrdski 8, 1000 Sofia, Bulgaria
idraganov@tu-sofia.bg

Abstract. The paper introduces and discusses the usability problem of text-based type of CAPTCHA. In particular, two types of text-based CAPTCHA, with text and with numbers, are in the focus. The usability is considered in terms of response time to find a solution for the two aforementioned types of CAPTCHA. To analyze the response time, an experiment is conducted on 230 Internet users, characterized by multiple features, like age, number of years of Internet use, education level, response time in solving text-based CAPTCHA and response time in solving text-number-based CAPTCHA. Then, association rules are extracted from the values of these features, by employing the Apriori algorithm. It determines a new and promising statistical analysis in this context, revealing the dependence of response time to CAPTCHA to the co-occurrence of the feature values and the strength of these dependencies by rule support, confidence and lift analysis.

Keywords: Text-based CAPTCHA · Text-number-based CAPTCHA · Association rules · Statistical analysis · Apriori algorithm

1 Introduction

CAPTCHA is an acronym for "Completely Automated Public Turing Text to tell Computers and Humans Apart". It is based on "Turing tests" which have to recognize the difference between humans and machine by the humans [13]. In fact, CAPTCHA is a program that uses reversed "Turing tests". Accordingly, it has to recognize the difference between humans and machine (computer). This presents the most intriguing part, because the computer has to have an ability to discriminate humans from the machine [11]. Essentially, it is a test evaluated by computers, which "only" humans can pass [14].

In some way, the CAPTCHA has analogies with cryptographic problem, in which humans have a key to decrypt the problem unlike the bots. If we take into account "Turing test", CAPTCHA has to fulfill the following conditions [11]: (i) it should be easy to generate many instances of the problem linked with

© Springer International Publishing Switzerland 2016
C. Dichev and G. Agre (Eds.): AIMSA 2016, LNAI 9883, pp. 78–88, 2016.
DOI: 10.1007/978-3-319-44748-3_8

their solutions, (ii) humans have to solve it easily, (iii) the most advanced bots, i.e. programs will fail to solve the problem, and (iv) the problem is succinctly explained to be easily understood by the humans. Also, it should not be forgotten that one of the most important CAPTCHA criteria is to be publicly available (open code) [11].

The CAPTCHA can be used for the following reasons [6]: (i) prevention and reduction of spams on forums, (ii) prevention of opening a large number of orders by users, (iii) protection of user accounts from bots attacks, and (iv) validation of online surveys by determining whether the humans or bot access them.

The CAPTCHA quality can be considered in the following directions [3]: (i) usability, (ii) security, and (iii) practicality. The usability considers difficulties of solving the CAPTCHA. It means that it quantifies the time to find the solution to the CAPTCHA. Security evaluates the difficulty of finding solutions to CAPTCHA by the computer (computer program, i.e. bot). The practicality refers to the way of creating and realizing the CAPTCHA by programmers tools.

In this study, the text-based type of CAPTCHA will be considered. Still, the only element that will be taken into account is its usability. Hence, we shall explore the complexity of solving CAPTCHA from the user's viewpoint. A similar statistical analysis was made with a population of 24 participants, which represents a very small population from the factor analysis point of view [8]. Our experiment will include a population of 230 different Internet users, which is a more representative population sample. Two types of text-based CAPTCHA will be under consideration: (i) with text only, and (ii) with numbers only. The tested Internet users vary by the following attributes: (i) age, (ii) years of Internet use, and (iii) educational level. The experiment will show the differences of using different text-based CAPTCHAs which include text or number according to its usability value. This approach is important for choosing adequate types of CAPTCHA and their implementation in different web environments. Also, it identifies the suitability of the CAPTCHA type to a certain group of Internet users. In this paper, the usability of the CAPTCHA will be evaluated by data mining tools like association rules, which is to the best of our knowledge for the first time used for analyzing the CAPTCHA in the known literature. The association rules method is primarily focused on finding frequent co-occurring associations among a collection of items. Hence, its goal is to find associations of items that occur together more often than you would expect from a random sampling of all possibilities. As a consequence to CAPTCHA time response analysis by association rules, the results will show the real dependence between different variables which are of great or small importance to the value of CAPTCHA users response time.

The paper is organized in the following manner. Section 2 describes the text-based CAPTCHA. Section 3 describes the experiment. It includes the data mining elements like association rules, which will be used as the main evaluation tool. Section 4 presents the results and gives the discussion. Section 5 draws the conclusions.

2   Text-Based CAPTCHA

First CAPTCHA was designed by Broder's team in 1997 for Altavista, to prevent automatic adding URL to a database of a web browser [9].

Text-based CAPTCHA is the most widespread type of CAPTCHA. It asks the user to decrypt the text which is usually distorted in some way [10]. The text-based scheme typically relies on sophisticated distortion of the text images rendering it unrecognizable to the state of the art. It is popular among programmers, because it can be easily created. Hence, it is characterized by high-raking practicality. Its security depends on the solution quality of different element combinations in text and its background. However, it is attacked by the most advanced types of bots, which include the OCR system in. Figure 1 shows samples of typical text-based type of CAPTCHA.

Fig. 1. The samples of the text-based CAPTCHA

Figure 2 illustrates the typical samples of text-based type of CAPTCHA, which include numbers, sometimes called number-based CAPTCHA.

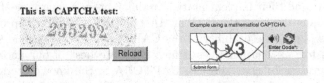

Fig. 2. The samples of the number-based CAPTCHA

Although the aforementioned text and number-based CAPTCHAs seem almost identical, the number-based CAPTCHA can include also some arithmetic operation. Hence, the user response to it can be different. However, from the user standpoint, it is extremely valuable to research the eventual response time differences in order to evaluate their suitability for certain group of Internet users. This is the main point that is going to be explored in the experimental part of our study.

3   Experiment

The experiment includes the testing of 230 Internet users, which have to solve given CAPTCHAs. Accordingly, each user is asked to solve a text and a number-based CAPTCHA. The value of 5 features including the following: (i) age, (ii) number of years of Internet use, (iii) education level, (iv) response time to solve text-based CAPTCHA, and (v) response time to solve number-based CAPTCHA is registered for each user, and then stored into a dataset. Hence, the dataset contains a total of 230 instances of 5 feature values for each Internet user.

The aim of the analysis is to find the association among the different feature values. In particular, we investigate if there is some implication between multiple feature values concerning the time spent by the user to efficiently solve text and number-based CAPTCHAs, and the strength of this implication. In fact, it is useful to understand in which users' conditions a given response time is obtained as well as the correlation between multiple feature values and the response time of the user. To pursue this goal, we find sets of feature values co-occurring frequently together, which are associated to a given response time of the user.

3.1   Association Rules

The most natural method to detect the frequent sets of feature values is the association rules one [1]. Next, we briefly recall the main concepts underlying the association rules and introduce the constraints for solving our task.

Let $T = T_1 \cup T_2 \cup, ..., \cup T_n$ be the set of items corresponding to all possible values of n features. In particular, let $T_i = \{t_i^1, t_i^2, ..., t_i^x\}$ be the set of x items representing all possible values of feature i. In our case, n is equal to 5 and T contains all the possible values of age, number of years of Internet use, education level, text-based CAPTCHA response time, and number-based CAPTCHA response time. A transaction I is a subset of n items extracted from T, such that $\forall t_i^j, t_k^h \in I$, $i \neq k$, and j and h define two possible values respectively of feature i and k. It indicates that items in I correspond each to a value of a distinct feature. In our case, given i.e. the possible values of "education level" feature, only one of them can be contained inside transaction I. Consequently, each transaction represents an instance of the aforementioned dataset.

Given these concepts, an Association Rule (AR) is defined as an implication $W \Rightarrow Z$, where W and Z are disjoint sets of items, representing respectively antecedent and consequent, characterized by *support* and *confidence*. Support measures how many times the items in $W \cup Z$ co-occur in the same transaction inside the dataset. It evaluates the statistical significance of AR. Confidence quantifies how many times the items in $W \cup Z$ co-occur in the transactions containing W. It is an estimate of the conditional probability of Z given W.

The aim is to detect the "meaningful" ARs from the dataset, having support $\geq minsupport$ and confidence $\geq minconfidence$, whose Z contains only items of text-based or number-based response time. It allows to analyze the co-occurrences of age values, number of Internet years use and education level, determining a given response time, and the strength of these co-occurrences, realizing a statistical analysis.

In order to detect the "meaningful" ARs, the well-known *Apriori* algorithm is employed [2]. It consists of two main steps: (i) detection of all the sets of items with frequency $\geq$ *minsupport* in the dataset; (ii) construction of the ARs from detected sets of items, considering the *minconfidence* threshold. The algorithm is based on the *anti-monotonicity* concept, for which if a set of items is unfrequent, also every its superset will be. For this reason, in each iteration, unfrequent sets of items are pruned from the algorithm. In the first step, the algorithm detects the sets of items of size k with frequency $\geq$ *minsupport*. Then, it enlarges them to size $k + 1$, by including the only sets of items of size 1 and frequency $\geq$ *minsupport*. These two steps are repeated starting from $k = 1$ to a certain value of k, for which no more sets of items of size $k + 1$ with frequency $\geq$ *minsupport* can be generated. In the second step, rules are generated for each frequent set of items F, finding all the subsets $f \subset F$ such that confidence value of $f \to F - f$ is $\geq$ *minconfidence*.

To evaluate the efficacy of predictability of the ARs, the *lift* measure is adopted [5]. *Lift* is the ratio between the confidence of the rule and the expected confidence, assuming that W and Z are independent [7]. If the lift takes a value of 1, then W and Z are independent. On the contrary, if the lift is bigger than 1, W and Z will co-occur more often than expected. This means that the occurrence of W has a positive effect on the occurrence of Z. Hence, the rule is potentially able to predict Z in different datasets.

4   Results and Discussion

All experimentation has been performed in MATLAB R2012a on a notebook quad-core at 2.2 GHz, 16 GB RAM and UNIX operating system.

In the aforementioned dataset, "age" can assume two different values, identifying users below 35 years, and users above 35 years. "Education level" has also two possible values, which are expressed as "higher education" and "secondary education". Furthermore, the values of the "number of years of Internet use" vary in the interval from 1 to 9 in the dataset. However, in order to improve the quality of the extracted ARs, this last feature has been regularly discretized by adopting an Equal-Width Discretization [12], to obtain only 3 intervals of interest, identified as "high Internet use" (> 6 years), "middle Internet use" (from 4 to 6 years), and "low Internet use" (< 4 years).

The two features, representing respectively the response time in solving text-based CAPTCHA and the response time in solving number-based CAPTCHA, have numerical values too, but corresponding to real numbers ranging from 0.0 to $+\infty$. A test using different discretization methods has been conducted on the numerical values and showed that clustering is the most suitable method for this task. Accordingly, response times have been discretized into 3 different intervals, by adopting the K-Medians clustering algorithm [4], whose advantage is the detection of natural groups of values based on their characteristics, overcoming the limitations of K-Means in managing the outliers. K-Medians determines the k value intervals, where $k = 3$ in this case, minimizing a function J, which quantifies the total sum of L_1 norm between each value in a given interval and its centroid. Given an interval, its centroid is the median over the values of the interval. Ten different executions of K-Medians algorithm have been performed on the response time values, and the set of intervals with the lowest value of J has been selected as the final solution. At the end, K-Medians determined 3 different intervals corresponding to "low response time" (< 20.09 s), "middle response time" (from 20.09 s to 39.97 s), and "high response time" (> 39.97 s).

Table 1 shows the possible values for each feature of the dataset. For discretized features, their corresponding intervals are reported.

Table 1. Possible values and corresponding intervals for each feature of the dataset

Features	Values	Interval
Age	Below 35	-
	Above 35	-
Education level	Higher education	-
	Secondary education	-
Number of years of Internet use	High Internet use	> 6 years
	Middle Internet use	4 - 6 years
	Low Internet use	< 4 years
Response time in solving text-based CAPTCHA	High response time	> 39.97 s
	Middle response time	20.09 s–39.97 s
	Low response time	< 20.09 s
Response time in solving number-based CAPTCHA	High response time	> 39.97 s
	Middle response time	20.09 s–39.97 s
	Low response time	< 20.09 s

An example of transaction obtained from the values, eventually discretized, of the features, corresponding to a row of the dataset, is given in Table 2.

Table 2. Example of transaction

Age	Educ. level	Num. years Internet use	Resp. time text	Resp. time text-number
Above 35	Higher educ	Middle Internet use	Middle resp. time	Middle resp. time

Furthermore, an example of AR defined over the values of the features "age", "number of years of Internet use", and "response time in solving text-based CAPTCHA", is given in Eq. (1).

$$\text{(age)} \quad \text{(Num. years Internet use)} \quad \text{(Resp. time text)}$$
$$above\ 35,\ middle\ Internet\ use \rightarrow middle\ response\ time \qquad (1)$$

The antecedent W of the AR is $\{above\ 35,\ middle\ Internet\ use\}$, while the consequent Z is $\{middle\ response\ time\}$. It can be observed that $W \cup Z$, which is $\{above\ 35,\ middle\ Internet\ use,\ middle\ response\ time\}$, is contained inside the transaction in Table 2. This AR expresses that a response time in solving text-based CAPTCHA between 20.09 s and 39.97 s is likely to occur in correspondence to users with age above 35 years and having an experience in Internet use varying from 4 to 6 years.

Tables 3-4 contain the ARs extracted from the dataset by the Apriori algorithm. For each AR, the antecedent W, the consequent Z, the support, confidence and *lift* are reported. Also, the thresholds of *minsupport* and *minconfidence* have been fixed respectively to 5 % and 50 %, because even if the sets of items are not so frequent in the dataset, the corresponding ARs are still interesting to analyze. It is worth to note that the *lift* is > 1 for all the ARs. This means that the antecedent and the consequent are always positively correlated.

Table 3 reports the ARs extracted from the dataset, whose consequent is a value associated to response time in solving text-based CAPTCHA. Looking at the ARs, some general considerations of interesting dependences can be performed. In particular, users with age below 35 years are almost always associated to a low response time in solving the text-based CAPTCHA. Also, users with age above 35 years are mostly associated to a middle-high response time in solving text-based CAPTCHA. The rule with the highest confidence of 0.90 is $\{below\ 35,\ higher\ educ.,\ middle\ Internet\ use \rightarrow low\ resp.\ time\}$. It indicates that in 90 % of transactions where users are: (i) below 35 years, (ii) have a higher education, and (iii) a middle experience in Internet use, the response time is low. The AR with the second highest confidence value of 0.8723 is $\{below\ 35,\ higher\ educ. \rightarrow low\ resp.\ time\}$. It denotes that in around 87 % of transactions where users are below 35 years having higher education, the response time to text-based CAPTCHA is low. Furthermore, it can be noted that the confidence of this rule is higher than the confidence of the rule $\{below\ 35,\ secondary : educ. \rightarrow low\ resp.\ time\}$. It indicates that in most of the cases when users are below 35 years and have a higher education, the response time is low. Also, it is more likely that young users with higher education level, rather than with secondary education level, provide a solution to text-based CAPTCHA in a reduced time. It is also interesting to observe that, when users with middle experience in Internet use are below 35 years, the response time is low with a confidence of 0.6364, while for users above 35 years, the response time is middle with the same confidence. It indicates that the age is discriminatory for obtaining a low response time, while the Internet use is not. Finally, a meaningful rule with lift value of 5.5740 is $\{above\ 35,\ low\ Internet\ use \rightarrow high\ resp.\ time\}$ indicating that users above 35 years with a low Internet use are strongly correlated to a high response time.

Table 4 shows the ARs extracted from the dataset, whose consequent is a value related to response time to solve a number-based CAPTCHA. It can be

Table 3. Association rules extracted from the dataset, for which the consequent is related to response time in solving text-based CAPTCHA

Antecedent	Consequent	Support	Confidence	Lift
Below 35, Higher educ., Middle internet use	Low resp. time	0.0786	0.9000	1.7466
Below 35, Higher educ.	Low resp. time	0.1790	0.8723	1.6929
Above 35, Low internet use	High resp. time	0.0524	0.7059	5.5740
Below 35	Low resp. time	0.4672	0.6687	1.2978
Below 35, High internet use	Low resp. time	0.1223	0.6667	1.2938
Below 35, Middle internet use	Low resp. time	0.1834	0.6364	1.2350
Above 35, Middle internet use	Middle resp. time	0.0917	0.6364	1.4870
Above 35, Higher educ., Middle internet use	Middle resp. time	0.0611	0.6364	1.4870
Below 35, Secondary educ., High internet use	Low resp. time	0.0742	0.6071	1.1783
Above 35, Higher educ.	Middle resp. time	0.1048	0.6000	1.4020
Below 35, Secondary educ.	Low resp. time	0.2882	0.5841	1.1335
Higher educ., High internet use	Low resp. time	0.0611	0.5833	1.1321
Higher educ.	Low resp. time	0.2183	0.5747	1.1153
High internet use	Low resp. time	0.1354	0.5741	1.1141
Secondary educ., High internet use	Low resp. time	0.0742	0.5667	1.0997
Above 35	Middle resp. time	0.1703	0.5652	1.3208
Below 35, Secondary educ., Middle internet use	Low resp. time	0.1048	0.5217	1.0125
Above 35, Secondary educ.	Middle resp. time	0.0655	0.5172	1.2086

noted that users with an age below 35 years and higher education level are mostly associated to a low response time in solving number-based CAPTCHA, independently from the experience in Internet use. Differently, users with an age above 35 years are strongly associated to a middle response time in solving number-based CAPTCHA, with lift values of around 2. In particular, the AR with the highest confidence of 1.00 is {$below\,35,\ higher\,educ. \rightarrow low\,resp.\ time$}. It indicates that, among the users below 35 years and with higher education level, the 100 % of them is able to quickly solve the number-based CAPTCHA. This means that an age below 35 years together with a higher education level are well-correlated to a low response time. However, also the users only having an age below 35 years or a higher education level are able to characterize the low response time, with smaller confidence values but higher support values

Table 4. Association rules extracted from the dataset, for which the consequent is related to response time in solving number-based CAPTCHA

Antecedent	Consequent	Support	Confidence	Lift
Below 35, Higher educ.	Low resp. time	0.2052	1.0000	1.4967
Below 35, Higher educ., Middle internet use	Low resp. time	0.0873	1.0000	1.4967
Below 35, Higher educ., High internet use	Low resp. time	0.0611	1.0000	1.4967
Below 35	Low resp. time	0.5851	0.8375	1.2535
Below 35, Middle internet use	Low resp. time	0.2402	0.8334	1.2473
Below 35, High internet use	Low resp. time	0.1528	0.8334	1.2473
Below 35, Secondary educ.	Low resp. time	0.3799	0.7699	1.1523
Below 35, Secondary educ., Middle internet use	Low resp. time	0.1528	0.7609	1.1388
Higher educ., High internet use	Low resp. time	0.0786	0.7500	1.1225
Below 35, Secondary educ., High internet use	Low resp. time	0.0917	0.7500	1.1225
High internet use	Low resp. time	0.1747	0.7407	1.1087
Secondary educ., High internet use	Low resp. time	0.0961	0.7333	1.0976
Higher educ.	Low resp. time	0.2664	0.7011	1.0494
Above 35, Higher educ., Middle internet use	Middle resp. time	0.0568	0.5909	2.1479
Above 35, Middle internet use	Middle resp. time	0.0830	0.5758	2.0928
Above 35, Higher educ.	Middle resp. time	0.1004	0.5750	2.0901
Above 35	Middle resp. time	0.1528	0.5072	1.8438

than in the case where both the features are considered. Again, users with: (i) an age below 35 years, (ii) a higher education level, (iii) a middle experience in Internet use, and (iv) a low response time in solving the number-based CAPTCHA, appear in around 9 % of transactions. It is a middle-low value, but with a confidence of 1.00. For this reason, the information retrieved from the AR {*below* 35, *higher educ., middle Internet use* → *low resp. time*} becomes more interesting. It indicates that in the 100 % of the cases where: (i) users are below 35 years, (ii) have a higher education level, and (iii) have a middle experience in Internet use, the number-based CAPTCHA is quickly solved. In contrast, among the users above 35 years, with higher education level and middle Internet use, a middle response time occurs in around 59 % of cases. For this reason, the connection between age above 35, higher education level and middle response time becomes promising. Finally, it is worth to note that a high Internet use is able to obtain a low response time, independently from education level.

In conclusion, from this analysis we observed that an age below 35 or a higher education level are strongly correlated to a low response time in solving the text-based CAPTCHA. Also, an age below 35 together with a higher education or a middle experience in Internet use are mostly correlated to a low response time in solving number-based CAPTCHA. Furthermore, comparison between text and number-based CAPTCHA shows that: (i) number-based CAPTCHA has higher support and confidence to be solved in low response time (below 35, higher educ.), and (ii) text-based CAPTCHA has higher support and confidence to be solved in the middle and high response time (above 35, low Internet use). Hence, number-based CAPTCHA should be easily solvable.

5   Conclusions

The study introduced a new statistical method for analysis of the response time of Internet users to solve the CAPTCHA. In particular, text and number-based CAPTCHAs were in the focus. Analysis consisted of investigating the dependence between the response time to solve CAPTCHA and typical features associated to the users, like age, education level and number of years of Internet use. It was statistically accomplished by adopting the ARs. Apriori algorithm was employed on the dataset of the users for obtaining the ARs together with their support, confidence and lift. Analysis of the obtained ARs revealed interesting co-occurrences of feature values and their association to a given response time to solve the text and number-based CAPTCHA. Also, it allowed to investigate on the strength of these dependencies by making considerations about the rule support, confidence and lift.

Future work will extend the method to different features, to other types of CAPTCHA and to alternative statistical measures. Also, other closed itemset mining algorithms in Java or C/C++ will be adopted and tested on the dataset.

Acknowledgments. The authors are fully grateful to Ms. Sanja Petrovska for the helpful support in collecting the data.

References

1. Agrawal, R., Imieliński, T., Swami, A.: Mining association rules between sets of items in large databases. In: Proceedings of the ACM SIGMOD International Conference on Management of Data - SIGMOD, pp. 207–216 (1993)
2. Agrawal, R., Srikant, R.: Fast algorithms for mining association rules in large databases. In: Proceedings of the 20th International Conference on Very Large Data Bases, VLDB, pp. 487–499 (1994)
3. Baecher, P., Fischlin, M., Gordon, L., Langenberg, R., Lutzow, M., Schroder, D.: CAPTCHAs: the good, the bad and the ugly. In: Frieling, F.C., (ed.) Sicherheit. LNI, Vol. 170, pp. 353–365 (2010)

4. Bradley, P.S., Mangasarian, O.L., Street, W.N.: Clustering via concave minimization. Adv. Neural Inf. Process. Syst. **9**, 368–374 (1997). MIT Press
5. Brin, S., Motwani, R., Ullman, J.D., Tsur, S.: Dynamic itemset counting and implication rules for market basket data. In: Proceedings of the ACM SIGMOD International Conference on Management of Data (ACM SIGMOD), pp. 265–276 (1997)
6. CAPTCHA. www.google.com, http://en.wikipedia.org/wiki/CAPTCHA
7. Hahsler, M.: A probabilistic comparison of commonly used interest measures for association rules (2015). http://michael.hahsler.net/research/association_rules/measures.html
8. Lee, Y.L., Hsu, C.H.: Usability study of text-based CAPTCHAs. Displays **32**(1), 81–86 (2011)
9. Lillibridge, M., Abadi, M., Bharat, K., Broder, A.: Method for selectively restricting access to computer systems. United States Patent 6195698, Applied 1998 and Approved 2001
10. Ling-Zi, X., Yi-Chun, Z.: A case study of text-based CAPTCHA attacks. In: Proceedings of International Conference on Cyber Enabled Distributed Computing and Knowledge Discover, pp. 121–124 (2012)
11. Naor, M.: Verification of a human in the loop or Identification via the Turing Test. Report, Weizmann Institute of Science (1996)
12. Sullivan, D.G.: Data mining V: preparing the data. http://cs-people.bu.edu/dgs/courses/cs105/lectures/data_mining_preparation.pdf
13. Turing, A.M.: Computing machinery and intelligence. Mind **59**, 433–460 (1950)
14. Von Ahn, L., Blum, M., Langford, J.: Telling humans and computers apart automatically. Commun. ACM **47**(2), 47–60 (2004)

New Model Distances and Uncertainty Measures for Multivalued Logic

Alexander Vikent'ev[1,2(✉)] and Mikhail Avilov[2]

[1] Sobolev Institute of Mathematics of the Siberian Branch of the Russian Academy of Sciences, Acad. Koptyug Avenue 4, 630090 Novosibirsk, Russia
vikent@math.nsc.ru
[2] Novosibirsk State University, Pirogova Str. 2, 630090 Novosibirsk, Russia
mikhail8avilov@gmail.com

Abstract. There is an increasing importance of problems regarding the analysis of propositions of experts and clustering of information contained in databases. Propositions of experts can be presented as formulas of n-valued logic L_n. This paper is concerned with defining metrics and degrees of uncertainty on formulas of n-valued logic. After metrics and degrees of uncertainty (as well as their useful properties) have been established, they are used for the cluster analysis of the sets of n-valued formulas. Various clustering algorithms are performed and obtained results are analyzed. Established methods can be further employed for experts propositions analysis, clustering problems and pattern recognition.

Keywords: Distances on formulas of n-valued logic · Metrics · Uncertainty measures · Cluster analysis · Pattern recognition

1 Introduction

The problem of ranking the statements of experts according to their informativeness and inducing the metric on the space of statements was introduced by Lbov and Zagoruiko in the early 1980s [5, 8, 12]. Expert statements were written in the form of logical formulas. Thus, the task of comparing and ranking the statements of experts turned into a task of comparing and ranking the logical formulas. In order to do this, the distance between the formulas and the uncertainty measure of the formulas were introduced.

Lbov and Vikent'ev used the normalized symmetric difference of the formulas models as a distance for the case of two-valued logic [3]. Then the model-theoretic approach to the analysis of multivalued formulas was proposed [1,2,4]. Formulas belong to the n-valued logic L_n [10]. Using the theory of models for n-valued formulas, the different versions of the distances and uncertainty measures were introduced and the properties of those quantities were established [4]. Then, to rank the statements of experts the clustering analysis of the finite sets of formulas of n-valued logic was performed based on the introduced distances and uncertainty measures. Clustering was performed for small n, and the utilized distance had constant weights [9].

© Springer International Publishing Switzerland 2016
C. Dichev and G. Agre (Eds.): AIMSA 2016, LNAI 9883, pp. 89–98, 2016.
DOI: 10.1007/978-3-319-44748-3_9

In this paper we introduce a new distance, which generalizes the distances used for multivalued logical formulas previously. The metric properties of the new distance are proven and it is shown that there is a continuum of such distances. A new uncertainty measure of multivalued logical formulas is also introduced. Using those new quantities the clustering analysis of finite sets of the formulas of n-valued logic L_n is performed. The results of various clustering algorithms are obtained for different dimensions n of logic L_n.

2   Definitions and Notations

In this section we define n-valued logic L_n and some useful model-theoretic properties.

Definition 1. *Propositional language L consists of the following propositional symbols:*
  1. x, y, z, ... – propositional variables;
  2. $\neg$, $\rightarrow$ – propositional logical connectives;
  3. $($, $)$ – auxiliary symbols.

Definition 2. *Formulas are the finite sequences of propositional symbols defined the following way:*
  1. x, y, z, ... – elementary formulas;
  2. If φ and ψ are formulas, then $\neg\varphi$, $\varphi \rightarrow \psi$ are formulas;
  3. No other finite sequences of propositional symbols, except those mentioned in 1, 2, are formulas.

Now we can introduce n-valued logic L_n.

Definition 3. *N-valued logic L_n is defined the following way:*
  $M_n = \langle V_n, \neg, \rightarrow, \{1\} \rangle$ *– n-valued Lukasiewicz matrix ($n \in N, n \geqslant 2$);*
  $V_n = \left\{ 0, \dfrac{1}{n-1}, ..., \dfrac{n-2}{n-1}, 1 \right\}$ *– set of truth values;*
  $\neg : V_n \rightarrow V_n$, *– unary negation operation;*
  $\rightarrow : V_n \times V_n \rightarrow V_n$ *– binary implication operation;*
  $\{1\}$ *– selected value of truth.*

Let us now introduce other logical connectives for the truth values of n-valued logic L_n.

Definition 4. *Logical connectives on V_n are defined from the input connectives:*
  $\neg x = 1 - x$ *– negation;*
  $x \rightarrow y = min\{1, 1 - x + y\}$ *– implication;*
  $x \vee y = (x \rightarrow y) \rightarrow y = max\{x, y\}$ *– disjunction;*
  $x \wedge y = \neg(\neg x \vee \neg y) = min\{x, y\}$ *– conjunction.*

We also formulate model-theoretic properties and notations that will be used further in this paper.

Definition 5. *Let Σ be the finite set of formulas of L_n. The set of propositional variables $S(\varphi)$ used for writing a formula φ of n-valued logic L_n is called the support of a formula ϕ.*

$S(\Sigma) = \cup_{\varphi \in \Sigma} S(\varphi)$ is called the support of the set Σ.

Definition 6. *A model M is a subset of attributed variables. $P(S(\Sigma))$ is the set of all models.*

We use the notation $\varphi_{\frac{k}{n-1}}$ if the formula φ has the value $\frac{k}{n-1}, k = 0, ..., n-1$ on a model M.

$M(\varphi_{\frac{k}{n-1}})$ is the number of models on which formula φ has the value $\frac{k}{n-1}$.

$M(\varphi_{\frac{k}{n-1}}, \psi_{\frac{l}{n-1}})$ is the number of models on which φ has the value $\frac{k}{n-1}$ and ψ has the value $\frac{l}{n-1}$.

Notes and Comments. The cardinality of the set of models $P(S(\Sigma))$ is

$$| P(S(\Sigma)) | = n^{|S(\Sigma)|}. \tag{1}$$

The proof of this fact as well as the definition of the logical connectives on models and other auxiliary statements are detailed in papers [1,4].

3   Model Distances and Uncertainty Measures

Let us show how to introduce a distance between formulas φ and ψ of n-valued logic L_n. It is natural to assume, that the less the absolute difference between the values of the formulas is, the closer those formulas are. So we will multiply the number of models with the same absolute difference values by the coefficient which considers the proximity of the values of the formulas. Those coefficients used to be precisely the truth values of L_n, so the model distance ρ_0 between formulas φ and ψ of n-valued logic L_n used to be defined as the normalized quantity $\tilde{\rho}_0$ [9]:

$$\rho_0(\varphi, \psi) = \frac{1}{n^{|S(\Sigma)|}} \cdot \sum_{k=0}^{n-1} \sum_{l=0}^{n-1} \frac{| k - l |}{n - 1} \cdot M(\varphi_{\frac{k}{n-1}}, \psi_{\frac{l}{n-1}}). \tag{2}$$

Example 1. Let $n = 5$. Then $\rho_0(\varphi \wedge \psi, \varphi \vee \psi) = 0.4$, $\rho_0(\varphi \wedge \psi \wedge \chi \wedge \omega, \varphi \to \omega) = 0.2576$.

The particularity of the quantity (2) is that its coefficients (weights) $\frac{|k-l|}{n-1}$ are constant. This particularity does not allow to adjust the weight of the quantity $M(\varphi_{\frac{k}{n-1}}, \psi_{\frac{l}{n-1}})$ (the number of models with the same absolute difference values) to be properly included in the final distance ρ_0.

The uncertainty measure I_0 of formula φ of n-valued logic L_n used to be defined as follows [9]:

$$I_0(\varphi) = \rho_0(\varphi, 1) = \sum_{i=0}^{n-2} \frac{n - 1 - i}{n - 1} \cdot \frac{M(\varphi_{\frac{i}{n-1}})}{n^{|S(\Sigma)|}}. \tag{3}$$

Example 2. Let $n = 5$. Then $I_0(\varphi \to \psi) = 0.2$, $I_0(\varphi \vee \psi \vee \chi \vee \omega) = 0.1416$.

The quantity (3) possesses the same particularity as the quantity (2): its weight $\frac{n-1-i}{n-1}$ is constant. This particularity also does not allow to adjust the weight of the quantity $M(\varphi_{\frac{i}{n-1}})$ to be properly included in the final uncertainty measure $I_0(\varphi)$.

Let us modify the quantities above to get rid of the particularities associated with the rigid structure of the weights. In order to do this, we substitute constant weights of the quantities ρ_0 and I_0 for arbitrary acceptable weights.

Definition 7. *The model distance between formulas φ and ψ of n-valued logic L_n, $S(\varphi) \cup S(\psi) \subseteq S(\Sigma)$ on $P(S(\Sigma))$ is called the quantity*

$$\rho(\varphi, \psi) = \frac{1}{n^{|S(\Sigma)|}} \cdot \sum_{k=0}^{n-1} \sum_{l=0}^{n-1} \lambda_{|k-l|} \cdot M(\varphi_{\frac{k}{n-1}}, \psi_{\frac{l}{n-1}}) \tag{4}$$

$$0 = \lambda_0 \leqslant \lambda_1 \leqslant ... \leqslant \lambda_{n-1} = 1;$$

$$n \geqslant 2.$$

Definition 8. *The uncertainty measure I of formula φ of n-valued logic L_n, $S(\varphi) \cup S(\psi) \subseteq S(\Sigma)$ on $P(S(\Sigma))$ is called the quantity*

$$I(\varphi) = \sum_{i=0}^{n-2} \alpha_i \cdot \frac{M(\varphi_{\frac{i}{n-1}})}{n^{|S(\Sigma)|}} \tag{5}$$

$$0 \leqslant \alpha_i \leqslant 1;$$

$$\alpha_k \geqslant \alpha_i \, \forall k \leqslant i;$$

$$\alpha_i + \alpha_{n-1-i} = 1 \, \forall i = 0, ..., \frac{n-1}{2}.$$

Notes and Comments. We got a new continuum of distances, possessing the properties of metrics.

In Definition 8 the coefficients α of the uncertainty measure I actually depend on the coefficients λ of the model distance ρ because the uncertainty measure is itself a distance. Moreover,

$$I(\varphi) = \rho(\varphi, 1). \tag{6}$$

So the given definition of the uncertainty measure for n-valued logic corresponds with the earlier ideas of Lbov and Bloshitsin for two-valued logic [8].

Let us now establish some of the properties of the model distance (4) and uncertainty measure (5). We start with the properties of the distance ρ introduced in Definition 7.

Theorem 1. *For any formulas φ, ψ, χ of Σ the following assertions hold:*

1. $0 \leqslant \rho(\varphi, \psi) \leqslant 1$;
2. $\rho(\varphi, \psi) = 0 \Leftrightarrow \varphi \equiv \psi$;
3. $\rho(\varphi, \psi) = \rho(\psi, \varphi)$;
4. $\rho(\varphi, \psi) \leqslant \rho(\varphi, \chi) + \rho(\chi, \psi)$;
5. $\varphi \equiv \varphi_1, \psi \equiv \psi_1 \Rightarrow \rho(\varphi, \psi) = \rho(\varphi_1, \psi_1)$.

Proof. Each of the assertions will be treated separately.

1. The distance calculation formula involves all models with coefficients from 0 to 1. $\rho(\varphi, \psi) = 0$ when $\varphi \equiv \psi$ and $\rho(\varphi, \psi) = 1$ when $\varphi \equiv \neg\psi$. φ and ψ take only the values 0 and 1 on the models involved. So $0 \leqslant \rho(\varphi, \psi) \leqslant 1$.

2. Necessity follows from the proof of the previous assertion. Sufficiency follows from that, given the definition of equivalence, if $\varphi \equiv \psi$ then the values of φ and ψ are the same on all models. Then for $k = l$ every $M(\varphi_{\frac{k}{n-1}}, \psi_{\frac{l}{n-1}})$ in the formula $\rho(\varphi, \psi)$ is multiplied by 0 hence $\rho(\varphi, \psi) = 0$.

3. Symmetrical pairs $M(\varphi_{\frac{k}{n-1}}, \psi_{\frac{l}{n-1}}) \neq M(\varphi_{\frac{l}{n-1}}, \psi_{\frac{k}{n-1}})$ are multiplied by the same coefficient. So $\rho(\varphi, \psi) = \rho(\psi, \varphi)$.

4. Follows from the model-theoretic properties given in Sect. 2 and the paper [9].

5. Follows from the definition of equivalence of the two formulas [11]. $\quad\square$

Notes and Comments. Assertions 2–4 are the properties of the metric. This means we can define a metric on the equivalence classes of the formulas of L_n.

Let us now establish the properties of the uncertainty measure I introduced in Definition 8.

Theorem 2. *For any formulas φ, ψ of Σ the following assertions hold:*

1. $0 \leqslant I(\varphi) \leqslant 1$;
2. $I(\varphi) + I(\neg\varphi) = 1$;
3. $I(\varphi \wedge \psi) \geqslant max\{I(\varphi), I(\psi)\}$;
4. $I(\varphi \vee \psi) \leqslant min\{I(\varphi), I(\psi)\}$;
5. $I(\varphi \wedge \psi) + I(\varphi \vee \psi) \geqslant I(\varphi) + I(\psi)$.

Proof. Each of the assertions will be treated separately.

1. $I(\varphi) = \rho(\varphi, 1)$ hence $0 \leqslant I(\varphi) \leqslant 1$.

2. $I(\varphi) + I(\neg\varphi) = \alpha_0 \cdot \dfrac{M(\varphi_0)}{n^{|S(\Sigma)|}} + \alpha_{n-1} \cdot \dfrac{M(\varphi_1)}{n^{|S(\Sigma)|}} + \displaystyle\sum_{i=1}^{n-2}(\alpha_i + \alpha_{n-1-i}) \cdot \dfrac{M(\varphi_{\frac{i}{n-1}})}{n^{|S(\Sigma)|}} =$

$$= \frac{|P(S(\Sigma))|}{n^{|S(\Sigma)|}} = 1.$$

3. $I(\varphi \wedge \psi) = \displaystyle\sum_{i=0}^{n-2}\alpha_i \frac{M((\varphi \wedge \psi)_{\frac{i}{n-1}})}{n^{|S(\Sigma)|}} =$

$$= \sum_{i=0}^{n-2}\alpha_i\left(\sum_{k=i}^{n-1}\left(\frac{M(\varphi_{\frac{k}{n-1}} \wedge \psi_{\frac{i}{n-1}})}{n^{|S(\Sigma)|}} + \frac{M(\varphi_{\frac{i}{n-1}} \wedge \psi_{\frac{k}{n-1}})}{n^{|S(\Sigma)|}}\right)\right) - \alpha_i \frac{M(\varphi_{\frac{i}{n-1}} \wedge \psi_{\frac{i}{n-1}})}{n^{|S(\Sigma)|}}.$$

$$I(\varphi) = \sum_{i=0}^{n-2} \alpha_i \sum_{k=0}^{n-1} \frac{M(\varphi_{\frac{i}{n-1}} \wedge \psi_{\frac{k}{n-1}})}{n^{|S(\Sigma)|}}$$

$$= \sum_{i=0}^{n-2} \alpha_i \left(\sum_{k=i}^{n-1} \frac{M(\varphi_{\frac{i}{n-1}} \wedge \psi_{\frac{k}{n-1}})}{n^{|S(\Sigma)|}} + \sum_{k=0}^{i} \frac{M(\varphi_{\frac{i}{n-1}} \wedge \psi_{\frac{k}{n-1}})}{n^{|S(\Sigma)|}} \right) - \alpha_i \frac{M(\varphi_{\frac{i}{n-1}} \wedge \psi_{\frac{i}{n-1}})}{n^{|S(\Sigma)|}}.$$

$$I(\varphi \wedge \psi) - I(\varphi) = \sum_{i=0}^{n-2} \sum_{k=0}^{i} (\alpha_k - \alpha_i) \frac{M(\varphi_{\frac{i}{n-1}} \wedge \psi_{\frac{k}{n-1}})}{n^{|S(\Sigma)|}} + \sum_{i=0}^{n-2} \sum_{k=i}^{n-1} \alpha_i \frac{M(\varphi_{\frac{i}{n-1}} \wedge \psi_{\frac{k}{n-1}})}{n^{|S(\Sigma)|}} \geqslant$$

$$\geqslant 0.$$

So $I(\varphi \wedge \psi) \geqslant I(\varphi)$. Similarly we have an evaluation for ψ: $I(\varphi \wedge \psi) \geqslant I(\psi)$. Hence $I(\varphi \wedge \psi) \geqslant max\{I(\varphi), I(\psi)\}$.

4. $I(\varphi \vee \psi) = \sum_{i=0}^{n-2} \alpha_i \frac{M((\varphi \wedge \psi)_{\frac{i}{n-1}})}{n^{|S(\Sigma)|}} =$

$$= \sum_{i=0}^{n-2} \alpha_i \left(\sum_{k=i}^{n-1} \left(\frac{M(\varphi_{\frac{k}{n-1}} \wedge \psi_{\frac{i}{n-1}})}{n^{|S(\Sigma)|}} + \frac{M(\varphi_{\frac{i}{n-1}} \wedge \psi_{\frac{k}{n-1}})}{n^{|S(\Sigma)|}} \right) \right) - \alpha_i \frac{M(\varphi_{\frac{i}{n-1}} \wedge \psi_{\frac{i}{n-1}})}{n^{|S(\Sigma)|}}.$$

$$I(\varphi) = \sum_{i=0}^{n-2} \alpha_i \sum_{k=0}^{n-1} \frac{M(\varphi_{\frac{i}{n-1}} \wedge \psi_{\frac{k}{n-1}})}{n^{|S(\Sigma)|}} =$$

$$= \sum_{i=0}^{n-2} \alpha_i \left(\sum_{k=i}^{n-1} \frac{M(\varphi_{\frac{i}{n-1}} \wedge \psi_{\frac{k}{n-1}})}{n^{|S(\Sigma)|}} + \sum_{k=0}^{i} \frac{M(\varphi_{\frac{i}{n-1}} \wedge \psi_{\frac{k}{n-1}})}{n^{|S(\Sigma)|}} \right) - \alpha_i \frac{M(\varphi_{\frac{i}{n-1}} \wedge \psi_{\frac{i}{n-1}})}{n^{|S(\Sigma)|}}.$$

$$I(\varphi) - I(\varphi \vee \psi) = \sum_{i=0}^{n-2} \sum_{k=0}^{i} \alpha_i \frac{M(\varphi_{\frac{i}{n-1}} \wedge \psi_{\frac{k}{n-1}})}{n^{|S(\Sigma)|}} - \sum_{i=0}^{n-2} \sum_{k=0}^{i} \alpha_i \frac{M(\varphi_{\frac{k}{n-1}} \wedge \psi_{\frac{i}{n-1}})}{n^{|S(\Sigma)|}} \geqslant$$

$$\geqslant \sum_{i=0}^{n-2} \sum_{k=0}^{i} \alpha_i \frac{M(\varphi_{\frac{i}{n-1}} \wedge \psi_{\frac{k}{n-1}})}{n^{|S(\Sigma)|}} \geqslant 0.$$

So $I(\varphi \vee \psi) \leqslant I(\varphi)$. Similarly we have an evaluation for ψ: $I(\varphi \vee \psi) \leqslant I(\psi)$. Hence $I(\varphi \vee \psi) \leqslant min\{I(\varphi), I(\psi)\}$.

5. Follows from the equalities used in the proof of assertions 3 and 4.   □

4   Clustering the Formulas of N-valued Logic

Clustering analysis is quite important while working with databases, statements of experts or performing statistical modeling [6]. For the sets of statements we know only the distances between the formulas and uncertainty measures of formulas. So two algorithms based on the distances were chosen for the clustering analysis – the hierarchic algorithm and the k-means algorithm. Those algorithms were adapted to work with multivalued formulas [7]. The complexity of computing the distance is exponential.

For the experiments below there was created a knowledge base, consisting of 300 multivalued formulas. The finite subsets of those formulas were randomly picked up for the clustering analysis. After that the value n for the logic L_n was chosen, the clustering algorithm was picked up and the weights λ were entered. Then the clustering analysis was performed utilizing either hierarchic algorithm or k-means algorithm. Both algorithms are based on new model distances and uncertainty measures introduced in Definitions 7 and 8 respectively. The results of the performed clusterings are presented in tables.

4.1 Hierarchic Algorithm, $n = 5$.

Let $n = 5$. Consider the set *test* 1, which consists of 8 formulas·of five-valued logic L_5: $\varphi_1 = x \to y$, $\varphi_2 = \neg(x \to y)$, $\varphi_3 = (x \lor z) \to y$, $\varphi_4 = \neg((x \land y) \lor z) \to w$, $\varphi_5 = y \to (x \land z)$, $\varphi_6 = (\neg y \lor (x \to z))$, $\varphi_7 = z \to (x \lor y)$, $\varphi_8 = \neg((z \land y) \to x$.

We perform clustering analysis of the set *test* 1 using hierarchic algorithm. We choose the following weights: $\lambda_0 = 0, \lambda_1 = \frac{1}{4}, \lambda_2 = \frac{2}{4}, \lambda_3 = \frac{3}{4}, \lambda_4 = 1$ (the standard weights). Based on the distances matrix the minimal distance is $\rho_{4,6} = 0,0510$, so the first cluster is $\varphi_{4,6}$. After six more iterations the results of the performed clustering are presented in Table 1.

Table 1. Hierarchic algorithm, *test* 1, $n = 5$

Iteration	Δ	Clusters
0	0,0000	$\varphi_1, \varphi_2, \varphi_3, \varphi_4, \varphi_6, \varphi_5, \varphi_7, \varphi_8$
1	0,0508	$\varphi_1, \varphi_2, \varphi_3, \varphi_{4,6}, \varphi_5, \varphi_7, \varphi_8$
2	0,1000	$\varphi_{1,3}, \varphi_2, \varphi_{4,6}, \varphi_5, \varphi_7, \varphi_8$
3	0,1376	$\varphi_{1,3}, \varphi_2, \varphi_{4,6,7}, \varphi_5, \varphi_8$
4	0,1376	$\varphi_{1,3}, \varphi_2, \varphi_{4,6}, \varphi_{5,8}$
5	0,2092	$\varphi_2, \varphi_{5,8}, \varphi_{1,3,4,6,7}$
6	0,2092	$\varphi_2, \varphi_{1,3,4,5,6,7,8}$

To stop our clustering algorithm we use the quantity Δ – the maximal value of uncertainty measure among all elements of every cluster. For instance, if we set $\Delta = 0,1500$, then the algorithm stops after fourth iteration and gives 4 clusters as a result: $\varphi_{13}, \varphi_2, \varphi_{46}, \varphi_{58}$. We set $\Delta = 0,2100$, so the algorithm stops after sixth iteration giving 2 clusters as a result: $\varphi_2, \varphi_{1,3,4,5,6,7,8}$. If we do not stop the algorithm, then after seven iterations all 8 formulas merge into a single cluster.

4.2 K-means Algorithm, $n = 5$.

Let $n = 5$. Consider the set *test* 1, which consists of 8 formulas of five-valued logic L_5: $\varphi_1 = x \to y$, $\varphi_2 = \neg(x \to y)$, $\varphi_3 = (x \lor z) \to y$, $\varphi_4 = \neg((x \land y) \lor z) \to w$, $\varphi_5 = y \to (x \land z)$, $\varphi_6 = (\neg y \lor (x \to z))$, $\varphi_7 = z \to (x \lor y)$, $\varphi_8 = \neg((z \land y) \to x$.

We perform clustering analysis of the set *test* 1 using k-means algorithm. Suppose we need 3 clusters as a result. We choose the following weights: $\lambda_0 = 0, \lambda_1 = \frac{1}{4}, \lambda_2 = \frac{2}{4}, \lambda_3 = \frac{3}{4}, \lambda_4 = 1$. Based on the matrix of distances we choose 3 centres: $\varphi_2, \varphi_4, \varphi_5$ (the centres are approximately equidistant from each other and the sum of distances between them is maximal). The remaining formulas are distributed into the source clusters according to those centres. This gives us the following clusters: $\varphi_2, \varphi_{1,3,4,6,7}, \varphi_{5,8}$. Then the algorithm calculates the centres of mass again and redistributes the formulas according to the renewed centres. After this we have the following clusters: $\varphi_2, \varphi_{1,3,4,6,7}, \varphi_{5,8}$. As we see, the clusters didn't change – this means the algorithm stops. The results of the performed clustering are presented in Table 2.

Table 2. K-means algorithm, *test* 1, $n = 5$

Iteration	Centres	Clusters
1	$\varphi_2, \varphi_4, \varphi_5$	$\varphi_2, \varphi_{1,3,4,6,7}, \varphi_{5,8}$
2	$\varphi_2, \varphi_3, \varphi_5$	$\varphi_2, \varphi_{1,3,4,6,7}, \varphi_{5,8}$

The algorithm stops after second iteration giving 3 clusters as a result: $\varphi_2, \varphi_{1,3,4,6,7}, \varphi_{5,8}$.

The results obtained after the clustering of the set *test* 1 utilizing the algorithms based on new distances and uncertainty measures are different from the results obtained after the clustering of the same set using the distances and uncertainty measures with rigid weights in [9]. This demonstrates the feasibility of using different distances in clustering algorithms.

4.3 Hierarchic Algorithm, $n = 9$.

Let $n = 9$. Consider the set *test* 2, which consists of 10 formulas of nine-valued logic L_9: $\varphi_1 = \neg(z \vee y)$, $\varphi_2 = (x \rightarrow y) \rightarrow w$, $\varphi_3 = \neg((x \rightarrow y) \wedge z)$, $\varphi_4 = (x \vee z) \wedge y$, $\varphi_5 = z \rightarrow (x \vee y)$, $\varphi_6 = (\neg y \vee (x \rightarrow z))$, $\varphi_7 = w \wedge (x \rightarrow z)$, $\varphi_8 = (y \vee z) \rightarrow (x \vee w)$, $\varphi_9 = z \rightarrow (x \wedge w)$, $\varphi_{10} = \neg(x \rightarrow y)$.

We perform clustering analysis of the set *test* 2 using hierarchic algorithm. We choose the following weights: $\lambda_0 = 0, \lambda_1 = \frac{1}{30}, \lambda_2 = \frac{1}{20}, \lambda_3 = \frac{1}{10}, \lambda_4 = \frac{1}{5}, \lambda_5 = \frac{3}{10}, \lambda_6 = \frac{2}{5}, \lambda_7 = \frac{3}{5}, \lambda_8 = 1$. The stopping criterion is $\Delta = 0,2000$. The results of the performed clustering are presented in Table 3.

Table 3. Hierarchic algorithm, *test* 2, $n = 9$

Iteration	Δ	Clusters
0	0,0000	$\varphi_1, \varphi_2, \varphi_3, \varphi_4, \varphi_5, \varphi_6, \varphi_7, \varphi_8, \varphi_9, \varphi_{10}$
1	0,0073	$\varphi_1, \varphi_{2,3}, \varphi_4, \varphi_5, \varphi_6, \varphi_7, \varphi_8, \varphi_9, \varphi_{10}$
2	0,0173	$\varphi_{2,3}, \varphi_4, \varphi_5, \varphi_{1,6,7}, \varphi_8, \varphi_9, \varphi_{10}$
3	0,0952	$\varphi_{2,3}, \varphi_4, \varphi_5, \varphi_{1,6,7}, \varphi_{8,9}, \varphi_{10}$
4	0,0952	$\varphi_{2,3}, \varphi_5, \varphi_{1,4,6,7}, \varphi_{8,9}, \varphi_{10}$
5	0,1907	$\varphi_{2,3}, \varphi_5, \varphi_{1,4,6,7}, \varphi_{8,9,10}$
6	0,1907	$\varphi_{2,3,5}, \varphi_{1,4,6,7}, \varphi_{8,9,10}$

The algorithm stops after sixth iteration giving 3 clusters as a result: $\varphi_{2,3,5}, \varphi_{1,4,6,7}, \varphi_{8,9,10}$.

4.4   K-means Algorithm, $n = 7$.

Let $n = 7$. Consider the set *test* 3, which consists of 15 formulas of seven-valued logic L_7: $\varphi_1 = y \rightarrow (x \wedge z)$, $\varphi_2 = \neg z \rightarrow w(x \wedge y)$, $\varphi_3 = z \rightarrow (x \vee y)$, $\varphi_4 = \neg((x \wedge y) \vee z) \rightarrow w$, $\varphi_5 = \neg(x \wedge z) \rightarrow y$, $\varphi_6 = (\neg y \vee (x \rightarrow z))$, $\varphi_7 = z \rightarrow (x \vee y)$, $\varphi_8 = \neg((z \wedge y) \rightarrow x$, $\varphi_9 = \neg z \rightarrow x$, $\varphi_{10} = \neg((x \wedge y) \vee z) \rightarrow w$, $\varphi_{11} = y \rightarrow (x \wedge w) \rightarrow z$, $\varphi_{12} = y \vee (x \rightarrow z)$, $\varphi_{13} = z \wedge (x \rightarrow y)$, $\varphi_{14} = (x \wedge z) \rightarrow w$, $\varphi_{15} = (x \vee w) \rightarrow y$.

We perform clustering analysis of the set *test* 3 using k-means algorithm. Suppose we need 4 clusters as a result. We choose the following weights: $\lambda_0 = 0, \lambda_1 = \frac{1}{6}, \lambda_2 = \frac{2}{6}, \lambda_3 = \frac{3}{6}, \lambda_4 = \frac{4}{6}, \lambda_5 = \frac{5}{6}, \lambda_6 = 1$. We also select the future centres of the clusters: $\varphi_2, \varphi_5, \varphi_7, \varphi_9$. During every iteration of the algorithm the formulas are distributed into the source clusters (according to the centres), then the centres of mass are calculated, and the resulting clusters are updated. The results of the performed clustering are presented in Table 4.

Table 4. K-means algorithm, *test* 3, $n = 7$

Iteration	Centres	Clusters
1	$\varphi_2, \varphi_5, \varphi_7, \varphi_9$	$\varphi_{1,2,3,8,10}, \varphi_{5,14}, \varphi_{6,7}, \varphi_{4,9,11,12}$
2	$\varphi_8, \varphi_5, \varphi_7, \varphi_9$	$\varphi_{2,3,8,10}, \varphi_{5,14}, \varphi_{6,7}, \varphi_{1,4,9,11,12}$
3	$\varphi_8, \varphi_5, \varphi_7, \varphi_4$	$\varphi_{2,3,8,10}, \varphi_{5,14}, \varphi_{6,7}, \varphi_{1,4,9,11,12}$

The algorithm stops after third iteration giving 4 clusters as a result: $\varphi_{2,3,8,10}, \varphi_{5,14}, \varphi_{6,7}, \varphi_{1,4,9,11,12}$.

5   Conclusions

In this paper, a continuum of new distances and uncertainty measures is offered for logical multivalued formulas. The new quantities are generalizations of the quantities that were used previously. Theorems 1 and 2 which respectively establish the properties of the new model distance and uncertainty measure are proven. Those new quantities can be used to analyse knowledge bases, to create expert systems, or to construct logical decision functions for the problems of recognition.

Based on new quantities the clustering analysis of multivalued logical formulas is performed. The formulas belong to the n-valued logic L_n. The software package for clustering analysis of sets of logical formulas, using the hierarchical and k-means algorithms is developed. The number of formulas, the dimension of logic L_n and the values of weights were chosen differently. The results are obtained for $n = 5, n = 7, n = 9$ (and more). A comparison of the clustering results with the results of previous works for the case $n = 5$ is also performed.

In the future we plan to use the new quantities for the analysis of the large sets of statements of experts. For this purpose the coefficient matrix composed of weights $\lambda_{|k-l|}$ will be used. This approach will explore the relationship between the selection of the optimal clustering and the properties of the coefficient matrix and multivalued logic.

Acknowledgments. This work is supported by the Russian Foundation for Basic Research, project nos. 10-0100113a and 11-07-00345a.

References

1. Vikent'ev, A.A.: Distances and degrees of uncertainty in many-valued propositions of experts and application of these concepts in problems of pattern recognition and clustering. Pattern Recogn. Image Anal. **24**(4), 489–501 (2014)
2. Vikent'ev, A.A.: Uncertainty measure of expert statements, distances in many-valued logic and adaptation processes. In: XVI International Conference "Knowledge-Dialogue-Solution" KDS-2008, Varna, pp. 179–188 (2008). (in Russian)
3. Vikent'ev, A.A., Lbov, G.S.: Setting the metrics and informativeness on statements of experts. Pattern Recogn. Image Anal. **7**(2), 175–183 (1997)
4. Vikent'ev, A.A., Vikent'ev, R.A.: Distances and uncertainty measures on the statements of N-valued logic. In: Bulletin of the Novosibirsk State University, Serious of Mathematics Mechanics, Computer Science, Novosibirsk, vol. 11, no. 2, pp. 51–64 (2011). (in Russian)
5. Lbov, G.S., Startseva, N.G.: Logical Solving Functions and the Problem on Solutions Statistical Stability. Sobolev Institute of Mathematics, Novosibirsk (1999). (in Russian)
6. Berikov, V.B.: Grouping of objects in a space of heterogeneous variables with the use of taxonomic decision trees. Pattern Recogn. Image Anal. **21**(4), 591–598 (2011)
7. Avilov, M.S.: The software package for calculating distances, uncertainty measures and clustering sets of formulas of N-valued logic. In: ISSC-2015, Mathematics, Novosibirsk, p. 6 (2015). (in Russian)
8. Lbov, G.S., Bloshitsin, V.Y.: On informativeness measures of logical statements. In: Lectures of the Republican School-Seminar "Development Technology of Expert Systems", Chiinu, pp. 12–14 (1978). (in Russian)
9. Vikent'ev, A.A., Kabanova, E.S.: Distances between formulas of 5-valued lukasiewicz logic and uncertainty measure of expert statements for clustering knowledge bases. In: Bulletin of the Tomsk State University, Tomsk, vol. 2, no. 23, pp. 121–129 (2013). (in Russian)
10. Karpenko, A.S.: Lukasiewicz Logics and Prime Numbers. Nauka, Moscow (2000). (in Russian)
11. Ershov, Y.L., Palutin, E.A.: Mathematical Logics. Fizmatlit, Moscow (2011). (in Russian)
12. Zagoruiko, N.G., Bushuev, M.V.: Distance measures in knowledge space. In: Data Analysis in Expert Systems, 117: Computing Systems, Novosibirsk, pp. 24–35 (1986). (in Russian)

Visual Anomaly Detection in Educational Data

Jan Géryk, Luboš Popelínský$^{(\boxtimes)}$, and Jozef Triščík

Knowledge Discovery Lab, Faculty of Informatics,
Masaryk University, Brno, Czech Republic
popel@fi.muni.cz

Abstract. This paper is dedicated to finding anomalies in short multivariate time series and focus on analysis of educational data. We present ODEXEDAIME, a new method for automated finding and visualising anomalies that can be applied to different types of short multivariate time series. The method was implemented as an extension of EDAIME, a tool for visual data mining in temporal data that has been successfully used for various academic analytics tasks, namely its Motion Charts module. We demonstrate a use of ODEXEDAIME on analysis of computer science study fields.

Keywords: Visual analytics · Academic analytics · Anomaly detection · Temporal data · Educational data mining

1 Introduction

Visual analytics [3,9,10,12,14] by means of animations is an amazing area of temporal data analysis. Animations allows us to detect temporal patterns, or better to say, patterns changing in time in much more comprehensive way than classical data mining or static graphs.

Motion Charts (MC) is a dynamic and interactive visualization method which enable analyst to display complex quantitative data in an intelligible way. The adjective dynamic refers to the animation of rich multidimensional data through time. Interactive refers to dynamic interactive features which allow analysts to explore, interpret, and analyze information hidden in complex data.

MC are very useful in analyzing multidimensional time-dependent data as it allows the visualization of high dimensional datasets. Motion Charts displays changes of element appearances over time by showing animations within a two-dimensional space. An element is basically a two-dimensional shape, e.g. a circle that represents one object from the dataset. The third dimension is time. Other dimension can be displayed inside circles e.g. in form of sectors or rings. The basic concept was introduced by Hans Rosling who popularized the Motion Charts visualization in a TED Talk[1]. MC enables exploring long-term trends which represent the subject of high-level analysis as well as the elements that form the

---

[1] http://www.ted.com/talks/hans_rosling_shows_the_best_stats_you_ve_ever_seen.html.

© Springer International Publishing Switzerland 2016
C. Dichev and G. Agre (Eds.): AIMSA 2016, LNAI 9883, pp. 99–108, 2016.
DOI: 10.1007/978-3-319-44748-3_10

patterns which represent the target analysis. The dynamic nature of MC allows a better identification of trends in the longitudinal multivariate data and enables the visualization of more element characteristics simultaneously [2]. E.g. in feature selection or mapping, it is visual analytics, and for time-dependent data even more animations, that can be helpful as a user is free to choose the feature selection according to his or her intentions and can see the results immediately.

Quite often we need not only to detect typical trends in time-dependent data but also to discover processes that differs from them the most significantly to find anomalous trends [1]. Naturally, a good feature selection significantly affect not only a detection of relationship but also of anomalies, the task that we try to solve here in collaboration of classical anomaly detection and visual analytics. In this paper we present a new tool ODEXEDAIME for anomaly detection in short series of time-dependent data. Its main advantage if compared with common anomaly detection methods is their comprehensibility and also their easy combination with visual analytics tool.

The paper is structured as follows. Section 2 contains a description of visual data mining tool EDAIME focusing on Motion Charts module. In Sect. 3 we gives an overview of the methods that we employed for outlier detection in time-dependent data focusing on short series. Section 4 describes ODEXEDAIME, a tool for outlier detection in short time series. Description of CS study fields dataset can be found in Sect. 5 and the results of experiments in Sect. 6. Discussion, conclusion and future work are presented at the end of the paper in Sect. 7.

2   Motion Charts in EDAIME

EDAIME [5–7], the tool for visual analytics in different kind of data has been addressed two main challenges. This tool enables visualization of multivariate data and the interactive exploration of data with temporal characteristics, actually, not only motion charts. EDAIME has been used not only for research purposes but also by FI MU management as it is optimised to process academic analytics (AA) [11]. For more information on properties and methods of EDAIME, see the demos

http://www.fi.muni.cz/~xgeryk/framework/video/clustering_of_elements.webm
http://www.fi.muni.cz/~xgeryk/framework/video/groups_of_elements.webm
http://www.fi.muni.cz/~xgeryk/framework/video/extending_animations.webm

X axis displays an average grade for each field (from 1.0 as Excellent to 4.0 as Failed), Y axis is an average number of the credits obtained (typically, 2 h course finished with exam is for 4 credits), the number in the bottom-right corner is the order of a semester. Green sectors means a fraction of successfully finished studies, red ones are for unsuccessful ones.

Menu Controls enables to control animation playback. Apart from play, pause, and stop buttons, there is also range input field which controls five levels of the animation speed. These controls facilitate the step-by-step exploration of the animation and allow functionality for transparent exploration of the data over the entire time span. Animation playback can be interactively changed by traversing mouse over semester number localised in right bottom of the EDAIME tool. Mouse-over element events trigger tooltip with additional element-specific information. One mouse click pauses animation playback and another one starts it again. Double-click restarts the animation playback. Cross axis can be activated to enable better reading values from axes and can be well combined with dimension distortion.

Menu Data mapping allows to map data into Motion Charts variables. The variables include average number of students, average number of credits, average grade, enrolled credits, obtained credits, completed studies, and incomplete studies. Controls for data selection are also particularly useful. Univariate statistical functions can be applied on any of the aforementioned variables. Bivariate functions are also available and can be applied on pairs of variables include enrolled and obtained credits, and complete and incomplete studies.

The main technical advantages over other implementations of Motion Charts are its flexibility, the ability to manage many animations simultaneously, and the intuitive rich user interface. Optimizations of the animation process were necessary, since even tens of animated elements significantly reduced the speed and contributed to the distraction of the analyst's visual perception. The Force Layout component of D3 provides the most of the functionality behind the animations, and collisions utilized in the interactive visualization methods. Linearly interpolated values are calculated for missing and sparse data.

3   Outlier Detection in Short Time Series

3.1   Basic Approach

Time series that we are interested in has three basic properties - (1) a fixed time interval between two observations, (2) same length, and (3) shortness of a time series. For the latter, we limit the length to be smaller than 15 what covers length of study (a number of semesters) of almost all students. We found that existing tools for multivariate time series are not appropriate mainly because of shortness of a time series in tasks that we focus on. We also tested methods for sequence mining [1], namely mining frequent subsequences but none of them displayed a good result. Actually, the time series under exploration lays somewhere between time series (but are quite short) and sequences. However, relation between sequence members look less important than dependence on time and moreover, anomalies in trend are important rather then point anomalies or subpart (subsequence) anomalies.

It was the reason that we decided to (1) transform each multivariate time series into a set of univariate ones, (2) apply to each of those series outlier detection method described bellow, and then (3) join the particular outlier detection factors into one for the original multivariate time series. We observed that this approach worked well, or even better, if compared with the state-of-the-art multivariate time series outlier detection methods.

Methods for anomaly detection in time series can be usually split into distance-based, deviation-based, shape-based methods (or its variant here, trend-based), and density-based (not used here) [1]. For all the methods below we checked two variants - original (non-normalised) data and normalised one - to limit e.g. an influence of a different number of students in the study fields.

3.2   Distance-Based Method

We employed two variants, *mean-based* method - mean M of a given feature is computed as an average of its values in all time series. Outlier factor is then computed as a distance of a given time series (actually its mean value m of the feature) from the mean M. The other method, called *distance-based* in the rest of this paper, computes euclidian (or Haming for non-numeric values) distance between two time series (two vectors). Outlier factor is computed as sum of distances from k nearest time series.

3.3   Trend-Based Method

This method computes how often the trend changed from increasing to decreasing or vice versa. Outlier factor is computed as difference of this value from mean value computed for all the rest of time seties in a collection.

3.4   Deviation-Based Method

This method compares difference of a feature value in two neighboring time moments for two time series. Difference of those two differences is taken as a distance. Rest is the same as for distance-based method.

3.5   Total Outlier Factor

For each dimension (i.e. for each dependent variables in an observation), and for a given basic method from the list above we compute a vector of length n where n is a number of dependent variables. Then we use LOF [4] (see also for formal definition of a local outlier factor) for computing the outlier factor for a given observation.

4   ODEXEDAIME

4.1   Algorithm

ODEXEDAIME (Outlier Detection and EXplanation with EDAIME), the tool for outlier detection in short multidimensional time series consists of four methods described above. We chose them because each of those method detect different kind of anomaly and we wanted to detect as wide spectre of anomalies as possible. The outlier detection method is unsuprevised, We do not have any

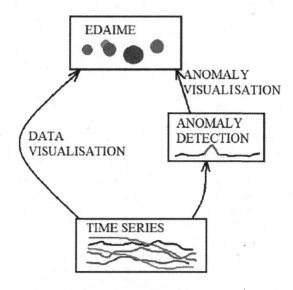

Fig. 1. ODEXEDAIME scheme

example of normal or abnormal anomalous series. The ODEXEDAIME algorithm can be split into five steps. In the first step, multivariate time series has been transformed into series of univariate, one-dimensional, time series. In the second step, an outlier factor has been computed for each univariate series and each of the four methods meanbased, distancebased, trendbased and deviation based. E.g. for our data where we analysed 7 features we obtain 28 characteristics for each multivariate time series. The outlier factors from the previous step are used for computing final outlier factor of the original multivariate time series. Local outlier factor LOF [4] has been used. The last step is visualisation. The scheme of ODEXEDAIME that has been implemented in Java can be seen in Fig. 1.

4.2 Visualisation

All the detected anomalous entities, e.g. a study field, are immediately visualised. Visualisation of anomalies is independent on features selected for visualisation. It means that features selected for anomaly detection can be different from features that has been chosen for visualisation. Layout of the ODEXEDAIME user interface can be seen on Fig. 2. The names of circles, actually CS study fields, are explained in the data section. A user select a use of EDAIME without or with anomaly detection. If the later was chosen, anomalous entities (circles) will be highlighted.

ODEXEDAIME can be found here
http://www.fi.muni.cz/~xgeryk/analyze/outlier/motion_chart_pie_anim_adv_
neobfus.pl

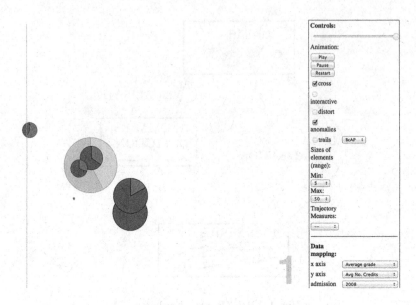

Fig. 2. ODEXEDAIME

Put the button *anomalies* on, to see the anomalous data. The acronym of a study field can be displayed after a pointer is inside a bubble. The outlying time series is/are that one(a) that is/are blinking.

5 Data

Data contains aggregated information about bachelor study fields at Faculty of Informatics, Masaryk University Brno. BcAP denotes Applied Informatics, PSK denotes Computer Networks and Communication, UMI denotes Artificial Intelligence and Natural Language Processing, GRA denotes Computer Graphics, PSZD denotes Computer Systems and Data Processing, PDS denotes Parallel and Distributed Systems, PTS is for Embedded systems, BIO denotes Bioinformatics, and MI denotes Mathematical Informatics. A field identifier is always followed by the starting year. E.g. BcAp (2007) concerns students of Applied Informatics that began their study in the year 2007. Data contains information on

- the number of students in every term;
- the average number of credits subscribed at the beginning of a term; and
- credits obtained at the end;
- a number of students that finished their study in the term; or
- moved to some other field; or
- changed at the mode of study (e.g. temporal termination); and also
- an average rate between 1 (Excellent) and 4 (Failed) for the study field in a term

6   Experiments and Results

We used all anomaly detection methods referred in Sect. 3 and then, for presentation in this Section, chose that ones with the highest local outlier factors where the maximal LOF was at least five-times higher than the minimum LOF for the chosen anomaly detection method.

For LOF parameter $k = 5$ (for k nearest neighbours) was used in all the experiments. We also checked smaller values (1..4) but the results were not better. For $k > 5$ the difference between the maximum and minimum value of LOF did not significantly change.

All the results obtained with ODEXEDAIME has been compared with anomaly detection performed by human (referred as an expert in this section) who can use only classical two-dimensional graphs.

Table 1. Distance-based outlier detection: applied informatics

	BcAP (2007)	PDS (2007)	BIO (2007)	PTS (2008)
LOF:	**23,10**	1,28	2,67	1,09
	GRA (2008)	**BcAP (2008)**	PSK (2007)	GRA (2007)
LOF:	0,99	**21,63**	0,91	0,91
	MI (2008)	PSZD (2007)	UMI (2008)	MI (2007)
LOF:	2,55	1,11	1,11	0,96
	BcAP (2006)	PSZD (2008)	PSK (2008)	UMI (2007)
LOF:	**18,07**	0,91	2,67	0,98

In Table 1, there are results for distance-based method when the euclidian distance was used. Similar results were obtained with Manhattan distance, only the difference between the highest value of LOF and the rest of values was slightly smaller, however still a magnitude higher for BcAP then for the other fields.

Table 2. Distance-based method after normalisation

	BcAP (2007)	**PDS (2007)**	BIO (2007)	PTS (2008)
LOF:	1,97	9,38	1,01	1,18
	GRA (2008)	BcAP (2008)	PSK (2007)	GRA (2007)
LOF:	1,0	1,04	0,96	1,02
	MI (2008)	PSZD (2007)	**UMI (2008)**	MI (2007)
LOF:	1,18	0,91	**3,03**	1,92
	BcAP (2006)	PSZD (2008)	PSK (2008)	UMI (2007)
LOF:	0,99	1,00	1,07	1,18

Several fields are massive, with tens or even hundreds students. To limit the influence of it, we normalised the data and again used distance-based method. After normalisation, see Table 2, we can observe that Parallel and distributed systems differs significantly, namely because of a grade and a number of credits (both subscribed and obtained). It is surprising that the second outlying filed in Artificial intelligence UMI. This field was not chosen as anomalous by an expert. However, both field are pretty similar w.r.t grades and numbers of credits, although for UMI the difference form the other fields is not so enormous. When looking for the same field one year sooner, there is no evidence for anomaly. We can conclude that for UMI it is just a coincidence.

Using trend-based method it is again PDS (2007) followed by MI (2008) (see Table 3) although with more than twice smaller outlier factor than PDS. Neither the latter was chosen by an expert. Possible explanation can be that both fields - PDS and MI - are more theoretical fields and are being chosen by good students but the values of features for MI do not differ so much from the rest of fields and are difficult to detect from two-dimensional graphs.

Table 3. Trend-based method after normalisation

	BcAP (2007)	**PDS (2007)**	BIO (2007)	PTS (2008)
LOF:	1,0	**6,25**	1,09	1,15
	GRA (2008)	BcAP (2008)	PSK (2007)	GRA (2007)
	1,0	1,0	1,03	1,0
	MI (2008)	PSZD (2007)	UMI (2008)	MI (2007)
LOF:	**3,75**	0,88	1,66	1,09
	BcAP (2006)	PSZD (2008)	PSK (2008)	UMI (2007)
	1,66	1,03	1,0	0,99

7   Conclusion and Future Work

We proposed a novel method for anomaly detection for short time series that employes anomaly detection and visual analytics, namely motion charts. We showed how this method can be used for analysis CS study fields.

There are many fields where ODEXEDAIME can be used, e.g. in analysis of trends in average salary or unemployment or in analysis of financial data. The current version transforms a multivariate time series into a set of univariate ones. For our task - analysis of Computer Scinece study fields - it is no disadvantage. However, it would be necessary to overcome this limit, as in general it may be not working. Limits of LOF are well-known - a user need to be careful when compares two values of LOFs. Again, here it was not a problem. In general, a probabilistic version of LOF probably need to be used.

There are several ways that should be followed to improve ODEXEDAIME. In the recent version results of different anomaly detection methods has been evaluated and then presented to a user separately. There is also possibility to use the method in supervised manner when normal and anomalous elements are available. Challenge is to use ODEXEDAIME for class-based outliers [8,13]. Actually explored study field are grouped into two study programs - Infromatics and Applied informatics. With these methods we would be able to find e.g. a study field from Informatics study program that is more close to the Applied Informatics study fields.

Acknowledgments. We thank to the members of Knowledge Discovery Lab at FIMU for their assistance and the anonymous referees for their comments. This work has been supported by Faculty of Informatics, Masaryk University.

References

1. Aggarwal, C.C.: Outlier Analysis. Springer, New York (2013)
2. Al-Aziz, J., Christou, N., Dinov, I.D.: Socr "motion charts": an efficient, open-source, interactive and dynamic applet for visualizing longitudinal multivariate data. J. Stat. Educ. **18**(3), 1–29 (2010)
3. Andrienko, G., Andrienko, N., Kopanakis, I., Ligtenberg, A., Wrobel, S.: Visual analytics methods for movement data. In: Giannotti, F., Pedreschi, D. (eds.) Mobility Data Mining and Privacy, pp. 375–410. Springer, Berlin (2008)
4. Breunig, M.M., Kriegel, H.-P., Ng, R.T., Sander, J.: LOF: identifying density-based local outliers. SIGMOD Rec. **29**(2), 93–104 (2000)
5. Géryk, J.: Using visual analytics tool for improving data comprehension. In: Proceedings for the 8th International Conference on Educational Data Mining (EDM 2015), pp. 327–334. International Educational Data Mining Society (2015)
6. Géryk, J., Popelínský, L.: Analysis of student retention and drop-out using visual analytics. In: Proceedings for the 7th International Conference on Educational Data Mining (EDM 2014), pp. 331–332. International Educational Data Mining Society (2014)
7. Géryk, J., Popelínský, L.: Towards academic analytics by means of motion charts. In: Rensing, C., Freitas, S., Ley, T., Muñoz-Merino, P.J. (eds.) EC-TEL 2014. LNCS, vol. 8719, pp. 486–489. Springer, Heidelberg (2014)
8. He, Z., Xu, X., Huang, J.Z., Deng, S.: Mining class outliers: concepts, algorithms and applications in CRM. Expert Syst. Appl. **27**(4), 681–697 (2004)
9. Keim, D.A., Andrienko, G., Fekete, J.-D., Görg, C., Kohlhammer, J., Melançon, G.: Visual analytics: definition, process, and challenges. In: Kerren, A., Stasko, J.T., Fekete, J.-D., North, C. (eds.) Information Visualization. LNCS, vol. 4950, pp. 154–175. Springer, Heidelberg (2008)
10. Keim, D.A., Mansmann, F., Schneidewind, J., Thomas, J., Ziegler, H.: Visual analytics: scope and challenges. In: Simoff, S.J., Böhlen, M.H., Mazeika, A. (eds.) Visual Data Mining. LNCS, vol. 4404, pp. 76–90. Springer, Heidelberg (2008)
11. Lauría, E.J., Moody, E.W., Jayaprakash, S.M., Jonnalagadda, N., Baron, J.D.: Open academic analytics initiative: initial research findings. In: Proceedings of the Third International Conference on Learning Analytics and Knowledge, LAK 2013, pp. 150–154, New York, NY, USA. ACM (2013)

12. Miksch, S., Aigner, W.: A matter of time: applying a data-users-tasks design triangle to visual analytics of time-oriented data. Comput. Graph. **38**, 286–290 (2014)
13. Nezvalová, L., Popelínský, L., Torgo, L., Vaculík, K.: Class-based outlier detection: staying zombies or awaiting for resurrection? In: Fromont, E., De Bie, T., van Leeuwen, M. (eds.) IDA 2015. LNCS, vol. 9385, pp. 193–204. Springer, Heidelberg (2015). doi:10.1007/978-3-319-24465-5_17
14. Tekusova, T., Kohlhammer, J.: Applying animation to the visual analysis of financial time-dependent data. In: 11th International Conference on Information Visualisation, pp. 101–108 (2007)

Extracting Patterns from Educational Traces via Clustering and Associated Quality Metrics

Marian Cristian Mihăescu[1], Alexandru Virgil Tănasie[1], Mihai Dascalu[2(✉)],
and Stefan Trausan-Matu[2]

[1] Department of Computer Science, University of Craiova, Craiova, Romania
mihaescu@software.ucv.ro, alexandru_tanasie@yahoo.com
[2] Computer Science Department, University Politehnica of Bucharest, Bucharest, Romania
{mihai.dascalu,stefan.trausan}@cs.pub.ro

Abstract. Clustering algorithms, pattern mining techniques and associated quality metrics emerged as reliable methods for modeling learners' performance, comprehension and interaction in given educational scenarios. The specificity of available data such as missing values, extreme values or outliers, creates a challenge to extract significant user models from an educational perspective. In this paper we introduce a pattern detection mechanism with-in our data analytics tool based on k-means clustering and on SSE, silhouette, Dunn index and Xi-Beni index quality metrics. Experiments performed on a dataset obtained from our online e-learning platform show that the extracted interaction patterns were representative in classifying learners. Furthermore, the performed monitoring activities created a strong basis for generating automatic feedback to learners in terms of their course participation, while relying on their previous performance. In addition, our analysis introduces automatic triggers that highlight learners who will potentially fail the course, enabling tutors to take timely actions.

Keywords: Clustering quality metrics · Pattern extraction · k-means clustering · Learner performance

1 Introduction

Finding educational patterns reflective of learner's performance represents a research question of particular interest that has been addressed from different perspectives in various contexts. As clustering has emerged as a de facto method for finding items that naturally group together, there is a wide range of algorithms that can be used to classify individual, differentiated mainly into three classes: hierarchical, non-hierarchical and spectral.

In terms of available data, items may be represented by actors (e.g., students or tutors) or learning assets (e.g., concepts, quizzes, assignments, etc.). Moreover, due to the variety of e-learning platforms, there is also a wide variety of structured and well defined input data, for example the PLSC DataShop [1] or the UCI Machine Learning Repository Student Performance Data Set [2]. However, designing and running a clustering process

© Springer International Publishing Switzerland 2016
C. Dichev and G. Agre (Eds.): AIMSA 2016, LNAI 9883, pp. 109–118, 2016.
DOI: 10.1007/978-3-319-44748-3_11

on a particular dataset in order to find meaningful educational insights represents a challenge.

In this paper we introduce an automatic feedback mechanism derived from pattern extraction data analytics applied on educational traces. Our approach for finding patterns is based on: (a) a continuous evaluation in terms of clustering quality metrics (CQM) applied on data models generated by the k-means algorithm and (b) the automatic highlight and manual removal by means of visual inspection of outliers and extreme values. From an educational perspective, generated patterns enable us to categorize new students based on their continuous performance evaluation, thus providing automatic personalized feedback. In other words, our main contributions cover the following aspects:

1. Define a custom data analytics pipeline that identifies relevant clusters by usage of clustering quality metrics (CQMs) in an educational context. The CQMs are implemented for Weka [3]. The current release does not support any standardized data models as outputs (i.e., cluster assignments based on standard CQMs).
2. Develop a software tool suited for finding educational patterns and providing automatic feedback to both learners and tutors.

The paper continues with a detailed state of the art, followed by methods and results. Afterwards, it discusses key findings derived from our dataset, points out strengths and limitations of our approach, and provides a roadmap for future work centered on increasing the efficacy of the educational tasks at hand.

2 Related Work

The quality of a clustering process is strongly related to the available dataset, the chosen clustering algorithm, as well as the implemented CQMs. The domain specific training dataset and the implemented quality metrics represent the building blocks for interpreting results in correspondence to the proposed learning analytics task. Within our experiments, we have opted to select a dataset from the Tesys e-learning platform [4]. As algorithm, we have used the k-means clustering [5] algorithm implemented in Weka [3]. Weka also implements Expectation Maximization [6] and COBWEB [7] for hierarchical clustering, as well as DBSCAN [8], but the selected algorithm was most adequate for our experiments in terms of visual representation.

However, the main drawback of Weka is the lack of implemented CQMs that would enable an easy and reliable analysis of the clustering outcomes for various clustering algorithms. Multiple CQMs are defined in literature and may be classified into various types, depending on the underlying approach: similarity metrics, clustering evaluations, clustering properties, distance measures, clustering distances, as well as cluster validity measures.

First, Jackson, Somers and Harvey [9] define similarity metrics as a way to compute coupling between two clusters. In contrast, Sneath and Sokal [10] cover the following distance, association and correlation coefficients: Taxonomic (distance), Camberra (distance), Jaccard (association), Simple matching (association), Sorensen-Dice (association), and Correlation (correlation).

Second, silhouette statistical analyses [11, 12] and visualization techniques have been successfully used by Jugo, Kovačić and Tijan [13] as main analytics engine for the cluster analysis of student activity in web-based intelligent tutoring systems.

Third, Hompes, Verbeek and van der Aalst [14] define specific clustering properties, i.e., cohesion, coupling, balance and clustering score. Their methodology evaluates weighted scoring functions that grade a clustering activity with a score between 0 (bad clustering) and 1 (good clustering).

Forth, the similarity measures between two clusters as defined by Meila et al. [15, 16] cover Clustering Error (CE), the Rand index (as a well-known representative of the point pair counting-based methods), and the Variation of Information (VI) [15]. Additional measures for comparing clusters were also defined: Wallace indices [17], Fowlkes-Mallows index [18], Jaccard index [19], or Mirkin metric [20]. These later distance measures are necessary to perform external cluster validation, stability assessments based on internal cluster validation, meta-clustering, and consensus clustering [16]. Classic cluster validity measures defined by Stein, Meyer zu Eissen and Wißbrock [21] include precision, recall, f-measure, Dunn index, Davies-Boulden, Λ and ρ measures. These internal measures determine the usefulness of obtained clusters in terms of structural properties.

From the information theory point of view, CQMs should meet the following primary axioms set by Kleinberg: (a) scale invariance, (b) consistency, and (c) richness. In addition, CQMs should follow the principles of relative point margin, representative set and relative margin [22]. However, aside from the wide variety of previously mentioned metrics, *SSE* (Sum of squared errors) is one of the most frequently used functions when inferring clusters in k-means algorithms. Minimization of this objective function represents the main goal and SSE may be regarded as a baseline quality threshold for most clustering processes.

From the application development perspective, many educational applications make intensive use of clustering processes. For example, Bogarín, Romero, Cerezo and Sánchez-Santillán [23] improve the educational process by clustering, Li and Yoo [24] perform user modeling via clustering, while Bian [25] clusters educational data to find meaningful patterns.

3   Method

Our learning analytics pipeline is built into two core components, one centered on data processing (Clustering and CQM identification) and the other focused on visual analytics. Data processing is performed at server side based on raw data from the e-learning platform. Visual analytics are performed in the web client using D3js (https://d3js.org/), a JavaScript library for manipulating network graphs. Figure 1 presents the high level workflow of our learning analytics pipeline.

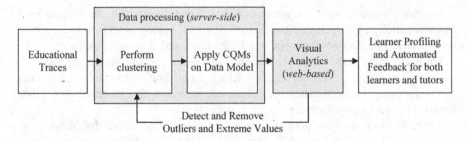

Fig. 1. Data analytics pipeline.

The input dataset has been obtained from logging activity performed during the academic year by students interacting with the Tesys e-learning platform [4]. The platform has specific logging mechanisms for extensively recording raw activity data that is divided into logging, messaging and quiz answering, indices described in detail in Sect. 4.

The data processing module creates the data model based on the input educational traces, the selected algorithm and the associated CQMs, generating in the end specific clusters and the computed quality metrics. Besides the data model itself, the data processing module is also responsible for identifying outliers and extreme values in accordance to the setup and the threshold values enforced by the data analyst.

The visual analytics module displays the generated clusters and marks outliers and extreme values as described by the data model. Its main goal is to properly present the model in a visual manner to the data analyst, facilitating the decision making process of manually removing outliers and extreme values under continuous evaluation of CQMs and centroid values. From this point forward, the design of our data analysis pipeline allows continuous trial and error sessions for the data analyst until relevant results from an educational perspective are obtained. The main characteristic of this module and of the entire pipeline is that it provides two types of relevant data: numerical information (by means of CQMs) and visual representations (by spatially generating presentations of clusters, centroids and outliers/extreme values). This enables the data analyst to properly decide if the current result is satisfactory or may potentially indicate that additional refinements need to be performed for improving the clustering results.

Once the patterns have been identified, the characterization of a new student may be performed. Assuming that three student clusters have been identified and manually labeled as "underperforming", "average" and "well-performing", new students may be characterized as belonging to a certain cluster via k-Nearest Neighbors algorithm. The labeling is performed by the tutor while taking into account the feature names and their intuitive interpretation. Proper usage of the CQMs in the labeling process is accomplished through the guidance of a data analyst. The overarching goal remains to improve each student's predicted knowledge level and to reduce as much as possible the number of underperforming students. Furthermore, parameters with a high variance need to be taken into consideration in order to determine the adequate course of action, which becomes in turn the automatic feedback provided by our system.

4 Results

The input dataset consisted of 558 students characterized by 17 attributes described in Table 1. Our initial run of the k-Means clustering algorithm produces the graphs depicted in Figs. 2 and 3 in terms of SSE, Silhouette, Dunn and Xi-Beni indices, as well as the centroids presented in Table 2.

Table 1. Initial classification attributes.

Attribute name	Description
A. Session parameters	
Total	Sessions started by the user
W[n]	Sessions in the n-th week, where n is takes values from 0 to 6
B. Message parameters	
TS	Total sent messages
TR	Total received messages
AS	Average duration between two consecutive sent messages
AR	Average duration between two consecutive received messages
WS	Words from sent messages
WR	Words from received messages
C. Self-assessment parameters	
TQ	Total answered questions
CQ	Correctly answered questions
PQ	Percentage of correctly answered questions

(a)   (b)

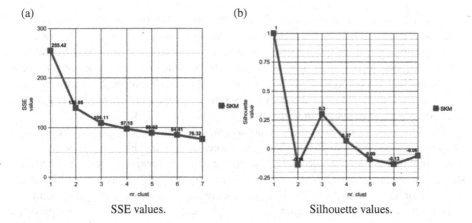

SSE values.   Silhouette values.

Fig. 2. (a) SSE values. (b) Silhouette values.

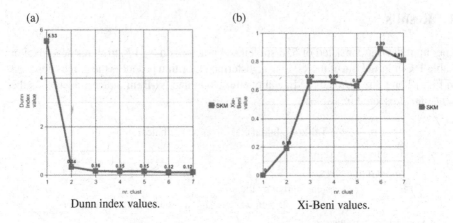

Dunn index values.   Xi-Beni values.

Fig. 3. (a) Dunn index values. (b) Xi-Beni values.

Table 2. Feature values for centroids.

	Total	W0	W1	W2	W3	W4	W5	W6
low	27.17	3.66	4.48	4.47	4.10	3.86	3.62	2.98
high	**75.21**	**11.98**	**12.31**	**1133**	**10.59**	**11.9**	**9.41**	**7.69**
average	31.33	4.1	4.58	4.84	5.00	5.09	4.10	3.61
	TS	TR	AS	AR	WS	WR	TQ	CQ	PQ
low	22.29	**9.92**	19.34	**319.70**	560.73	**231.28**	*151.51*	*58.42*	*21.10*
high| **50.2**	6.76	**23.93**	144.35	**1022.76**	155.88	**1540.35**	**1042.16**	**68.40**
average	*9.99*	*2.50*	*5.20*	*75.86*	*234.72*	*78.23*	215.76	82.50	32.88

Figure 2a shows an elbow for *k* equals three clusters. For two clusters, a significant decrease is observed −45.2 % from the SSE maximum value. While moving forward to three clusters, there is also an important decrease of 22 % from the previous SSE value. The subsequent decreases in SSE values are smaller than 10 %, thus leading to the conclusion that the optimal number of clusters in terms of the steepest descent may be two or three. Decision among these values is a matter of domain knowledge, but additional CQMs provide insight in terms of the optimal selection.

The silhouette values from Fig. 2b are a clear indicator of the quality of generated clusters. A value between .25 and .5 reflects a week structure, while a value between .5 and .7 highlights a reliable structure. Due to the nature of analysis, we do not expect strong structures since we are monitoring activities performed by learners. The .3 value obtained for three clusters is by far the highest score and indicates an appropriate number of clusters, with a reasonable structure. All other values are close to zero or negative, meaning that no substantial structure could be identified.

Dunn index values (see Fig. 3a) lays between zero and ∞ and should be maximized as higher values indicate the presence of compact and well separated clusters. The obtained index values show that the clusters are not well defined and this is usually due to the presence of noise in the input dataset. For our particular experiment, this index yields to the conclusion that a high percentage of extreme values and outliers are found in the dataset.

The Xi and Beni index (see Fig. 3b) is significant when a value is a minimum of its neighbors. Therefore, two and five clusters represent minimum values while disregarding the zero value for one cluster. Still, the index value for three clusters is 4.5 % higher that the index value for five clusters; thus, from an educational perspective, we may consider that three clusters are more representative than five.

Based on all previous CQM analyses, we can conclude that the optimal number of clusters for our analysis is three. Starting from this categorization, Table 2 presents each centroid's feature values. After color coding each minimum, maximum and middle values, we can easily observe that each centroid becomes representative for a corresponding learner level ("under-performing", "average" and "well-performing"), highlighting also that the obtained centroids have educational significance.

In addition, a Principal Component Analysis was applied in order to better represent our data and to reduce the dimensionality of the initial classification attributes. The *Total* attribute was eliminated due to multicollinearity to W[0–6], while the *AS* attribute was disregarded due to low communality. In the end, 4 principal components were identified as having corresponding eigenvalues of 1 or greater and accounting for 90.13 % variance. The identified components represent a refinement of the initial classification, as messages are now split into in-degree and out-degree while relating to each student's activity (Table 3).

Table 3. Rotated component matrix using varimax with Kaiser normalization.

Classification attribute	Component 1	2	3	4
1. Logging activity				
W[0]	.923			
W[1]	.941			
W[2]	.943			
W[3]	.944			
W[4]	.940			
W[5]	.916			
W[6]	.888			
2. Testing activity				
TQ		.967		
CQ		.967		
PQ		.872		
3. In-degree messaging activity				
TR			.971	
AR			.864	
WR			.942	
4. Out-degree messaging activity				
TS				.975
WS				.985

For visualization purposes, three emerging components –¹ *Logging, Testing* and *Messaging activities* (aggregation of both in-degree and out-degree) – were used to create dedicated views that facilitate the identification of outliers and extreme values. Figure 4 presents a print screen from our visual analytics application in which the centroids are marked with a large circle and the assigned items have similar colors. Three clusters (orange, green and blue) were annotated as representing "underperformingl", "average" and "well-performing" learners. As the coordinate axes represent Testing and Messaging activities, it becomes obvious that students who were most engaged and performed best at their tests were clustered into the well-performing group, while the ones with the lowest participation defined the underperforming cluster.

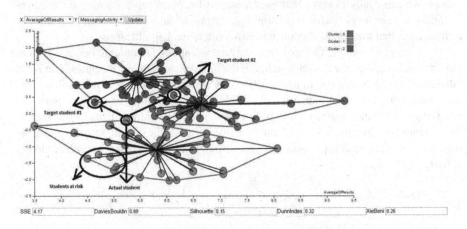

Fig. 4. Clustering visualization with corresponding CQMs.

Once students have been assigned to a cluster, they may select target students with better performance from another cluster or from their own cluster and just-in-time recommendations are offered as a means to improve their activity (e.g., X relevant additional messages should be posted, or Y supplementary tests should be taken). Moreover, the tutor is presented with a list of at-risk students who represent a subset of

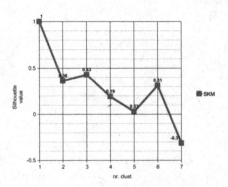

Fig. 5. Silhouette values after the removal of extreme values and outliers.

underperforming students with the lowest activity level. For example, 5 students were marked as "at-risk" in Fig. 4 and will receive personalized feedback from the tutor, encouraging them to become more actively involved.

Starting from the initial distribution, outliers and extreme values are removed using Interquartile Range with an extreme values factor set to 1.5 and an outlier factor set to 1.1. The outlier and extreme value removal led to a Silhouette value of .43 (see Fig. 5). This significant increase from .3 highlights a more solid structure of clusters, thus creating a reliable baseline for providing feedback to learners.

5   Conclusions and Future Work

Our approach consists of a learning analytics solution that integrates clustering algorithms, PCA transformations, visual analytics and clustering quality metrics. In order to obtain a detailed insight on the quality of the clustering process, several CQMs were implemented in order to facilitate the categorization of learners. Outlier and extreme values removal produced more robust structures, effective for running learning analytics and feedback mechanisms. From an educational perspective, reference learners were clustered and used for inferring educational traces corresponding to newcomers in the course.

Future work includes the integration of additional features to represent learners, implementation of other CQMs for a better evaluation of the obtained models, as well as comparative clustering algorithms which might be more appropriate for specific learning analytics tasks. A timeline view focused on displaying the knowledge level of students or centroids for each successive week will be introduced in order to provide a more fine-grained perspective of each learner's engagement.

Acknowledgements.   The work presented in this paper was partially funded by the FP7 2008-212578 LTfLL project and by the EC H2020 project RAGE (Realising and Applied Gaming Eco-System) http://www.rageproject.eu/ Grant agreement No. 644187.

References

1. Koedinger, K.R., Baker, R., Cunningham, K., Skogsholm, A., Leber, B., Stamper, J.: A data repository for the EDM community: the PSLC DataShop. In: Romero, C., Ventura, S., Pechenizkiy, M., Baker, R. (eds.) Handbook of Educational Data Mining. CRC Press, Boca Raton (2010)
2. Cortez, P., Silva, A.: Using data mining to predict secondary school student performance. In: 5th FUture BUsiness TEChnology Conference (FUBUTEC 2008), Porto, Portugal, pp. 5–12 (2008)
3. Hall, M., Frank, E., Holmes, G., Pfahringer, B., Reutemann, P., Witten, I.H.: The WEKA data mining software: an update. SIGKDD Explor. **11**(1), 10–18 (2009)
4. Burdescu, D.D., Mihaescu, M.C.: TESYS: e-learning application built on a web platform. In: International Conference on e-Business (ICE-B 2006), Setúbal, Portugal (2006)

5. MacQueen, J.B.: Some methods for classification and analysis of multivariate observations. In: 5th Berkeley Symposium on Mathematical Statistics and Probability, pp. 281–297. University of California Press, Berkeley (1967)

6. Dempster, A.P., Laird, N.M., Rubin, D.B.: Maximum likelihood from incomplete data via the EM algorithm. J. Roy. Stat. Soc.: Ser. B (Methodol.) **39**(1), 1–38 (1977)

7. Dasgupta, S., Long, P.M.: Performance guarantees for hierarchical clustering. J. Comput. Syst. Sci. **70**(4), 555–569 (2005)

8. Ester, M., Kriegel, H.-P., Sander, J., Xu, X.: A density-based algorithm for discovering clusters in large spatial databases with noise. In: International Conference on Knowledge Discovery and Data Mining (KDD-96), pp. 226–231. AAAI Press (1996)

9. Jackson, D.A., Somers, K.M., Harvey, H.H.: Similarity coefficients: measures of co-occurrence and association or simply measures of occurrence? Am. Nat. **133**(3), 436–453 (1989)

10. Sneath, P.H.A., Sokal, R.R.: Principles of Numerical Taxonomy. W.H. Freeman, San Francisco (1963)

11. Rousseeuw, P.J.: Silhouettes: a graphical aid to the interpretation and validation of cluster analysis. J. Comput. Appl. Math. **20**, 53–65 (1987)

12. Kaufman, L., Rousseeuw, P.J.: Finding Groups in Data. An Introduction to Cluster Analysis. Wiley-Interscience, New York (1990)

13. Jugo, I., Kovačić, B., Tijan, E.: Cluster analysis of student activity in a web-based intelligent tutoring system. Sci. J. Maritime Res. **29**, 75–83 (2015)

14. Hompes, B.F.A., Verbeek, H.M.W., van der Aalst, W.M.P.: Finding suitable activity clusters for decomposed process discovery. In: Ceravolo, P., Russo, B., Accorsi, R. (eds.) SIMPDA 2014. LNBIP, vol. 237, pp. 32–57. Springer, Heidelberg (2015). doi: 10.1007/978-3-319-27243-6_2

15. Meilă, M.: Comparing clusterings by the variation of information. In: Schölkopf, B., Warmuth, M.K. (eds.) COLT/Kernel 2003. LNCS (LNAI), vol. 2777, pp. 173–187. Springer, Heidelberg (2003)

16. Patrikainen, A., Meilă, M.: Comparing subspace clusterings. IEEE Trans. Knowl. Data Eng. **18**(7), 902–916 (2006)

17. Wallace, D.L.: Comment. J. Am. Stat. Assoc. **383**, 569–576 (1983)

18. Fowlkes, E.B., Mallows, C.L.: A method for comparing two hierarchical clusterings. J. Am. Stat. Assoc. **383**, 553–569 (1983)

19. Rand, W.M.: Objective criteria for the evaluation of clustering methods. J. Am. Stat. Assoc. **66**, 846–850 (1971)

20. Mirkin, B.: Mathematical Classification and Clustering. Kluwer Academic Press, Boston (1996)

21. Stein, B., Meyer zu Eissen, S., Wißbrock, F.: On cluster validity and the information need of users. In: 3rd IASTED International Conference on Artificial Intelligence and Applications (AIA 2003), Benalmádena, Spain, pp. 404–413 (2003)

22. Ben-David, S., Ackerman, M.: Measures of clustering quality: a working set of axioms for clustering. In: Neural Information Processing Systems Conference (NIPS 2008), pp. 121–128 (2009)

23. Bogarín, A., Romero, C., Cerezo, R., Sánchez-Santillán, M.: Clustering for improving educational process mining. In: 4th International Conference on Learning Analytics and Knowledge (LAK 2014), pp. 11–15. ACM, New York (2014)

24. Li, C., Yoo, J.: Modeling student online learning using clustering. In: 44th Annual Southeast Regional Conference (ACM-SE 44), pp. 186–191. ACM, New York (2006)

25. Bian, H.: Clustering student learning activity data. In: 3rd International Conference on Educational Data Mining, Pittsburgh, PA, pp. 277–278 (2010)

Natural Language Processing
and Sentiment Analysis

Classifying Written Texts Through Rhythmic Features

Mihaela Balint[1], Mihai Dascalu[1(✉)], and Stefan Trausan-Matu[1,2]

[1] Computer Science Department, University Politehnica of Bucharest,
Bucharest, Romania
{mihaela.balint,mihai.dascalu,
stefan.trausan}@cs.pub.ro
[2] Research Institute for Artificial Intelligence of the Romanian Academy,
Bucharest, Romania

Abstract. Rhythm analysis of written texts focuses on literary analysis and it mainly considers poetry. In this paper we investigate the relevance of rhythmic features for categorizing texts in prosaic form pertaining to different genres. Our contribution is threefold. First, we define a set of rhythmic features for written texts. Second, we extract these features from three corpora, of speeches, essays, and newspaper articles. Third, we perform feature selection by means of statistical analyses, and determine a subset of features which efficiently discriminates between the three genres. We find that using as little as eight rhythmic features, documents can be adequately assigned to a given genre with an accuracy of around 80 %, significantly higher than the 33 % baseline which results from random assignment.

Keywords: Rhythm · Text classification · Natural language processing · Discourse analysis

1 Introduction

Rhythm refers to the quest for harmonious proportions in all creative acts, which is essential for both human emotion and cognition. Rhythm brings thoughts and feelings to resonance, and facilitates understanding, remembering, and learning [1]. A creative piece is built as an ensemble of identical and different units, and rhythm emerges as a particular succession of these units. Examples of units are musical beats, linguistic phonemes, or colors and shapes used in paintings.

Text classification or categorization is the task of assigning a written document to a class from a set of predefined classes. The increasing importance of this task follows the increasing amount of textual information available online, and the need to efficiently index and retrieve such information. Researchers have approached the problem using statistical methods and machine learning, with the latter attaining accuracies comparable to the human expert standard. In machine learning, the distinctive features of individual classes are learned from a set of pre-classified documents. Preferred features include single words, syntactic phrases (two or more words plus the syntactic relationship between them), or n-grams [2]. A high number of words in the vocabulary leads to a high

© Springer International Publishing Switzerland 2016
C. Dichev and G. Agre (Eds.): AIMSA 2016, LNAI 9883, pp. 121–129, 2016.
DOI: 10.1007/978-3-319-44748-3_12

number of features, difficult or impossible to handle by classifiers. Even in the simplest case of single words, the resulting high dimensional feature space requires efficient algorithms of feature selection prior to entering inductive learning algorithms [3].

Our hypothesis is that the communicative purpose of a text influences significantly the rhythm of that text; thus, rhythmic features would become predictors for text categorization. The purpose of this work is to test this hypothesis, by evaluating how well rhythmic features extracted from already categorized text function as predictors. Section 2 presents relevant studies in rhythm analysis. Section 3 describes the first two steps of our method, namely the proposed set of features, and the feature extractor (together with the three corpora selected to demonstrate its use). The third step, namely feature selection, is discussed in Sect. 4, followed by the results of the classification using the selected features. Section 5 is dedicated to conclusions and future work.

2   Related Work

There are multiple perspectives on what constitutes linguistic rhythm analysis, and, most of the time, metrical phonology is implied. Phonology is the branch of linguistics that investigates the systematic organization of sounds in languages. Metrical phonology uses syllabification (at word level) and constituency parsing (at sentence level) to create a hierarchy of stresses inside clauses. The stress phenomenon refers to the relative emphasis placed on a syllable (word level) or syntactic category (sentence level). Rules for stress assignment in English are presented in seminal works written by Chomsky and Halle [4], and Liberman and Prince [5], while an analysis on French Literary text is performed by Boychuk et al. [6].

Several works compare rhythmic behavior in language and music, from which the concept of rhythm is derived. Jackendoff and Lerdahl [7] carry out a complete grammatical parallel between the tree structures used to represent rhythm in language and music. This profound similarity could be explained by Barbosa and Bailly's [8] theory that humans have an internal clock which needs to synchronize with the external clock of the stimulus (the meter in language, or the beat in music). The internal clock hypothesis is in accordance with Beeferman's [9] study, which demonstrates that sentences with a higher probability of occurrence, i.e. sentences that are actually preferred by writers, are more rhythmical. For this result, he uses a corpus of over 60 million syllables of Wall Street Journal text, in both its original form and in a second form, altered to randomize word order inside sentences. He finds that the stress entropy rate is higher in the second case. The rhythm of language appears to be culturally regulated. Galves et al. [10] extract streams of stresses from corpora of newspaper articles written in both European and Brazilian Portuguese, and use Variable Length Markov Chains [11] to model rhythmic realization in the two corpora, arriving at different final models. Where cultural background influences linguistic rhythm, it similarly influences musical rhythm, as shown by Patel and Daniele [12]. As a tool of comparison, they use the normalized Pairwise Variability Index (nPVI), introduced by Grabe and Low [13] to capture the difference in duration between successive vocalic intervals. Patel and Daniele contrast the nPVI's of spoken English and French with the nPVI's computed from English and French instrumental music scores, and obtain statistically significant differences (in the same direction for both language and music, albeit smaller in music). Their conclusion is strengthened by London and Jones's [14] refined method to compute the nPVI of music.

However, linguistic rhythm does not have to be restricted to metrics. According to Boychuk et al. [6], a high degree of rhythmization is achieved whenever there are elements with a high frequency of occurrence and the occurrences are close to each other. They build a tool for the French language, with the option of highlighting repetitions of specific words, vowels, consonants, or phonemic groups. Other features include detection of coordinated units, same-length units, or affirmative, interrogative, exclamatory, and elliptical sentences. In our research, we adopt this more general view of rhythm as repetition and alternation of linguistic elements.

3   Method Description

This section describes our method for the rhythmic evaluation of texts. We model a text as a sequence of elementary units. To separate units, we use the loci where readers naturally insert pauses, and we obtain two kinds of units: sentences (separated by sentence boundaries), and punctuation units (separated by punctuation markers in general). For example, there are four punctuation units in the sentence "Shall we expand[1], be inclusive[2], find unity and power[3]; or suffer division and impotence[4]". Rhythmic features will characterize individual units (e.g. the length of a unit in syllables) or interactions between neighboring units (e.g. the anaphora phenomenon – two or more units which start with the same sequence of words). Subsection 3.1 presents the pre-classified data chosen as ground truth for our model, while Subsect. 3.2 describes the full set of rhythmic features, prior to the step of feature selection.

3.1   Data Collection

In our method, any text corpus can be a data source. We opted for the comparison of three corpora, chosen to exhibit various degrees of rhetoric: a corpus of famous speeches (extracted from http://www.famous-speeches-and-speech-topics.info/famous-speeches/), student essays from the Uppsala Student English (USE) corpus (http://ota. ox.ac.uk/desc/2457), and the raw texts from the RST-DT corpus of Wall Street Journal articles [15]. Table 1 presents the relevant properties of the three datasets. In order to obtain accurate corpus statistics, the full datasets underwent feature extraction. Subsequently, when we evaluated the relevance of rhythmic features in text classification, we balanced the data, by keeping the longest (in number of sentences) 110 documents from each category. This does not eliminate imbalance pertaining to age, gender, or nationality, but this is a pilot study created to demonstrate the strength and scalability of our model. The model as it is now can be further used to look for significant differences in rhythmicity according to age, gender, or nationality.

Our feature extractor was implemented in the Python programming language, using the NLTK package for natural language processing (http://www.nltk.org/) and the SQLite3 package for interfacing with SQL databases (https://www.sqlite.org/). We

Table 1. Statistics of the three datasets.

Dataset	# of documents	# of sentences
Speeches	110	14,111
RST-DT	380	8,281
USE	1,266	49,851

loaded the raw content of documents into three distinct databases (one for each corpus), that were subsequently filled with the extracted document features.

3.2 Rhythmic Features

The analysis presented in this paper relies on five main categories of features: organizational, lexical, grammatical, phonetical, and metrical. They are refined versions of the features we introduced in previous work [16].

Organizational features include the average word length, the length of units in either words or syllables, and patterns of length variation along sequences of units. The number of syllables is computed using the CMU Pronouncing Dictionary (http://www.speech.cs.cmu.edu/cgi-bin/cmudict). Rhythm can occur from a particular alternation of long and short units. We count rising (successive units keep getting longer), falling (they keep getting shorter), alternating (shorter and longer units alternate), or repetitive (same-length) patterns, and the maximum length of such patterns. For frequent words in a document, we determine how often they occur in the beginning (first third), at the middle (second third), or at the end (last third) of units.

Lexical features refer to types of lexical repetition. Words or n-grams in a document are considered frequent if their number of occurrences exceeds the value *(text_length * threshold/n_gram_length)*, where the threshold can be varied. Stop words are eliminated in the detection of frequent words, but accepted inside n-grams which contain at least two non stop words.

In the case of frequent words or n-grams, there is no restriction on the maximum distance between successive occurrences. We use a variable parameter *delta* to impose this kind of restriction when counting *duplicated units* (several identical units), *anaphora* (several units starting with one or more identical words), *epistrophes* (several units ending in the same word(s)), *symploces* (several units presenting a combination of the anaphora and epistrophe phenomena), *anadiploses* (a second unit starting the way a first unit ends), *epanalepses* (single units starting and ending with the same word(s)). We consider only the maximal and non-redundant occurrences of these phenomena. Therefore, if *n* neighboring units have the same start, that is considered to be a single anaphora. If they share *w* words, the anaphora is counted only once, not once for every initial substring of the maximal one.

Grammatical features consider the frequencies of parts-of-speech, commas, and types of sentence boundaries (full-stops, question marks, exclamation marks) in each document. Each sentence is parsed using the Stanford Parser (http://nlp.stanford.edu) and the resulting trees of constituents are used to detect syntactic parallelism between neighboring sentences (located within a given distance of each other). Parallelism can be checked either for the entire or only up to a given depth of the tree. With the obvious exception of terminal nodes (corresponding to actual words), nodes in equivalent positions should be labelled with the same main part-of-speech category. Another kind of noun, verb, adjective, etc. is allowed in place of a kind of noun, verb, adjective, but a noun cannot be in place of a verb, for example. Figure 1 illustrates this point using an excerpt from Jesse Jackson's speech "Common ground and common sense". Non-identical nodes which still fulfill the standard for syntactic parallelism are shown in boldface.

```
(ROOT
  (S
   (NP (PRP We))
   (VP (VBP have) (NP (JJ public) (NNS accommodations)))
   (. .)))
  (ROOT
   (S (NP (PRP We)) (VP (VBP have) (NP (JJ open) (NN housing))) (. .)))
```

Fig. 1. Two syntax trees marked for syntactic parallelism.

Phonetical features refer to phonetical repetition, in much the same way that lexical features refer to lexical repetition The representative phenomena are the ones of *assonance* (the repetition of a vocalic phoneme over a small amount of text), *alliteration* (the same for consonants), and *rhyme* (defined here as the repetition of the same phonemic sequence, not necessarily at the end of words).

To compute *metrical* features, for each syllabified document the complete stream of stresses (primary, secondary, or no-stress) is extracted. We record the frequencies of units built with an odd number of syllables, and of units ending in a stressed syllable. For the latter feature, the stress from monosyllabic stop words is removed, because, in practice, monosyllabic words are sometimes stressed, sometimes not, and the CMU dictionary does not handle this problem satisfactorily.

4   Feature Selection, Results and Discussions

This section describes our approach to feature selection and testing for feature relevance, together with corpus statistics and classification results for the refined set of features.

Discriminant Function Analysis (DFA) is a statistical analysis which predicts a categorical dependent variable (a class from a set of predefined classes) from the behavior of several independent variables (called predictors). Performing a DFA over a given dataset requires that independent variables respect a normal distribution, and that no discriminating variables be a linear combination of other variables. These requirements guide the reduction of the feature space described in the previous section. First, we remove all features which demonstrate non-normality. Second, we assess multicollinearity based on pair-wise correlations with a correlation coefficient $r > .70$, and filter multicollinear features to keep only the feature with the strongest effect in the model (see Table 2 for the final list of rhythmic features and their descriptive statistics).

The results indicate rhetorical preferences. Conciseness and fluency are achieved through the usage of short words and the alternation of long and short units. The main themes of a document, captured in frequent words, tend to occupy the middle of units, with the beginning and end of units functioning as background and elaboration. Essays contain more frequently used words and fewer commas, which might be explained by the lower English proficiency of their authors. Speeches do not repeat many words, but they make the most use of figures of speech based on repetition, especially anaphora. Anaphora in reference to punctuation units, not sentences, are particularly indicative of a document's genre.

Table 2. General statistics of rhythmic features - $M(SD)$.

Rhythmic feature	Article	Essay	Speech
# of syllables per word	1.576 (0.081)	1.444 (0.088)	1.475 (0.094)
% of rising word-length patterns	0.171 (0.052)	0.176 (0.051)	0.190 (0.057)
% of falling word-length patterns	0.182 (0.052)	0.173 (0.045)	0.170 (0.042)
% of repetitive word-length patterns	0.035 (0.031)	0.042 (0.029)	0.032 (0.027)
longest rising word-length sequence	2.140 (0.807)	2.340 (0.805)	2.640 (1.002)
longest falling word-length sequence	2.150 (0.826)	2.330 (0.692)	2.410 (0.881)
longest repetitive word-length sequence	0.840 (0.614)	1.060 (0.529)	1.030 (0.642)
% of falling syllable-length patterns	0.187 (0.052)	0.179 (0.044)	0.172 (0.045)
longest rising syllable-length sequence	2.050 (0.806)	2.350 (0.840)	2.620 (0.967)
longest repetitive syllable-length sequence	0.570 (0.582)	0.860 (0.438)	0.850 (0.618)
% of frequent words at the beginning of sentences	0.203 (0.096)	0.172 (0.061)	0.143 (0.092)
% of frequent words at the end of sentences	0.199 (0.086)	0.220 (0.068)	0.174 (0.103)
% of frequent words at the beginning of punctuation units	0.210 (0.080)	0.167 (0.056)	0.149 (0.078)
% of frequent words at the end of punctuation units	0.285 (0.089)	0.297 (0.065)	0.308 (0.088)
# of words deemed frequent	41.33 (20.26)	53.47 (19.70)	39.34 (22.44)
normalized # of sentence anaphora	0.005 (0.003)	0.006 (0.003)	0.007 (0.004)
normalized # of punctuation unit anaphora	0.005 (0.003)	0.008 (0.004)	0.013 (0.006)
normalized # of commas	0.060 (0.015)	0.041 (0.015)	0.060 (0.015)
% of sentences with an odd # of syllables	0.507 (0.068)	0.509 (0.062)	0.497 (0.068)

Table 3 denotes the features that vary significantly between the three datasets, in descending order of effect size, determined through a multivariate analysis of variance (MANOVA) [17, 18]. There is a significant difference among the three datasets in terms of rhythmic features, Wilks' $\lambda = 0.259$, $F(28, 628) = 21.635$, $p < .001$ and partial $\eta^2 = .491$.

We predict the genre of a given text using a stepwise Discriminant Function Analysis (DFA) [19]. Only eight variables from Table 2 (marked with italics) are deemed significant predictors, denoting complementary features of rhythmicity: the number of syllables per word, the normalized number of falling syllable-length patterns, the percentage of frequent words located at the end of sentences, the percentage of frequent words located at the beginning of punctuation units, the number of words deemed frequent, the normalized number of sentence anaphora, the normalized number of punctuation unit anaphora, and the normalized number of commas. Figure 2 depicts the two retained canonical discriminant functions ($\chi^2(df = 7) = 171.773$, $p < .001$).

Table 3. Tests of between-genre effects for significantly different rhythmic features.

Rhythmic feature	df	F	p	η^2 partial
normalized # of punctuation unit anaphora	2	96.433	<.001	.371
# of syllables per word	2	68.483	<.001	.295
normalized # of commas	2	55.41	<.001	.253
% of frequent words at the beginning of punctuation units	2	20.335	<.001	.111
# of words deemed frequent	2	14.84	<.001	.083
% of frequent words at the beginning of sentences	2	13.968	<.001	.079
longest rising syllable-length sequence	2	11.826	<.001	.067
longest repetitive syllable-length sequence	2	9.885	<.001	.057
longest rising word-length sequence	2	9.077	<.001	.053
normalized # of sentence anaphora	2	8.441	<.001	.049
% of frequent words at the end of sentences	2	7.646	.001	.045
longest repetitive word-length sequence	2	4.6	.011	.027
% of rising word-length patterns	2	3.909	.021	.023
% of repetitive word-length patterns	2	3.255	.040	.020

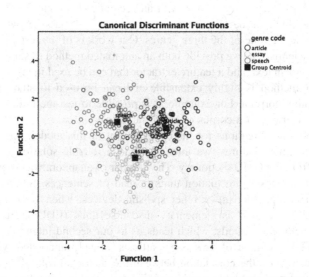

Fig. 2. Separation of genres based on canonical discriminant functions derived from rhythmic features.

The results presented in Table 4 show that the DFA based on these eight features correctly allocated 269 out of 330 texts, for an accuracy of 81.51 % (the chance level for this analysis being 33.33 %). Using leave-one-out cross-validation (LOOCV), the DFA achieved an accuracy of 79.69 % (see the confusion matrix in Table 4 for detailed results). The resulting weighted Cohen's Kappa of 0.723 demonstrates substantial agreement between the actual genre and the genre assigned by the model.

Table 4. Confusion matrix for DFA classifying texts pertaining to different genres.

	Genre	Predicted Group Membership		
		Article	Essay	Speech
Original	Article	99	7	4
	Essay	14	84	12
	Speech	8	16	86
Cross-validated	Article	97	8	5
	Essay	16	81	13
	Speech	8	17	85

5   Conclusions and Future Work

The main purpose of this paper was to test the ability to predict the genre of a given document based on rhythmic features. We used a dataset of 330 documents, equally distributed between three genres: famous speeches, student essays, and newspaper articles. A Discriminant Function Analysis based on the most predictive eight features of our model performed classification with an accuracy of around 80 %, significantly higher than the 33.33 % baseline represented by a trivial classifier which randomly assigns a document to one of the three genres. Our work is of interest to both linguists and computer scientists, as we provide both an automated method to study the rhythmic properties of English text, and a feature extractor that can be used in text categorization. Moreover, our method is highly extensible and can be used to study the rhythmic properties of other corpora. For example, it is possible to test the intuition that words are longer, on average, in a corpus of scientific articles.

We consider two directions for the development of this model. First, in terms of refining our rhythmic features, we intend to find a reliable solution to characterize words absent from the CMU dictionary. The number of anaphora was shown to greatly vary when calculated on punctuation units instead of sentences. Similar results may occur for syntactic parallelism or other stylistic devices, when we experiment with other types of units, such as elementary discourse units (EDUs). EDUs are units separated on rhetorical grounds, which leads us to our second intended development. Using the RST-DT corpus of newspaper articles, already annotated with rhetorical relations, we can study the correlation between the rhetorical role of an EDU and its rhythmic properties, with viable applications in rhetorical relation labelling.

Acknowledgements. The work presented in this paper was partially funded by the EC H2020 project RAGE (Realising and Applied Gaming Eco-System) http://www.rageproject.eu/ Grant agreement No 644187.

References

1. Lefebvre, H.: Rhythmanalysis: Space. Time and Everyday Life. Continuum, London (2004)
2. Fürnkranz, J.: A study using n-gram features for text categorization. Austrian Research Institute for Artificial Intelligence, Wien (1998)
3. Yang, Y., Pedersen, J.O.: A comparative study on feature selection in text categorization. In: 14th International Conference on Machine Learning (ICML 1997), pp. 412–420. Morgan Kaufmann Publishers Inc., San Francisco (1997)
4. Chomsky, N., Halle, M.: The Sound Pattern of English. Harper & Row, New York (1968)
5. Liberman, M., Prince, A.: On stress and linguistic rhythm. Linguist. Inq. **8**(2), 249–336 (1977)
6. Boychuk, E., Paramonov, I., Kozhemyakin, N., Kasatkina, N.: Automated approach for rhythm analysis of french literary texts. In: 15th Conference of Open Innovations Association FRUCT, pp. 15–23. IEEE, St. Petersburg (2014)
7. Jackendoff, R., Lerdahl, F.: A grammatical parallel between music and language. In: Clynes, M. (ed.) Music, Mind, and Brain, pp. 83–117. Springer, Heidelberg (1982)
8. Barbosa, P., Bailly, G.: Characterisation of rhythmic patterns for text-to-speech synthesis. Speech Commun. **15**(1–2), 127–137 (1994)
9. Beeferman, D.: The rhythm of lexical stress in prose. In: 34th Annual Meeting of the Association for Computational Linguistics (ACL). ACL, Santa Cruz (1996)
10. Galves, A., Galves, C., Garcia, J., Garcia, N., Leonardi, F.: Context tree selection and linguistic rhythm retrieval from written texts. Ann. Appl. Stat. **6**(1), 186–209 (2012)
11. Buhlmann, P., Wyner, A.J.: Variable length Markov chains. Ann. Stat. **27**(2), 480–513 (1999)
12. Patel, A.D., Daniele, J.R.: An empirical comparison of rhythm in language and music. Cognition **87**(1), B35–B45 (2003)
13. Grabe, E., Low, E.L.: Durational variability in speech and the rhythm class hypothesis. In: Gussenhoven, C., Warner, N. (eds.) Papers in Laboratory Phonology, pp. 515–546. Mouton de Gruyter, Berlin (2002)
14. London, J., Jones, K.: Rhythmic refinements to the nPVI measure: a reanalysis of Patel & Daniele (2003a). Music Percept. Interdisc. J. **29**(1), 115–120 (2011)
15. Carlson, L., Marcu, D., Okurowski, M.E.: Building a discourse-tagged corpus in the framework of Rhetorical Structure Theory. In: 2nd SIGdial Workshop on Discourse and Dialogue (SIGDIAL 2001), vol. 16, pp. 1–10. Association for Computational Linguistics, Stroudsburg (2001)
16. Balint, M., Trausan-Matu, S.: A critical comparison of rhythm In music and natural language. Ann. Acad. Rom. Scientists Ser. Sci. Technol. Inf. **9**(1), 43–60 (2016)
17. Stevens, J.P.: Applied Multivariate Statistics for the Social Sciences. Lawrence Erblaum, Mahwah (2002)
18. Garson, G.D.: Multivariate GLM, MANOVA, and MANCOVA. Statistical Associates Publishing, Asheboro (2015)
19. Klecka, W.R.: Discriminant Analysis. Quantitative Applications in the Social Sciences Series, vol. 19. Sage Publications, Thousand Oaks (1980)

Using Context Information for Knowledge-Based Word Sense Disambiguation

Kiril Simov$^{(\boxtimes)}$, Petya Osenova, and Alexander Popov

Institute of Information and Communication Technology, BAS,
Akad. G. Bonchev. 25A, 1113 Sofia, Bulgaria
{kivs,petya,alex.popov}@bultreebank.org

Abstract. One of the most successful approaches to Word Sense Disambiguation (**WSD**) in the last decade has been the knowledge-based approach, which exploits lexical knowledge sources such as Wordnets, ontologies, etc. The knowledge encoded in them is typically used as a sense inventory and as a relations bank. However, this type of information is rather sparse in terms of senses and the relations among them. In this paper we present a strategy for the enrichment of WSD knowledge bases with data-driven relations from a gold standard corpus (annotated with word senses, syntactic analyses, etc.). We focus on English as use case, but our approach is scalable to other languages. The results show that the addition of new knowledge improves the accuracy of WSD task.

Keywords: Knowledge-based word sense disambiguation · Inference of semantic relations · Context representation

1 Introduction

The recent success of *knowledge-based Word Sense Disambiguation* (**KWSD**) approaches depends on the quality of the *knowledge graph* (**KG**) — whether the knowledge represented in terms of nodes and relations (arcs) between them is sufficient for the algorithm to pick the correct senses of the ambiguous words. Several extensions of the KG constructed on the basis of WordNet have been proposed and already implemented. The solutions to Word Sense Disambiguation (**WSD**) related tasks usually employ lexical databases, such as WordNets and ontologies. However, lexical databases suffer from sparseness with respect to the availability and density of relations. One approach towards remedying this problem is the BabelNet [1], which relates several lexical resources — WordNet[1] [2], DBpedia[2], Wiktionary[3], etc. Although such a setting takes into consideration the role of lexical and world knowledge, it does not incorporate contextual knowledge learned from actual texts (such as collocational patterns, for example). This happens because the knowledge sources for WSD systems usually capture only a

---

[1] https://wordnet.princeton.edu/.
[2] http://wiki.dbpedia.org/.
[3] https://en.wiktionary.org/wiki/Wiktionary:Main_Page.

C. Dichev and G. Agre (Eds.): AIMSA 2016, LNAI 9883, pp. 130–139, 2016.
DOI: 10.1007/978-3-319-44748-3_13

fraction of the relations between entities in the world. Many important relations are not present in the ontological resources but could be learned from texts.

Here we present approaches towards the enrichment of WSD knowledge bases with context information, represented as relations over semantically annotated corpora. These context relations are taken from gold standard corpora. We focus on English (*SemCor* [3]) as use case, but our approach is scalable to other languages as well. Such an approach is justified by the fact that the lexical databases are sparse with respect to the available knowledge, its density and appropriateness. Also, the predominance of paradigmatic knowledge (synonymy, hypernymy, etc.) is balanced by the addition of syntagmatic relations (valency) — see [4]. From the perspective of knowledge representation lexical databases contain terminological knowledge (T-Box in terms of KL-One) and semantically annotated corpora contain world knowledge (A-Box in terms of KL-One). The current paper demonstrates that adding such context information improves the accuracy of Knowledge-based WSD.

The structure of the paper is as follows: the next section discusses the related work on the topic. Section 3 presents the manually annotated with senses resource — *SemCor*. Section 4 introduces the knowledge-based tool for WSD. Section 5 describes the creation of a new knowledge graph on the basis of gloss logical form encoded in eXtended WordNet (XWN). Section 6 demonstrates two approaches to encode sentences as context relations. Section 7 reports on the performed experiments. Section 8 concludes the paper.

2   Related Work

Knowledge-based systems for WSD have proven to be a good alternative to supervised systems, which require large amounts of manually annotated data. Knowledge-based systems require only a knowledge base and no additional corpus-dependent information. An especially popular knowledge-based disambiguation approach has been the use of popular graph-based algorithms known under the name of "Random Walk on Graph" [5]. Most approaches exploit variants of the PageRank algorithm [6]. Agirre and Soroa [7] apply a variant of the algorithm to WSD by translating WordNet into a graph in which the synsets are represented as nodes and the relations between them are represented as arcs. The resulting graph is called a *knowledge graph* in this paper. Calculating the PageRank vector **Pr** is accomplished through solving the equation:

$$\mathbf{Pr} = cM\mathbf{Pr} + (1 - c)\mathbf{v} \tag{1}$$

where M is an $N \times N$ transition probability matrix (N being the number of nodes in the graph), c is the damping factor and $\mathbf{v}$ is an $N \times 1$ vector. In the traditional, static version of PageRank the values of $\mathbf{v}$ are all equal ($1/N$), which means that in the case of a random jump each vertex is equally likely to be selected. Modifying the values of $\mathbf{v}$ effectively changes these probabilities and thus makes certain nodes more important. The version of PageRank for which the values in $\mathbf{v}$ are not uniform is called *Personalized PageRank* (PPR). The words in the text that are to be disambiguated are inserted as nodes in the KG and are connected to their potential senses via directed arcs. These newly introduced nodes serve to inject initial probability mass (via the vector **v**) and thus to make their associated sense nodes especially relevant in the *knowledge graph*. Applying the PPR algorithm iteratively over the resulting graph determines the most appropriate sense for each ambiguous word. Montroyo et al. [8] present a combination of knowledge-based and supervised systems for WSD, which demonstrate that the two approaches can boost one another, due to the fundamentally different types of knowledge they utilise (paradigmatic vs. syntagmatic). They explore a knowledge-based system that uses heuristics for WSD depending on the position of word potential senses in the WordNet knowledge graph (**WN**). In terms of supervised machine learning based on an annotated corpus, it explores a Maximum Entropy model that takes into account multiple features from the context of the to-be-disambiguated word. This earlier line of research demonstrates that combining paradigmatic and syntagmatic information is a fruitful strategy, but it does so by doing the combination in a postprocessing step, i.e. by merging the output of two separate systems; also, it still relies on manually annotated data for the supervised disambiguation.

The success of KWSD approaches apparently depends on the quality of the knowledge graph – whether the knowledge represented in terms of nodes and relations between them is sufficient for the algorithm to pick the correct senses of ambiguous words. An approach similar to ours is described in [9], which explores the extraction of syntactically supported semantic relations from manually annotated corpora: *SemCor*. *SemCor* was processed with the MiniPar parser and the subject-verb and object-verb relations were extracted. The new relations were represented on several levels: as word-to-class and class-to-class relations. The extracted selectional relations were then added to WordNet. The main difference with our approach is that the set of relations used in our work is larger, including whole sentences in which different n-ary relations are encoded such as subject-verb-object-indirect-object relations, adjective-noun-noun relations, etc.). In our case we have added much more relations.

3   The Sense Annotated Resources: *SemCor* and eXtended WordNet

As it was stated above, our goal is to experiment with different kinds of semantic relations. The relations missing in WordNet are the syntagmatic ones. As sources of such types of relations we consider semantically annotated resources extended with syntactic information. In this case we are able to extract syntagmatic relations between semantic classes of syntactically related words. For English we use a parsebank created over the texts in *SemCor* and XWN which is annotated also with syntax and logical forms. Since *SemCor* has been exploited for extracting relations, it was divided into test and training parts in ratio of one-to-three.

The sense annotations in *SemCor* were also performed manually on the base of WordNet. It comprises texts from Brown corpus[4] which is a balanced corpus. In this respect *SemCor* contains really diverse types of texts. We use *SemCor* in two ways: first, for testing the WSD for English; and second, as a source for extracting of new semantic relations. To achieve this, we parsed *SemCor* with a dependency parser included in the IXA pipeline[5]. Then we divided the corpus into a proportion one-to-three: first part comprises 49 documents (from br-a01 to br-f44) and it was used as a test set in the experiments reported below in the paper. The rest of the documents formed the training set from which the new relations were extracted. The new semantic relations were extracted on the basis of the syntactic relations in the dependency parses of each sentence in the training part of *SemCor*.

4 Knowledge-Based Tool for the WSD

The experiments that serve to illustrate the outlined approaches were carried out with the UKB[6] tool, which provides graph-based methods for WSD and measuring lexical similarity. The tool uses a set of random walk on graph algorithms, described in [7]. The tool builds a knowledge graph over a set of relations that can be induced from different types of resources, such as WordNet or DBPedia; then it selects a context window of open class words and runs the algorithm over the graph. We have used the UKB default settings, i.e. a context window of 20 words that are to be disambiguated together, and 30 iterations of the PPR algorithm. The UKB tool requires two resource files to process the input file. One of the resources is a dictionary file with all lemmas that can be possibly linked to a sense identifier. In our case WordNet-derived relations were used for our knowledge base; consequently, the sense identifiers are WordNet IDs. For instance, a line from the dictionary extracted from WordNet looks like this:

```
predicate 06316813-n:0 06316626-n:0 01017222-v:0
           01017001-v:0 00931232-v:0
```

First comes the lemma associated with the relevant word senses, after the lemma the sense identifiers are listed. Each ID consists of eight digits followed by a hyphen and a label referring to the POS category of the word. Finally, a number following a colon indicates the frequency of the word sense, calculated on the basis of a tagged corpus. When a lemma from the dictionary has occurred in the analysis of the input text, the tool assigns all the associated word senses to the word form in the context and attempts to disambiguate its meaning among them.

The second resource file required for running the tool is the set of relations that is used to construct the knowledge graph over which algorithms are run.

---

[4] http://clu.uni.no/icame/manuals/BROWN/INDEX.HTM.

[5] http://ixa.si.ehu.es/Ixa.

[6] http://ixa2.si.ehu.es/ukb/.

As an initial knowledge graph we are using the resource files for version 3.0, distributed together with the tool, have been used in our experiments. The distribution of UKB comes with a file containing the standard lexical relations defined in WordNet, such as hypernymy, meronymy, etc., as well as with a file containing relations derived on the basis of common words found in the synset glosses, which have been manually disambiguated. The format of the relations in the KG is as follows:

```
u:SynSetId01 v:SynSetId02 s:Source d:w
```

where `SynSetId01` is the identifier of the first synset in the relation, `SynSetId02` is the identifier of the second synset, `Source` is the source of the relation, and `w` is the weight of the relation in the graph. In the experiments reported in the paper, the weight of all relations is set to 0.

This tool is used for performing all the experiments reported in the next section. The goal in this paper is to investigate the impact of the different sets of relations over the knowledge graph.

5 New Knowledge Graph from Logical Form

Here we present an approach towards the enrichment of WSD knowledge bases with relations from gold standard corpora. In our previous work we focused on Bulgarian (*BTB* [10]) and English (*SemCor* [3]) corpora as use cases and as sources of new semantic relations. The extraction of new semantic relations from gold corpora is a mechanism for balancing the predominance of paradigmatic knowledge (synonymy, hypernymy, etc.) by the addition of syntagmatic relations.

The new relations are extracted from eXtended WordNet (XWN) by using the logical form of glosses in WordNet. This corpus was already used for the extraction of semantic relations from the co-occurrences of the synset concept and the concepts assigned during the annotation to the words in the gloss. For example, the synset {disyllable, dissyllable} — 06290539-n, is defined by "a word having two syllables." After the analysis, the following synsets are selected: 06286395-n — *word*, 06304671-n — *syllable*, 02203362-v — *have*. Each of these synsets is related to the synset which the gloss belongs to[7]:

```
u:06290539-n v:06286395-n
u:06290539-n v:06304671-n
```

The logical form for this gloss in XWN is the following

```
disyllable:NN(x1) ->
    word:NN(x1) have:VB(e1, x1, x2)
    two:JJ(x2) syllable:NN(x2)
```

---

[7] In the knowledge graph constructed in this way and distributed with the UKB system, the relation between the noun synset and the verb synset for *have* is not presented.

In our opinion, each predicate that originates from a verbal, adjectival, adverbial, or prepositional lemma expresses an event. In the example, `have:VB(e1, x1, x2)` denote the event of "holding" of object denoted by `x2` by the object denoted by `x1`. Both of these objects are participants of the event of "holding" `e1`. From this we extract the following relations:

```
u:02203362-v v:06286395-n
u:02203362-v v:06304671-n
u:06286395-n v:06304671-n
```

In the case of {ice-cream cone} defined by "ice cream in a crisp conical wafer" the following logical form is presented:

```
ice-cream_cone:NN(x1) ->
    ice_cream:NN(x1) in:IN(x1, x2)
    crisp:JJ(x2) conical:JJ(x2)
    wafer:NN(x2)
```

From it we have extracted relations between "ice cream" and "wafer" on the basis of the predicate `in:IN(x1, x2)`, also between "crisp" and "wafer", "conical" and "wafer", and between "crisp" and "conical" in the appropriate senses. This set of relations forms a knowledge graph which we denote as **WNGL** — knowledge graph constructed on the basis of the logical form of the glosses in WordNet.

6   Semantically Annotated Sentences as Context for WSD

We consider each sentence in a semantically annotated corpora as representation of a context in which the WSD is already performed by an expert. These contexts are similar to the context created by the UKB system during the WSD task. In our view this explains the good results reported below. In this section we present two approaches to represent such contexts as knowledge graphs for WSD.

The extraction of contexts from manually annotated corpora with word senses can be performed in different ways. The connection between the nodes that are semantically annotated can be determined at least in two ways:

– As a sequence of nodes corresponding to the order of the words in the sentence;
– As a syntactic structure of nodes corresponding to a parse of the sentence.

The second approach is demonstrated on the basis of SemCor. The training part of SemCor was parsed with the dependency parser from the IXA pipeline. The set of relations presented here is based on the dependency tree for each sentence. Each node in the dependency tree corresponds to a new node for the relevant word. Then it is related to the head node. An additional relation points to the node corresponding to the WordNet Synset. Since SemCor consists of text fragments, the sentences that are from the same fragment are connected via relations between the roots of the dependency trees. The root of the second tree is related to the root of the first sentence, the root of the third sentence to the root of the second sentence, etc.

For example, the sentence "Evidence that other sources of financing are unavailable must be provided." is analyzed with a dependency parser of IXA pipeline. From this analysis we construct a set of relations corresponding to the syntactic tree. Figure 1 depicts the top

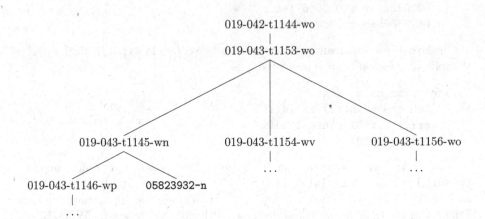

Fig. 1. The top fragment of the dependency tree

where 019-043-t****-** represents the nodes in the dependency tree, the first three digits represent: the number of the file from which the sentence is selected, the number of the sentence and the number of tokens in the sentence. Nodes like 05823932-n ("Evidence") are from the knowledge graph of WordNet V3.0. There are nodes of the syntactic tree that are not mapped to a synset, because not each word in the sentence is mapped to a synset. The relation between 019-043-t1153-wo and 019-042-t1144-wo is the relation between the root of the sentence and the root of the previous sentence. The set is called **GraphRelSC**. Note that the parsing of the sentences is done automatically. Thus, there might be errors.

The first approach mentioned above on the representation of context is illustrated on the basis of glosses in XWN. For each ⟨gloss⟩ element in XWN we consider the element ⟨wsd⟩, containing the words of the gloss with assigned synset id from WordNet V2.0:

```
<wsd>
<wf>a</wf>
<wf wn20="ENG20-05501538-n" wnsn="1">kind</wf>
<wf>of</wf>
<wf wn20="ENG20-02650459-n">artificial heart</wf>
<wf>that</wf>
<wf wn20="ENG20-02139918-v">has</wf>
```

```
<wf wn20="ENG20-02526983-v">been</wf>
<wf wn20="ENG20-01123102-v">used</wf>
<wf>with</wf>
<wf>some</wf>
<wf wn20="ENG20-06869923-n">success</wf>
</wsd>
```

We have performed the following operations:

- The synset id for synset of the gloss is added as a first element;
- The WordNet V2.0 ids are converted to WordNet V3.0 ids as they are used in UKB knowledge graphs;
- For each word annotated with WordNet id we created a node which connects to the node of the corresponding WordNet synset and to the node of the preceding word annotated with WordNet synset id in the gloss.

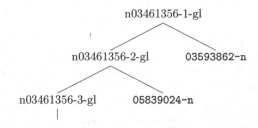

Fig. 2. The beginning of graph for the sequence of words in the gloss and their relation to synsets nodes.

Having performed this procedure on the above gloss example we created a set of relations depicted in Fig. 2. The nodes with labels `n03461356-*-gl` are nodes corresponding to the word in the gloss and the other nodes are corresponding to the synsets in WordNet V3.0. We call the resulting set of relations **WN30glCon**.

7   Experiments

The experiments that illustrate the outlined approaches were carried out with the UKB tool, which provides graph-based methods for WSD and measuring lexical similarity. We have performed experiments with two algorithms implemented within UKB: Static and PPRw2w — see [7]. We have selected these two because the latter has shown the best results during our previous experiments and the first one is the fastest one. As it was mentioned, we exploit the semantically annotated corpus *SemCor*. As baselines we consider the results achieved with the standard knowledge graphs distributed within UKB system: **WN** for the relations in WordNet and **WNG** for the relations extracted from **XWN** on the basis of co-occurrences.

The new knowledge graphs are: **WNGL** — a knowledge graph based on analysis of the logical forms of the glosses in **XWN**; **GraphRelSC** — a knowledge graph comprising sentences from *SemCor*; and **WN30glCon** — a knowledge graph comprising sentences from **XWN**. The results for these knowledge graphs are compared to the baselines and also to some combinations of them. The results show improvement on *SemCor* for both algorithms — Table 1.

Table 1. Comparison of the results for the standard knowledge graphs with the newly constructed knowledge graphs.

KG	Static	PPRw2w
WN	56.60	56.35
WNG	56.00	57.33
WN + WNG	**59.55**	**62.24**
WNGL	60.46	60.35
WN + WNGL	66.61	67.19
WN + WN30glCon	67.00	66.42
WN + GraphRelSC	67.04	65.97
WN + GraphRelSC + WNGL	68.41	68.51
WN + WN30glCon + GraphRelSC	68.74	68.15
WN + WN30glCon + GraphRelSC + WNGL	**68.77**	68.48
WN + WNG + WN30glCon + GraphRelSC + WNGL	68.39	**68.59**

The results show the following important facts: (1) the combination of relations from different sources might improve the results significantly[8]; (2) the improvement is not monotonic with respect to the number of the relations. Obviously the topology of the graph plays an important role for the Random Walk on Graphs algorithms. Also in current experimental setup only local context in the text is considered. Thus, if two senses share local connectivity in the text, they will be hard for disambiguation even when more relations are added. This problem will be studied in our future work.

8   Conclusion

The experiments with adding various bundles of relations from WordNet and from syntactically and semantically annotated corpora for English have shown several directions to be considered in our future work.

---

[8] This result for English is far from state-of-the-art, but it is based only on 25 % of *SemCor*. Also, our goal here is only to compare the various knowledge graphs.

First of all, the addition of syntagmatic syntactic-based relations in form of context improves the results of KWSD task, since they balance the paradigmatic lexical relations. Then, the accuracy depends also on the integrity of the domain – in more homogeneous domains the accuracy is more stable and increases, while in more heterogeneous domains the accuracy drops. We consider the accuracy as a measure of quality of the knowledge graph with respect to the KWSD task. The conclusion is that adding important linguistic and world knowledge in form of relations between lexical concepts does not necessarily improve the quality of the knowledge graph.

Another issue is the differing impact of the various relations on the knowledge graph. Since the quantity of the added information is huge, our idea was to reduce it through the selection of the contributing relations without losing the quality of the result. This strategy is not trivial. It requires a lot of sets of experiments as well as new mechanisms for evaluating the graph and optimizing the algorithm.

Acknowledgements. This research has received partial support by the EC's FP7 project: "QTLeap: Quality Translation by Deep Language Engineering Approaches" (610516).

References

1. Navigli, R., Ponzetto, S.P.: BabelNet: the automatic construction, evaluation and application of a wide-coverage multilingual semantic network. Artif. Intell. **193**, 217–250 (2012). Elsevier
2. Fellbaum, C. (ed.): WordNet an Electronic Lexical Database. The MIT Press, Cambridge (1998)
3. Miller, G.A., Leacock, C., Tengi, R., Bunker, R.T.: A semantic concordance. In: Proceedings of HLT 1993, pp. 303–308 (1993)
4. Cruse, D.A.: Lexical Semantics. Cambridge University Press, Cambridge (1986)
5. Agirre, E., de López, O., Soroa, A.: Random walks for knowledge-based word sense disambiguation. Comput. Linguist. **40**(1), 57–84 (2014)
6. Brin, S., Page, L.: The anatomy of a large-scale hypertextual web search engine. Comput. Netw. **56**(18), 3825–3833 (2012)
7. Agirre, E., Soroa, A.: Personalizing PageRank for word sense disambiguation. In: Proceedings of 12th Conference of the European Chapter of the ACL (EACL 2009), pp. 33–41 (2009)
8. Montoyo, A., Suárez, A., Rigau, G., Palomar, M.: Combining knowledge-and corpus-based word-sense-disambiguation methods. J. Artif. Intell. Res. (JAIR) **23**, 299–330 (2005)
9. Agirre, E., Martinez, D.: Integrating selectional preferences in WordNet. In: Proceedings of 1st International WordNet Conference (2002)
10. Popov, A., Kancheva, S., Manova, S., Radev, I., Simov, K., Osenova, P.: The sense annotation of BulTreeBank. In: Proceedings of TLT13, pp. 127–136 (2014)

Towards Translation of Tags in Large Annotated Image Collections

Olga Kanishcheva[1(✉)], Galia Angelova[2], and Stavri G. Nikolov[3]

[1] National Technical University "Kharkiv Polytechnic Institute",
21 Kirpichov Street, Kharkiv 61002, Ukraine
kanichshevaolga@gmail.com
[2] Institute of Information and Communication Technologies,
Bulgarian Academy of Sciences, 25A Acad. G. Bonchev Street,
1113 Sofia, Bulgaria
galia@lml.bas.bg
[3] Imagga Technologies Ltd., bul. Cherni Vrah 47A, floor 4, 1407 Sofia, Bulgaria
stavri.nikolov@imagga.com

Abstract. This paper presents an approach for translation of tags in professional and social image databases, using an original lexical resource extracted from Wikipedia. The translation integrates a tag sense disambiguation algorithm based on WordNet and Wikipedia (as external resources defining word senses). Our disambiguation technique uses the Lesk algorithm, extended gloss overlaps and similarity measures in order to achieve successful resolution of lexical ambiguity and accurate translation of tags. We show how to involve Wikipedia as a source of translation correspondences since open WordNets are not available for most languages. Experimental results and performance evaluation show 97 % accuracy for professional images and 86 % accuracy for social images from Flickr. This translation technique can be applied by auto-tagging programs and information retrieval systems.

Keywords: Multilingual language processing · Tag translation · Word sense disambiguation · Annotation disambiguation · Image auto-tagging

1 Introduction

Today many tasks require multilingual or multimodal processing. We find multiple views of the same entity across various modalities and languages. For example, news articles which are published in multiple languages are essentially different views of the same facts. Similarly, video, audio and images present multiple views of the same movies. Therefore transformation (translation) of keywords into another language is needed, for example, to automatically translate tags of a travel-related image or ask natural language questions about videos and images. New approaches to language and speech processing need to be developed and integrated with computer vision.

Image tag translation is a novel shared task aimed at the generation of image description in a target language, given an image and one or more descriptions in a different (source) language. Figure 1 shows interfaces of auto-tagging programs which allow translating image tags. One of the key issues is tag sense disambiguation (TSD).

© Springer International Publishing Switzerland 2016
C. Dichev and G. Agre (Eds.): AIMSA 2016, LNAI 9883, pp. 140–150, 2016.
DOI: 10.1007/978-3-319-44748-3_14

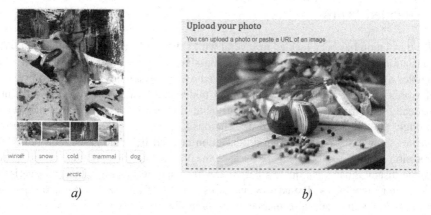

a)   b)

Fig. 1. Auto-tagging by: (a) Clarifai (www.clarifai.com) and (b) Imagga (www.imagga.com)

Tag ambiguity leads to inaccuracy and misunderstandings. For example, Fig. 2 represents the automatically generated tags in English and Russian for the image in Fig. 1(b). This tagset contains the keyword "*bulb*" that is translated as "*лампа*" (*lamp*) while it had to be translated as "*луковица*" (*a rounded underground plant part*). Therefore resolving the ambiguity of tags will help to improve the accuracy of keyword machine translation. However, a major problem for the automatic TSD is the limited context of isolated keywords in the tagsets. It is well known that the successful resolution of lexical ambiguity exploits observations of the various entities that occur around the text unit to be disambiguated but text is often missing in our task.

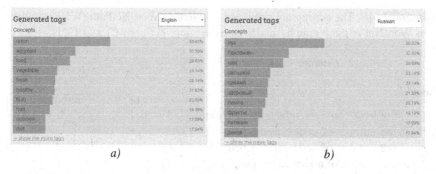

a)   b)

Fig. 2. Imagga's English and Russian tags generated for the image in Fig. 1(b)

The paper is organized as follows: Sect. 2 considers related work and introduces notions and algorithms used for TSD. In Sect. 3 we describe our approach for translation of image tags. Section 4 presents the experimental results and their evaluation (the accuracy achieves 97 % for images from *Professional image marketplace*); examples of correct and wrong translations are discussed as well. Finally, in Sect. 5 we briefly sketch future work and present the conclusion.

2 Related Work

Tag translation has been tackled in several works. Etzioni et al. [1, 2] implement lexical translation via a "translation graph", a massive lexical resource where each node denotes a word in some language and each edge denotes a word sense shared by a pair of words. Their graph contains 1,267,460 nodes and 2,315,783 edges. It was automatically constructed from machine-readable dictionaries and Wiktionaries. Paths through the graph suggest word translations. A probabilistic inference procedure enables to quantify confidence in a translation derived from the graph, and thus trade precision against recall. Using the graph, the PANIMAGES cross-lingual image search engine disambiguates the queries before sending them to Google. Thus, for queries in 33 "minor" languages, PANIMAGES increases correct results by 75 % on the first 15 pages (270 results), and also increases the precision by 27 %.

In [3] the author mapped noun tags of images from an annotated collection to the word senses of a foreign language lexical resource (via the bilingual correspondences of an existing bilingual dictionary). He used the English-Russian dictionary by V.K. Mueller [4] to enhance the *Yet Another RussNet* synsets with *Flickr* photos.

Stampouli A. et al. proposed a mashup that combines data and functionality of Flickr and Wikipedia and aims at enhancing content retrieval for web users [5].

Tag disambiguation is also tackled in [6, 7] which are not directly related to image processing. In [6] the authors use Wikipedia to define Neighbor tags to help users understand the meaning of each tag in a folksonomy with informal, uncontrolled, and personalized vocabulary. In [7] all the annotations involving a target tag are modelled with a 3-order tensor; spectral clustering over the hypergraph induced from it helps to discover the clusters representing the senses of the target tag within a folksonomy.

We shall use the following notions and algorithms:

Definition 1. [8] The conceptual similarity *ConSim*, or WordNetPath (WUP) similarity when WordNet is the hierarchy under consideration, between C_1 and C_2 is

$$ConSim(C_1, C_2) = 2 \cdot N_3 / (N_1 + N_2 + 2 \cdot N_3)$$

where

- C_3 is the least common subsumer of C_1 and C_2;
- N_1 is the number of nodes on the path $C_1 \rightarrow C_3$;
- N_2 is the number of nodes on the path $C_2 \rightarrow C_3$;
- N_3 is the number of nodes on the path from C_3 to the top.

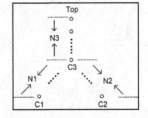

Fig. 3. A conceptual hierarchy

Thus $ConSim(C_1, C_2) \in [0,1]$ and 1 indicates the highest possible similarity score for a pair of concepts or words (Fig. 3).

The classical **Lesk algorithm** [9] for word sense disambiguation (WSD) matches the text contexts of an ambiguous word w to its sense definitions. Given a target word w, the following score is computed for each sense S of w:

$$scoreLeskVar(S) = |context(w) \cap gloss(S)|,$$

where $context(w)$ is the bag of all content words in a context window around the target word w. For our task, given an ambiguous keyword w, all other tags for the same image will be viewed as the context of w use [11]. List of keyword senses are taken either from WordNet or from Wikipedia.

The **Extended Gloss Overlap** approach [10] is based on the Lesk mapping but expands the glosses of the words being compared to include glosses of *related* words/concepts; here we shall consider the relation *Hyponyms* and *Hypernyms* of the words being compared. This helps to tackle longer contexts and to overcome the significant limitation of short glosses (e.g. the ones in WordNet have average length of 7 words). As in the Lesk algorithm, the synset having the largest weight is chosen:

$$score_{ExtLesk}(S) = \sum_{S':\ S \xrightarrow{rel} S'\ or\ S \equiv S'} |context(w) \cap gloss(S')|,$$

where $context(w)$ is the bag of all content words in a context window around the target word w and $gloss(S')$ is the bag of words in the textual definition of a sense S' which is either S itself or related to S through a hypernym/hyponym relation rel. The overlap scoring mechanism is also parametrized and can be adjusted to take into account gloss length (i.e. normalization) or to include function words.

In this paper we consider image tags as sets of keywords and try to translate them from English into Russian after TSD. When selecting a tag assigned to an image, we match the other tags of the same image to the entries of specific lexical resources [11]. One of them is WordNet; another original resource is based on Wikipedia. So we use either the WordNet hierarchy or the one imposed by the linked Wikipedia disambiguation pages.

3 Our Approach to Tag Translation

In [11] we proposed a TSD algorithm given an image I with tags that have one or multiple senses in WordNet. Ambiguity is resolved for each tag separately. Briefly, for a tag t with multiple senses $t_1, t_2, ..., t_n$, we build a similarity matrix M with n rows corresponding to the t's senses and $k+1$ columns where k is the total number of all WordNet senses for the tags $\mathcal{JT} = \{p_1, p_2, ..., p_m\}$ of I, $p_j \neq t$ for $j = 1, ..., m$. Thus all senses of the $\mathcal{JT}$ tags, enumerated in a list L, provide disambiguation context C for t. In each cell a_{ij} of M we calculate the WUP similarity between the i-th sense of t and the j-th sense in the list L. After summing up all row values $a_{ij}, j=1,...,k$ in the cell $a_{i,k+1}$ of M, we select the row q, $1 \leq q \leq n$ with the maximal score in the column $k+1$ of M and consider the meaning t_q as the t's sense that is meant in the annotation of I. In addition, for each sense of every $p \in \mathcal{JT}$ we calculate its similarity to the WordNet definitions of $t_1, t_2, ..., t_n$ using the algorithms of Lesk and the Extended Gloss overlap, and show how to aggregate the latter measures with the M values in specific vectors. The experiments show TSD accuracy higher than 90 % [11]. It turned out that considering contexts of all tag senses gives better results than using only tags with a single WordNet sense.

Searching for publicly available multilingual vocabulary, we build a special lexicon based on Wikipedia. For each English tag with many Wikipedia senses we collect the corresponding Russian translations and fragments of the Wikipedia definitions in English (Table 1). We store the initial 1–3 sentences from the Wikipedia article in order to construct longer contexts for further tag disambiguation.

Table 1. Senses of keyword "*spring*" in the English Wikipedia and their Russian translations

English tag	Russian translation	Definition in the English Wikipedia
spring	*весна*	S_1: *spring is one of the four conventional temperate seasons following winter and preceding summer there are various technical definitions of spring but local usage of the term varies according to local climate cultures and customs*
spring	*пружина*	S_2: *a spring is an elastic object used to store mechanical energy springs are usually made out of spring steel*
spring	*родник*	S_3: *a spring is any natural situation where water flows from an aquifer to the earth's surface it is a component of the hydrosphere*

Figure 4 sketches the pipeline of tag processing components. Annotated images arrive to our system. Their keywords are assigned manually or automatically (in this experiment, they come from the Imagga's auto-tagging program). At the preprocessing phase, for each image we split the tags into *(i)* "*original*" (assigned automatically by the original Imagga's platform, where the annotation is based on WordNet synsets for recognized objects and thus each tag's sense is internally fixed) and *(ii)* "*additional*" ones (derived by Imagga from similar images from large and diverse photo collections in order to extend the description of the image content). After this we refine the "*additional*" tags associated with images, for example, find plural forms and transform them to singular [12].

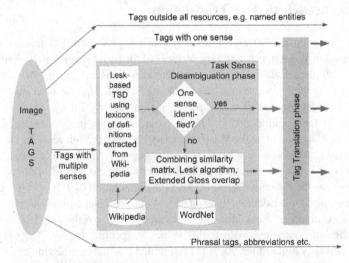

Fig. 4. A Pipeline for tag sense disambiguation and Translation

Our goal is to translate all image tags from English into Russian. The focus is particularly on the "*additional*" tags because they might be ambiguous, but the "*original*" tags with one sense are very useful as disambiguation context. Tag translation is a subsequent step after all tag senses are justified.

Initially we justify the number of tag senses. If a tag has one sense in Wikipedia we directly translate it. Note that a tag can have one sense in WordNet but more in Wikipedia which happens for terminology. Then we apply our algorithm [11].

Figure 4 shows a core processing step: to apply the simple Lesk algorithm for TSD using our lexicon based on Wikipedia which normally contains longer definitions than WordNet. Longer definitions provide more vocabulary that supports resolution of ambiguity. If we find a correct sense that we can translate this tag (Example 1).

Fig. 5. *{water, spring, grass, rock}*

Example 1. Consider the image in Fig. 5 and the tag *spring*. We use the Lesk algorithm and match the definitions of senses (Table 1) to the tagset {*water, grass, rock*}. Then the word *water* is met once in the gloss of the sense S_3. Therefore, the weight of S_3 sense is 1, and the senses S_1 and S_2 have weights zero. Therefore the tag *spring* is translated as родник.

If no decision is made using the Lesk-based filter, we continue with the TSD approach in [11]. Then we have to calculate the WUP similarity and also construct the Extended Glosses. The latter allows extending the disambiguation context and we can try the Lesk algorithm once again (Example 2 and *plate* definitions in Table 2).

Fig. 6. *{plate, salad, meat, tomato}*

Example 2. The *plate* tag annotates the image in Fig. 6. Our TSD algorithm [11] finds in Word-Net that here *plate* has the sense: *(n) plate (dish on which food is served or from which food is eaten).*

Then the nouns from the WordNet gloss are mapped by the Lesk algorithm to the definitions in Table 2. Thus, we find the word *food* in sense S_1 and obtain the correct sense (S_1) and the corresponding translation тарелка.

4   Experiments and Results

Our approach was evaluated manually on 60 images from *Professional image marketplace* and 60 images from *Flickr*, with 7,651 tags in total (see Table 3). We evaluate how often the correct tag translation is provided using the Wikipedia resource.

All professional images were classified manually by human annotators into three categories with informal names "*Food*", "*Scene*", and "*Children*". Social images from

Table 2. Senses of keyword *"plate"* in the English Wikipedia and their Russian translations

English tag	Russian translation	Definition from English Wikipedia
plate	тарелка	S_1: *a plate is a broad concave but mainly flat vessel on which food can be served a plate can also be used for ceremonial or decorative purposes*
plate	тектоника плит	S_2: *plate tectonics from the late Latin tectonicus from the Greek pertaining to building is a scientific theory that describes the large-scale motion of earth's lithosphere this theoretical model builds on the concept of continental drift which was developed during the first few decades of the 20th century*
plate	фотопластинка	S_3: *photographic plates preceded photographic film as a capture medium in photography the light-sensitive emulsion of silver salts was coated on a glass plate typically thinner than common window glass instead of a clear plastic film*

Table 3. Tags in the test datasets

Tags	Professional image marketplace	Flickr
Original tags (with one sense)	*350*	*505*
Additional tags / from them ambiguous	*4,334 / 3,765*	*2,462 / 2,198*
Total	**4,684**	**2,967**

Flickr were classified also into three categories *"Scene"*, *"Flowers"*, and *"Animals"*. For the images taken from *Professional image marketplace*, some 1,210 tags out of the 4,334 *"additional"* tags were translated using Google translate (free multilingual statistical machine translation service provided by Google), and 3,124 tags (72 %) have been translated with our approach.

For the social images, 936 tags out of the 2,462 *"additional"* tags were translated using Google translate, and 1,526 tags (62 %) have been translated with our approach.

All translations were analyzed manually by annotators, where the humans were asked to compare the results received by Google translate and by our approach. In addition, the human annotators had access to bilingual dictionaries. The tricky issue in the comparison is that actually we do not know the exact intended meaning communicated by the photo authors so benchmarking is difficult. The accuracy for the translation of professional tags is presented in Fig. 7a. It is seen that our method produces more correct translations (on average 2 % improvement). This is mostly due to the longer definitions of Wikipedia that provide more vocabulary for checking mappings to the tagset keywords.

The accuracy for tag translation of *Flickr* images is lower than the one for professional images (about 86 %, Fig. 7(b). Social images are often low-quality photos without clearly focused objects, which is somewhat problematic for the auto-tagging platforms. Therefore these images are annotated by fewer tags with lower relevance.

In Fig. 8 the green line columns show the quantity of tags which have been translated using the Lesk algorithm with lexicons based of definitions extracted from

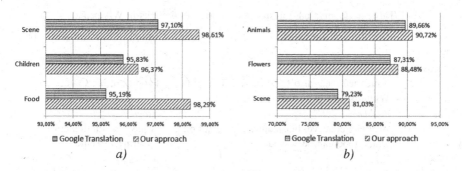

Fig. 7. Evaluation of translation accuracy per category. (a) *Professional image marketplace,* (b) *Flickr*

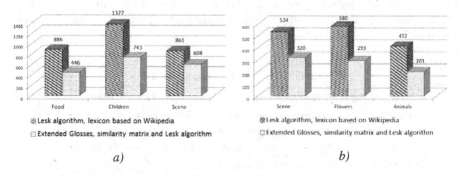

Fig. 8. Efficiency of components: (a) *Professional image marketplace,* (b) *Flickr*

Wikipedia. Purple dot columns show the number of tags translated by calculation of similarity matrix, Lesk algorithm and Extended Gloss overlap as suggested in [11].

We present some examples to illustrate the results. Distinguishing between senses in WordNet and Wikipedia is not easy. E.g. the noun tag *caucasian* from the category "*Children*" has 2 senses in WordNet and 3 senses in Wikipedia:

WordNet: ● <u>S:</u> (n) White, White person, **Caucasian** (a Caucasian)
● <u>S:</u> (n) **Caucasian**, Caucasian language (a number of languages spoken in the Caucasus that are unrelated to languages spoken elsewhere)

Wikipedia: ● <u>S:</u> (n) Peoples of the Caucasus
● <u>S:</u> (n) Languages of the Caucasus
● <u>S:</u> (n) Caucasian race

Due to these reasons we start the analysis by using Wikipedia's definition because Wikipedia is the resource that will deliver the translations we need. WordNet glosses are analyzed at the next step when the related words are taken into consideration. Table 4 shows some examples of translations based on Wikipedia.

Table 4. Correct tag translations using the definitions in Wikipedia

English tag	Our Russian translation	Google's Russian translation
caucasian	*европеоидная раса* Caucasian race (also Caucasoid or Europid). White people in general	*кавказец* Peoples of the Caucasus
plate	*тарелка* A broad, concave, but mainly flat vessel on which food can be served	*пластина* A piece of Earth's crust and uppermost mantle
table	*стол* An item of furniture with a flat top	*таблица* A table of structured data

The lexical resource based on Wikipedia allows getting more meaningful and literary translations including terminological ones as well as translations of words in phrases. However there are also incorrect translations. Tables 5, 6 and 7 present examples.

Table 5. Meaningful tag translations

English tag	Our Russian translation	Google's translation
parsley	*петрушка кудрявая*/curled parsley	*петрушка*/parsley
snack	*легкая закуска*/light snack	*закуска*/snack
olive	*олива европейская*/european olive	*оливковый*/olive *adj*

Table 6. Translations in phrasal tags: similar performance as Google translate

English tag	Our Russian translation	Google's translation
mid adult	*середине взрослых*	*середине взрослых*
caucasian appearance	*кавказской внешности*	*кавказской внешности*

Table 7. Incorrect tag translations – *home* from the *Flickr* category "*Children*"

English tag	Correct translation	Our translation	Google's translation
home	*дом, семья* home, family	*главная* main page	*главная* main page

5   Conclusions

Here we present an approach for automatic translation of tags that annotate images in large public collections. The main advantage of the suggested algorithm is the combination of solutions for tag sense disambiguation and tag translation. The novel aspect is the involvement of Wikipedia's definitions which provide contexts that might help to better disambiguate tag senses. Using Wikipedia looks a reasonable approach because its content constantly grows and articles in new languages are included every day. Especially valuable is the content in the so called minor languages for which few linguistic resources exit. We show how to borrow from Wikipedia translation equiv-alences. Wikipedia also allows thinking about automatic translation of the terminology and the named entities that occur as image tags. Another advantage of our approach is that it can be used for all language pairs provided by Wikipedia (given that there are no open WordNets with sufficient size for most languages). Up to our knowledge, tag translation in the context of image auto-tagging is considered for the first time.

As future work we plan to study better the mismatch between senses in WordNet and Wikipedia and to assess more deeply how it influences our decision making in regard to the selection of "correct" tag translation. These experiments will require a more comprehensive evaluation. We also plan to use further disambiguation techniques e.g. unsupervised approaches (but the lack of large image collections with multilingual annotations remains a major obstacle to test a variety of tools). Finally, the Extended Gloss overlap can be applied not only for the *Hyponyms* and *Hypernyms* of the tag under consideration but also for other glosses of words linked to the focal one with relations like *part_of, attribute*, etc.

Acknowledgements. This research is partially supported by the EC FP7 grant 316087 AComIn *"Advanced Computing for Innovation"*, 2012–2016. It is also related to the COST Action IC1307 *"Integrating Vision and Language (iV&L Net)"*. The authors are thankful to Imagga's technical team for their comments, recommendations and experimental datasets.

References

1. Etzioni, O., Reiter, K., Soderland, S., Sammer, H.: Lexical translation with application to image search on the web. In: Macgoad, B. (ed.) Proceedings of Machine Translation Summit XI, pp. 175–182 (2007)
2. Reiter, K., Soderland, S., Etzioni, O.: Cross-lingual image search on the web. In: Proceedings of the IJCAI 2007 Workshop on Cross-Lingual Information Access (2007). http://search.iiit.ac.in/CLIA2007/papers/CLIS.pdf
3. Ustalov, D.: TagBag: Annotating a foreign language lexical resource with pictures. In: Khachay, M.Y., Konstantinova, N., Panchenko, A., Ignatov, D.I., Labunets, V.G. (eds.) AIST 2015. CCIS, vol. 542, pp. 361–369. Springer, Heidelberg (2015). doi:10.1007/978-3-319-26123-2_35
4. Mueller, V.K.: English-Russian Dictionary. http://mueller-dict.sourceforge.net/cgi-bin/dict.cgi?word=00-database-info;dict=mueller-dict;scroll=on#defs
5. Stampouli, A., Giannakidou, E., Vakali, A.: Tag disambiguation through Flickr and Wikipedia. In: Yoshikawa, M., Meng, X., Yumoto, T., Ma, Q., Sun, L., Watanabe, C. (eds.) DASFAA 2010. LNCS, vol. 6193, pp. 252–263. Springer, Heidelberg (2010)
6. Kangpyo, L., Hyunwoo, K., Hyopil, S., Hyoung-Joo, K.: Tag sense disambiguation for clarifying the vocabulary of social tags. In: Proceedings of the International Conference on Computational Science and Engineering, CSE 2009, vol. 4, pp. 729–734 (2009)
7. Kaipeng, L., Binxing, F., Weizhe, Z.: Unsupervised tag sense disambiguation in folksonomies. J. Comput. 5(11), 1715–1722 (2010)
8. Wu, Z., Palmer, M.: Verb semantics and lexical selection. In: Proceedings of the 32nd Annual Meeting of the Association for Computational Linguistics, pp. 133–138 (1994)

9. Lesk, M.: Automatic sense disambiguation using machine readable dictionaries: how to tell a pine cone from an ice cream cone. In: Proceedings of the 5th Annual International Conference on Systems Documentation, SIGDOC 1986, pp. 24–26. ACM, USA (1986)

10. Banerjee, S., Pedersen, T.: Extended gloss overlaps as a measure of semantic relatedness. In: Proceedings of the 18th International Joint Conference on AI (IJCAI), pp. 805–810 (2003)

11. Kanishcheva, O., Angelovaou, G.: About sense disambiguation of image tags in large annotated image collections. In: Margenov, S., Angelova, G., Agre, G. (eds.) Innovative Approaches and Solutions in Advanced Intelligent Systems. Studies in Computational Intelligence, vol. 648, pp. 133–149. Springer, Heidelberg (2016)

12. Kanishcheva, O., Angelova, G.: A pipeline approach to image auto-tagging refinement. In: Bădică, et al. (eds.) Proceedings of the 7th Balkan Conference on Informatics. ACM, New York (2015)

Linking Tweets to News: Is All News of Interest?

Tariq Ahmad$^{(\boxtimes)}$ and Allan Ramsay

School of Computer Science, The University of Manchester, Oxford Road,
Manchester M13 9PL, UK
tariq.ahmad@postgrad.manchester.ac.uk, allan.ramsay@cs.man.ac.uk

Abstract. In a world where news is being generated almost continuously by many different news providers on many different platforms, it would be useful in certain industries to be able to determine how much of that news is actually being read, which news items are not interest generating or, indeed, if there are topics being discussed on Twitter that have not even been reported in the news. Twitter generates vast numbers of Tweets daily and has a massive active user base, so it is ideal as a way of gauging what news people are, or are not, interested in. This paper proposes a technique to efficiently relate Tweets to news articles and then to determine which news articles are of interest, which are not, and what is being discussed on Twitter that is not even in the news.

Keywords: Twitter · News articles · Similarity · TF-IDF

1 Introduction

Twitter generates 500 million Tweets daily [4] and has a 320 million active monthly user base [8]. Many of these Tweets will talk about a current news topic, many of them will refer to something that has not appeared in the mainstream news yet and many will be "noise". If we could link news articles to Tweets we could perhaps determine which news items are generating interest and which are not. Tweets are random in nature, in that they may contain unstructured text, abbreviations, slang, acronyms, emoticons, incorrect grammar and incorrect spelling [1]. When we took individual Tweets and tried to link them to news articles using simple word matching this produced inaccurate matches. This is because individual Tweets do not contain enough information to perform this task. When we tried to match 10,000 Tweets to 11,888 news articles, using simple word matching alone (excluding hashtags, URLs and usernames) we had 9 % returned as matching. However when we looked at the results, a Tweet such as "@TekelTowers @mygransfortea not mock", is clearly not related to the suggested news article headlined "Donald Trump says he did not mock New York Times reporter's disability". Finding Tweets that are actually related to news stories, rather than finding ones that share low frequency terms with them, is a challenging task.

The problem is that individual Tweets are very short, and hence TF-IDF vectors for them will be heavily swayed by specific terms. The aim of the work

© Springer International Publishing Switzerland 2016
C. Dichev and G. Agre (Eds.): AIMSA 2016, LNAI 9883, pp. 151–161, 2016.
DOI: 10.1007/978-3-319-44748-3_15

reported here is to group Tweets into larger collections, which we will refer to as "stories", in order to downplay the effects of individual terms. If the Tweet above about going to gran's for tea had been linked into the larger body of material about daily life which is posted by *@TekelTowers* then the presence of the word *'mock'* would have been masked by information about Scottish football, and this Tweet would not have been linked to material relating to Donald Trump.

Tweets include two devices for grouping them – usernames and hashtags. Usernames are helpful if there is a particular individual whose opinions you care about, but as [2], among others, has noted this does not tend to lead to thematically linked material. Hashtags, however, highlight the topics in Tweets [2], e.g. *"@Vistaprint are worse then @Ryanair with their shopping cart experience #tiring"* and hence two or more Tweets sharing the same hashtag should be related. We therefore use Twitter hashtags and usernames to group Tweets into "topics that people are talking about". We refer to these as "Tweet stories".

News articles tend not to have such strong indicators. They do, however, tend to contain large numbers of words, and hence TF-IDF vectors do provide a robust way of grouping them into stories, particularly if we also make use of the time-stamps associated with entries on RSS feeds.

The final step in the work reported here is to link Tweet stories to RSS stories to establish which news stories are being talked about on Twitter, and by elimination which news stories are **not** being talked about on Twitter and which Twitter stories are not showing up in the news. Tweet stories that are not linked or loosely linked to news stories are deemed as "things that people are talking about, that are not in the news". This exercise will yield a number of different outcomes:

1. Topics that appear in news articles and are being discussed on Twitter
2. Topics that appear in news articles but are not being discussed on Twitter
3. Topics that do not appear in news articles but are being discussed on Twitter

There are a number of different players such as news providers, governments, politicians or data analysts who could potentially benefit from knowing about the relationship between news stories and what is being discussed on Twitter.

News providers would be interested in knowing what topics are in the news that people are not talking about on Twitter because that might draw their attention to the fact that they may be spending substantial resources on producing content that, actually, is not of interest to anyone. They might also be interested in knowing that there are things that people are discussing on Twitter that they are not reporting on. For example, if there was a lot of discussion on Twitter about the weakness of Putin but this was not reported in the news, this might be of considerable interest to certain types of people. In general, as discussed by Howard et al. [3], countries with repressive state-controlled media would certainly be interested to know that there are things being discussed on Twitter that might precede major events on the ground.

Twitter also moves faster than platforms for traditional news dissemination. If we could find patterns in Tweet stories that indicated that they were likely to turn into news stories, we could act more quickly in the face of, for instance, epidemics, which might show up on Twitter as informal observations about symptoms and absence from work well before there was reliable epidemiological evidence.

2   Data Collection

We need a large number of Tweets. Given that we need to use time-stamps as part of the process of linking individual Tweets and articles into Tweet stories and news stories, and that we then need to use this same information to constrain the links between Tweet and news stories, it is easier to construct a corpus with the required characteristics than trying to find existing corpora that cover the same time span and hence can reasonably be expected to contain linked stories.

We therefore gathered Tweets originating in the UK, with no restriction on topic, and news articles from RSS feeds available from a number of major news providers with UK outlets, namely Google, BBC, The Independent, The Guardian, The Daily Express, Sky and CNN, over a period of 30 days. The final dataset contained 693,527 Tweets and 11,896 news article webpages.

We did not restrict the news articles or Tweets to any specific topic, keyword or news event. The aim of the project is to find out what kinds of Twitter stories are under- or over-represented in the traditional news media: making any kind of prejudgement about what area we want to look at would inevitably compromise this goal. We may have some initial thoughts about the likely outcome of this investigation, but it would be a mistake to build those thoughts into the data-gathering process.

3   Stories

Documents that are published in the same time interval have a larger chance of being similar that those that are not chronologically close [2]. However, we cannot simply assume that documents are similar based on their timestamp.

3.1   RSS Stories

We use cosine similarity to group our news articles. Guo et al. [2] represent a news article using its *summary* and its title, however we represent a news article by using its *content* and its title.

The advantage of using the summary is that it is an easily identifiable piece of text, without extraneous links to other articles, adverts and external websites, and without a large amount of embedded HTML. The disadvantage is that it is, indeed, a summary, and hence contains less information than the full article.

Extracting the content from the content of an article poses a number of challenges, because RSS feeds tend not to follow a consistent format, and the content of the article is often buried inside a large amount of irrelevant material.

However, a reasonable rule of thumb is that consecutive passages of material separated by plain <p> tend to be content, whereas things like adverts, links to other pages, and side-bars tend be contained inside other kinds of tags. Not all news providers follow this pattern, but for the ones noted above this is a reliable way of finding the content.

We construct stories as follows. Initially, we have no stories. We examine the first document D1 and perform cosine similarity against all the other documents that are grouped into stories that were last updated in the last two days, and look for the story that has the highest similarity score above a similarity threshold. Since D1 is the first document we have ever seen, we do not have any matching stories. Therefore this document forms a story of its own containing just the one document D1. When we add a document to a story we update the story TF matrix that we will use to compare subsequent documents. We now look at the next document D2 and again compare it to all documents that are grouped into stories, using the story TF matrix. If D2 is similar enough to S1, then D2 is added to story S1, otherwise it forms a new story S2. This process is repeated until all documents for Day 1 have been classified into stories or they are left unclassified because the similarity measure did not exceed the threshold value for any story.

For Day 2 we follow the same procedure, but with one crucial difference. As well as comparing documents from Day 2 with stories created on Day 2, we also compare documents from Day 2 with stories created on the previous day (i.e. Day 1), this gives us the concept of "running stories", i.e. stories about the same topic that appeared in the news over multiple continuous days.

The effectiveness of this strategy depends on choosing a suitable threshold for deciding whether an article should be included in a running story or should form the basis of a new story by itself. If the threshold is too high, every article forms a story by itself, if it is too low then articles that are not linked get grouped into the same story. Given that we are interested in assessing the importance of the news stories that are and are not linked to Twitter stories, it is important to get this right, since the most obvious measure of the importance of a story is the amount of space that is devoted to reporting it, and we can only measure that accurately if we have indeed grouped articles appropriately.

To determine a trustworthy, reliable threshold we conducted an online survey. We selected 100 random news article pairs with cosine similarities in the range 0.5–1.0, ensuring that we had 20 pairs with similarity 0.5–0.6, 20 pairs with similarity 0.6–0.7 and so on. Furthermore, we ensured that the survey presented pairs that had been answered the least first to ensure that we had answers for as many of the pairs as possible. The survey presented four users with the headline and short description from two related articles and asked them to say if the two were related or unrelated. The results are presented in Table 1.

The results show that when the cosine similarity was from 0.9–1.0, 92 % of the time the users agreed with our classification that the articles were related. However, as discussed, too high a threshold is problematic. When the cosine similarity was from 0.8–0.9, 70 % of the time the users agreed with our classification.

Table 1. Results from survey

Pair cosine similarity	Total questions answered	Unrelated (%)	Related (%)
0.9–1.0	79	7.59	92.41
0.8–0.9	78	29.49	70.51
0.7–0.8	78	55.13	44.87
0.6–0.7	78	53.85	46.15
0.5–0.6	78	66.67	33.33

Below 0.8 we see that our classification was correct less than 50 % of the time, so it is not sensible to use a threshold less than this as this would lead to unrelated articles being grouped into stories.

0.8 would seem to be the best value to use. However, to ensure that this inter-annotator agreement was not simply due to chance we used Fleiss' kappa to get a kappa value of 0.69, which according to the commonly cited scale as described in [6] means that there is "Substantial agreement" between our annotators.

We therefore decided that 0.8 was a reliable value to use as our threshold.

3.2 Tweet Stories

We use hashtags to group our Tweets. The concept is similar to that employed for RSS stories, in that we create stories and then turn them into running stories as appropriate. Even though we have vastly larger amounts of data than for our RSS stories, this process is much more efficient and can process large numbers of Tweets easily. One slight difference in the creation of Tweet stories, is that we do not allow Tweet stories to contain only one Tweet. This is because, as we have seen, it is very easy for an individual Tweet to get falsely linked to a story on the basis of a few keywords, but it is much harder for a Tweet story to falsely get related to a news story. That is not to say it will **never** happen, just that it is much harder for it to happen. As in [10], any Tweets not grouped with other Tweets are discarded. We find that these are typically short, single, random Tweets such as *"Just spilt milk on my laptop!!! Help!!!"* or *"@StephenChamber8 no I love them"*. This is not the content "that does not appear in news articles but is being discussed on Twitter", these are single, random Tweets.

4   Linking Tweet Stories and RSS Stories

We have now managed to group together Tweets into Tweet stories and RSS news articles into RSS stories. The numbers of Tweets has been reduced by approximately 33 % but the number of RSS documents remains unchanged because we allow RSS news stories to contain one or more items.

We now want to link the Tweet stories to the RSS stories. We would like to do this using cos_{TF-IDF}. However some stories run for days which leads to

Table 2. Numbers of documents before and after grouping into stories

	Initial	After grouping	Stories
RSS	11,896	11,896	11,457
Tweet	693,527	462,212	102,938

the construction of TF matrices which contain thousands of words. This gives us a problem in that we find that as the number of words increases, the cosine similarity algorithm gets slower (Table 3).

Table 3. Cosine similarity computation comparison

Entity	Average words	Time (s)
Tweet	12	0.00003
Document	392	0.0002
Story	3371	0.001

In order to find the best match for each of our Tweet stories in the set of RSS stories we would need to perform $102,938 \times 11,457$ cosine similarity computations. This takes of the order of tens of hours to compute. As others have discovered [7], this is impractical, even if we restrict attention to stories that have overlapping times (remember that Tweet stories often start before the corresponding news stories, so we cannot simply focus on pairs that cover the same time periods).

4.1 Top 5 Words

We know that there will be lots of stories that will be unrelated to each other. We would like to exclude these from the computation-intensive cosine similarity part of our process because that would be wasting time and effort. Ideally, we need a quick way of finding stories that **might** be linked, and then we can use cosine similarity to confirm or refute that.

The main idea in this paper is that we create cosine similarity "candidates" by determining the top five words from each story using TF-IDF and then simply linking stories that share one or more top words. If two or more stories mention the same important keywords then we hypothesise that they might be describing the same event, and then we use cosine similarity on these to determine whether these stories are really related or not.

One might think that because stories share keywords that they would almost certainly be linked, but this is not the case. A typical example of this scenario is where the word "quiz" is flagged as a top five word in a Twitter story and in an RSS story, but upon closer inspection we see that the Tweets are to do with *"Jimmy Carrs Big Fat Quiz of the Year"* whereas the RSS story is about a suspect who was "quizzed" by the police. Clearly these two stories should not be linked, cosine similarity will ensure that is the case. This process gave us 2,536 candidates between Twitter and RSS stories. We then performed cosine similarity on these to get a measure of how similar the pair is.

5 Results

5.1 Tweet Stories Not Related to News Stories

One of the immediately obvious results is that the number of Tweet stories has been reduced from 102,938 to 2,536 stories that relate to an RSS news story.

From the original 102,938 stories we find that 72,191 are stories that were created using usernames alone (e.g. *@LevityMusic*). These stories contain a varying number of Tweets, from 2–259, with only 3,688 stories containing ten or more Tweets. These are discarded as discussed.

The remaining 28,211 stories are those for which we are unable to find related RSS stories - these are the topics that are being discussed on Twitter but do not appear in the news (Table 4).

Table 4. Tweet stories results summary

Result	%
Based on username	70.1
Not related to a news article	27.4
Related to a news article	2.5

The 28,211 stories that are not related to news articles also have varying numbers of Tweets ranging from 2–2,319, with only 2,212 stories containing ten or more Tweets. The story with 2,319 Tweets was about *#Sherlock*. Other unmatched stories containing large numbers of Tweets were about *#HappyNewYear*, *#EthanAndGraysonTo1Mil*, *#LoveTheDarts*, *#BoxingDay* and *#2016*.

Some of the Tweet stories that contained 100s of Tweets were about things like *#corrie*, *#SutthakornTour*, *#Lotto5Millionaires* and *#2015Wipe*.

At the bottom end of the scale where we have stories containing less than ten Tweets we see hashtags like *#10SpinHitsRewind*, *#hungryeyes*, *#band*, *#mmromance* and *#independentwomanbeautyclinic*.

The stories with high numbers of Tweets seem to be about topics (Darts, Sherlock, New Year) that we would think, and probably are, being reported on in news articles. However, the *#EthanAndGraysonTo1Mil* hashtag looks like exactly the kind of thing we are looking for - it has a high number of Tweets but a Google search on *"Ethan And Grayson To 1 Million"* reveals only one relevant news article.

5.2   News Stories Related to Tweet Stories

From the numbers of stories listed in Table 2, we only manage to produce 2,536 supposedly related stories. However, the cosine similarities of these are 0.24 or below. We expected to get more closely matched stories due to the steps we performed above. We know that Tweets are short, in comparison to RSS news articles, therefore Tweet stories are also short in comparison to RSS news stories. In our dataset, on average, a Tweet contains 12 words whereas a news article contains 770 words.

Extrapolating this we can see that this mismatch in numbers of words increases for stories. We know cosine similarity uses the concept of treating words as vectors, if vectors are close to each other then they are similar. The reason our similarity measures are small is because our RSS stories have a high number of dimensions compared to our Tweet stories and, although the RSS vector will be far away from the corresponding Tweet vector, the keywords we have identified will help to ensure that we get at least some sort of positive similarity measure.

We expect that the items we found will be closely related, however on first inspection this does not appear to be the case. The supposed best match we found (with a similarity score of 0.24) was a false-positive between a news article titled *"Obama confident executive action on gun control is constitutional"* and Tweets like *"Liberty London ??? #libertylondon #beautiful #london @Liberty London"*. These are clearly not related. However, the next best score of 0.21 is between an article titled *"No reason to resign - Van Gaal"* and Tweets like *"Lous van Gaal sat motionless, what an absolute shock! #mufc #GiggsIn #LVGOut"*. These, clearly, are related.

5.3   News Stories Not Related to Tweet Stories

Recall, that originally we had 11,896 RSS news articles. These were then grouped into 11,457 stories and then further processed into 2,536 RSS-Tweet candidate pairs. From this process we find that there are 158 RSS stories that did not get matched to a Tweet story as part of the "create candidate pairs" process. Out of these there are 43 that contain more than two documents. If we look at the content of some these we see headlines such as *"Paris attacks: 5000 rounds fired in St-Denis raids, prosecutor says"*, *"Global climate march 2015: hundreds of thousands march around the world"*, *"Jihadi John'dead': MI5 on alert amid fears of Isil revenge attack"* and *"Failed flood defences cast doubt on UK readiness for new weather era"*.

These are definitely stories that we would reasonably expect to see discussed on Twitter. We can perhaps surmise that we failed to match these because of our mismatch in collecting worldwide articles against UK Tweets.

Alternatively, as mentioned by Zhao et al. [9], this could be because Twitter users show relatively low interest in world news.

Gold Standard. We selected 400 spread out pairs from the upper end of our results and hand marked them as "Correctly identified" or "Incorrectly identified" to get Precision and Recall graphs.

The Precision graph shows that the pair with the highest score was actually a mismatch, then between a score of 0.10 and 0.22 we see that a high proportion of the matches we are seeing are correct. Between 0.15 and 0.10 we see the matches we find are correct only 50 % of the time. Beyond a score of 0.10, as we might expect with such low scores, there are fewer and fewer correct matches.

Our Recall graph shows that beyond a certain small score there are no further matches to be found.

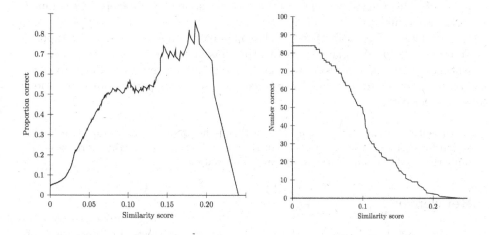

Fig. 1. Precision   **Fig. 2.** Recall

6 Related Work

There are many techniques that have been proposed for grouping "documents" (i.e. news stories, Tweets or indeed any kind of body of text).

Kaur and Gelowitz [5] proposed using cosine similarity to group Tweets. Although this approach grouped Tweets effectively, they only used a relatively small number of Tweets (3,500). For the numbers we used, cosine similarity quickly became impractical as we saw. Furthermore, they remove punctuation, URLs, usernames and also make use of a stoplist. We prefer to let the mathematics of TF-IDF take care of common words like "the" and "and" that do not tell us anything.

In [10], Petrović et al. also link Tweets to news articles using cosine similarity but they do not use a threshold as we do.

Similar to our work, Sankaranarayanan et al. [11] also use cosine similarity in conjunction with publication time - but they use the mean publication time of Tweets in the story.

The work done by Guo et al. [2] is also a similar concept in that they attempt to link Tweets to news, but they try to predict the URL referred news article based on the text in each Tweet.

We also find, as discussed by Zhao et al. [9], that Twitter is a good source of topics that have low coverage in the traditional news media.

7 Conclusion

It has been shown that by creating cosine similarity candidates by using the top five words in each story, we can filter out many of the stories that were unlikely to match. This then allows us to perform cosine similarity in an efficient manner on the remaining pairs.

At a later stage, in another work, we plan to find informal sentiment expressions in Tweets. We plan to do that by looking for orthodox expressions or sentiment markers in news sources and looking for over-representation of particular terms in Tweet stories. This project also serves as part of our motivation for that project.

Acknowledgments. Tariq would like to thank the Qatar National Research Foundation for their financial support. Allan Ramsay's contribution to this work was also supported by the Qatar National Research Foundation.

References

1. Albogamy, F., Ramsay, A.: POS tagging for Arabic tweets. In: Proceedings of the International Conference Recent Advances in Natural Language Processing, Hissar, pp. 1–8. INCOMA Ltd., Shoumen (2015). http://www.aclweb.org/anthology/R15-1001
2. Guo, W., Li, H., Ji, H., Diab, M.T.: Linking tweets to news: a framework to enrich short text data in social media. In: ACL (1), pp. 239–249. Citeseer (2013)
3. Howard, P.N., Duffy, A., Freelon, D., Hussain, M.M., Mari, W., Mazaid, M.: Opening closed regimes: what was the role of social media during the Arab spring? In: Social Science Research Network (2011)
4. internetlivestats: Twitter usage statistics. http://www.internetlivestats.com/twitter-statistics. Accessed Mar 2016
5. Kaur, N., Gelowitz, C.M.: A tweet grouping methodology utilizing inter and intra cosine similarity. In: 2015 IEEE 28th Canadian Conference on Electrical and Computer Engineering (CCECE), pp. 756–759. IEEE (2015)
6. Landis, J.R., Koch, G.G.: The measurement of observer agreement for categorical data. Biometrics **33**(1), 159–174 (1977)
7. Petrović, S., Osborne, M., Lavrenko, V.: Streaming first story detection with application to Twitter. In: Human Language Technologies: The 2010 Annual Conference of the North American Chapter of the Association for Computational Linguistics, HLT 2010, pp. 181–189. Association for Computational Linguistics, Stroudsburg, PA, USA (2010)
8. Twitter: About Twitter. https://about.twitter.com/. Accessed Mar 2016

9. Zhao, W.X., Jiang, J., Weng, J., He, J., Lim, E.-P., Yan, H., Li, X.: Comparing Twitter and traditional media using topic models. In: Clough, P., Foley, C., Gurrin, C., Jones, G.J.F., Kraaij, W., Lee, H., Mudoch, V. (eds.) ECIR 2011. LNCS, vol. 6611, pp. 338–349. Springer, Heidelberg (2011)

10. Petrović, S., Osborne, M., McCreadie, R., Macdonald, C., Ounis, I., Shrimpton, L.: Can Twitter replace newswire for breaking news? In: Proceedings of the Seventh International AAAI Conference on Weblogs and Social Media (2013)

11. Sankaranarayanan, J., Samet, H., Teitler, B.E., Lieberman, M.D., Sperling, J.: Twitterstand: news in tweets. In: Proceedings of the 17th ACM SIGSPATIAL International Conference on Advances in Geographic Information Systems, pp. 42–51 (2009)

A Novel Method for Extracting Feature Opinion Pairs for Turkish

Hazal Türkmen, Ekin Ekinci, and Sevinç İlhan Omurca[✉]

Computer Engineering Department, Kocaeli University, İzmit, Kocaeli, Turkey
hazalturkmen91@gmail.com,
{ekin.ekinci, silhan}@kocaeli.edu.tr

Abstract. Reviews made by online users, are one of the most important sources for consumers who give importance them during decision process and for companies which benefit from them during development process. Since internet has become a part of our daily lives, the number of reviews expands; it is getting difficult day by day to obtain a comprehensive view of user opinions from these reviews manually. Thus, sentiment analysis becomes an indispensable task for analyzing user reviews automatically. Recently, feature-based opinion mining methods are gaining importance in terms of fine-grained sentiment analysis. In this paper, we propose a Push Down Automata (PDA) based Feature-Opinion Pair (FOP) extraction for Turkish hotel reviews. At first, context free grammars are proposed by using Turkish linguistic relations then PDA is applied for extracting FOPs. Experimental results are showed that the proposed approach provides an efficient solution for discovering accurate FOPs.

Keywords: Feature based sentiment analysis · Push Down Automata · Context free grammar · Turkish language

1 Introduction

Online review websites have showed itself as one of the most popular channels with millions of users who express their feelings, opinions, manners, appraisals, evaluations on individuals, events or properties [1]. Through these mediums consumers' opinions about products and services can be easily obtained. As a result of constant and unavoidable increase in this online word-of-mouth behavior, an important source for data analysis can be provided and online user reviews has been studied intensively because of its powerful impact on consumers [2]. Users have started using this content while they are giving decision on variety of subjects. Although these millions of comments written by a variety of people can be seen as a mentor to users to make decisions, it's getting extremely difficult to obtain the sense on the feature they are interested in. Therefore, an automatic system is needed.

Sentiment analysis, which has been studied on since 1990's and in recent years has increasingly gained importance, is identified as extracting information, which is related to opinion, from document in hand [3]. In general, sentiment analysis can be document

---

In an earlier version of this paper, the acknowledgement was missing. An acknowledgement has now been added. An erratum to this chapter is available at https://doi.org/10.1007/978-3-319-44748-3_39

C. Dichev and G. Agre (Eds.): AIMSA 2016, LNAI 9883, pp. 162–171, 2016.
DOI: 10.1007/978-3-319-44748-3_16

level, sentence level or feature level. Document or sentence level sentiment analysis is applied in many studies, also for fine-grained sentiment analysis on online customer reviews feature-based opinion mining methods are gaining importance. In document or sentence level analysis, general opinions about the product are obtained whereas feature based sentiment analysis gives sentiments separately for the product features. Therefore sentiment analysis based on feature is a much more convenient approach. For instance, when reviews about a hotel are examined, instead of learning general opinions (good or bad) about the hotel, determining opinions about variety of aspects such as stuff or customer relations is more precious.

While features are called as titles which express opinions [4], opinions are positive or negative feelings about the features and reviews are consist of combinations of features and opinions. After the step of determining features and opinions in the reviews, the main step which aims to determine FOPs is realized. After determining FOPs, polarity of opinions about a specific feature can be determined also; correspondingly the feature based sentiment analysis can be achieved more accurately.

In the literature there are few representative works about FOPs extraction. In both studies Hu and Liu [5, 6] utilized association rule mining to extract product features and for these features the nearest opinion words in the sentence were taken. These opinion words were feature specific so features and nearest opinion words constituted the feature opinion pairs (FOPs). Chan and King [7], proposed a Feature-Opinion Association algorithm, which was a corpus-based approach to find relation between noun features and opinions. In this study, nearest opinion word, co-occurrence frequency, co-occurrence ratio, likelihood-ratio test and combined method were used as relation functions to calculate relation between FOPs. Yin, Wang and Guo [8], developed a domain ontology-based method to extract FOPs from Chinese reviews. To initialize domain ontology, concepts and relations were used and mapping product features and opinions to the conceptual space of domain ontology the FOPS were obtained. Kamal [9] combined supervised machine learning and rule-based approaches to determine FOPS from subjective sentences. Actually, the extraction of FOPs was carried out in the rule-based approach which is based on linguistic and semantic analysis of texts. Zhou, Lou and Qin [10], enacted an approach consists of three steps which were preprocessing, lexical base building and extraction of FOPs. In lexical base building features were extracted based on rules and in the last step a rule based method were applied to Chinese restaurant reviews for extracting FOPs. Quan and Ren [11] extracted FOPS by using dependency parse and dependency distance were calculated based on direct and indirect dependency. After that, candidate opinion words were filtered by using PMI-TF IDF measure. Lakkaraju, Socher and Manning [12] devised hierarchical deep learning model for extracting single aspect-sentiments and multi aspect-sentiments from two review datasets. Marx and Yellin-Flaherty [13] implemented recurrent and hierarchical frameworks which are based on neural networks and called as Joint Aspect Sentiment Model and Separate Aspect Sentiment Model for extracting single and multiple aspect sentiment pair extraction. Che et al. [14] proposed a step, which is called Sent_Comp, to facilitate FOP extraction. Sent_Comp is a sentence compression method used for removing sentiment-unnecessary information.

In this study, we present a novel PDA based feature opinion pair extraction method from Turkish user reviews in hotel domain for the purpose of realizing a feature based sentiment analysis system. Our method differs from existing methods in terms of proposed automata solution by using Turkish linguistic rules. PDA is chosen for the solution due to its convenience to a rule based linguistic analysis problem. From this aspect, it can be a pioneering study with regard to extracting FOPs especially for Turkish language. Experimental results obtained from hotel reviews show that F-score rate of the model is 0.85.

The rest of the paper is organized as follows: Sect. 2 gives the theoretical background of our work and also proposed method is described in detailed in the same section. The dataset description and the results evaluation are presented in Sect. 3. Finally we conclude with a summary.

2 Extracting FOP with PDA

Sentiment analysis contains a lot of interesting and challenging task in itself. It's known that extracting FOPs is also a fundamental phase of sentiment analysis. The main idea of the proposed approach in this paper is that FOPs are to be extracted with PDA from users' reviews by using Turkish syntactic relations. When the studies in literature are examined, it's possible to find studies which were done for Chinese and English languages [14, 15]. However, there is no any study for Turkish language. Turkish language has displayed extremely different characteristic structure in terms of syntactic than other languages, mostly studied on [16]. In the developed system, PDA was applied on extracting FOPs and the performance of the proposed approach evaluated by using hotel reviews.

2.1 Syntactic Language Analysis

As in the previous studies, in the syntactic language analysis, adjectives are accepted as opinions o; nouns and phrases are accepted as features f [6, 17]. The set of all user reviews about a specific product is indicated as $R = \{R_1, R_2, \ldots, R_n\}$. If we assume that any review R_i within the R is comprised of f and o belonging to a given product, then R_i can be illustrated as $R_i = \{\ldots, a_n, \ldots, s_n, \ldots\}$. Accordingly, f and o are labeled on the user reviews to implement the proposed system. These labels are represented by a JSON data structure [4].

In FOPs extraction, the syntactic relation between nouns and adjectives is analyzed and it is seen that nouns and adjectives have two basic syntactic relations in Turkish. These are illustrated with two examples in Fig. 1. In Fig. 1(a) indicating the first syntactic relation, the word 'nice' modifies the word 'pool' by following it. In Fig. 1(b) indicating the second syntactic relation, the word 'clean' modifies the word 'room' by coming before it. By using the coordination relation (COO) of features or opinions, these two basic rules are extended as seven rules by identifying f and o sequencing which can occur in a sentence according to Turkish language rules. The extended rules are given in Table 1.

As it is understood from the rules, FOPs extraction which is one of the basic tasks of feature based sentiment analysis depends on the syntactic relation between nouns and adjectives (Fig. 2).

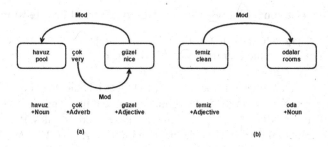

Fig. 1. Syntactic structures

Table 1. Rules used in FOP extraction

Rule	Sentence	Explanation of rule
1	$R = \{f_1, o_1, \ldots, o_n\}$	more than one o modifies one f.
2	$R = \{o_1, f_1, \ldots, f_n\}$	one o modifies more than one f.
3	$R = \{f_1 o_1, \ldots, f_n o_n\}$	Respectively each and every o modifies the previous f coming before itself.
4	$R = \{f_1, \ldots, f_n, o_1, \ldots, o_n\}$	the whole of o modifies the whole of previous f.
5	$R = \{o_1, \ldots, o_n, f_1, \ldots, f_n\}$	the whole of o modifies the whole of following f.
6	$R = \{o_1 f_1, \ldots, o_n f_n\}$	Respectively each and every o modifies the following f coming after itself.
7	$R = \{f_1 o_1, \ldots, o_n f_i o_1 \ldots o_n\}$	The whole of the o modifies a previous f coming before itself.

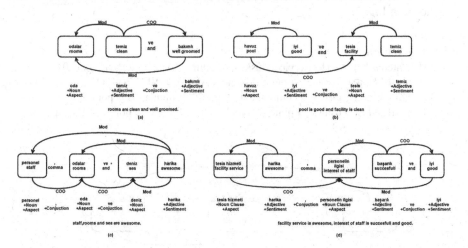

Fig. 2. Example of syntactic structure rules for FOPs extraction only four of seven of the syntactic structures mentioned above is showed here. Respectively stem, type and label of the words (if it is a feature or opinion) have been given under each box. (a) is syntactic structures equals to Rule 1, (b) is Rule 2, (c) is Rule 4 and (d) is Rule 7. According to constitutive relationship mentioned in Fig. 1, Rule 2, 5 and 6 is the opposite of Rule 1, 4 and 3 respectively. Since Rule 7 is free from other rules, it is dealt separately and with-the-rule-syntactic-structure is shown with (d).

2.2   Syntactic Relations' Identification with Using Context Free Grammar

The aim of this paper is to present a method that uses context-free grammars as models which extract FOPs from the user reviews. This is done by discovering FOPs with PDA which accepted by a given context-free grammar.

Context Free Grammar which is accepted by PDAs, as it is shown in Eq. 1, consists of 4-tuples. V_N is a finite set of variables, V_T is finite set of terminals and V_N is disjoint from V_T, S is the initial symbol and $S \in V_N$, P is the rewrite or produce rules. P, $A \Rightarrow \beta$ and it means A can be replaced with β. If we define A and β here; $A \in N$, $V = N \cup T$, $\beta \in V^*$.

$$G = (V_N, V_T, P, S) \tag{1}$$

The string of nonterminals and terminals produced at each step of zero or more derivation from S is called a sentential form $\Rightarrow^*$. The language generated by a context-free grammar represented by L(G) which considers the set of all the terminal strings that can be derived from S using the rules of L(G):

$$L(G) = \left\{ w \mid w \in V_T^*, S^* \Rightarrow w \right\} \tag{2}$$

In the study, each of seven rules specified for sequences contains f and o are dealt with separate rules such as $L(Rule_1)$, $L(Rule_2)$, ..., $L(Rule_n)$. In this case, $L(Rule_n)$ stands for the language selected for Rule n. f and o labels in the sentence represent the V_N set.

If we explain Rule 1 by handling a sample review sentence as "service is good and fast"; the Json data structure for this sentence is as following and syntactic structure of it is shown in Fig. 3.

Json version:
[{"sentence":[{"word":"service","usage":"feature","type":"NOUN","pol":0},
{" word ":"good","usage":"opinion","type":"ADJECTIVE","pol":1}
{" word ":"and","usage":" conjunction ","type":" conjunction ","pol":0},
{" word ":"fast","usage":"opinion","type":" ADJECTIVE ","pol":1}]}]"}]

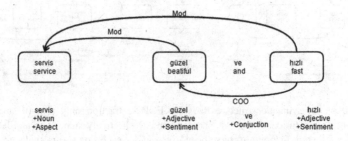

Fig. 3. Syntactic structure for "service is good and fast"

This sample sentence indicates Rule 1 and $L(Rule_1)$ language is given below.

$$L(Rule_1) = (V_N = \{S, B, f, o\}, V_T = \{f, o\}, P = \{S \rightarrow fB, B \rightarrow o|oB\}, S)$$
$$S \rightarrow fB$$
$$B \rightarrow o|oB$$

String produced as a result of Rule 1 is: $S \rightarrow fB \rightarrow foB \rightarrow foo$ and extracted FOPs are: <service, good> <service, fast>.

Table 2. The simulated PDAs for FOP extraction.

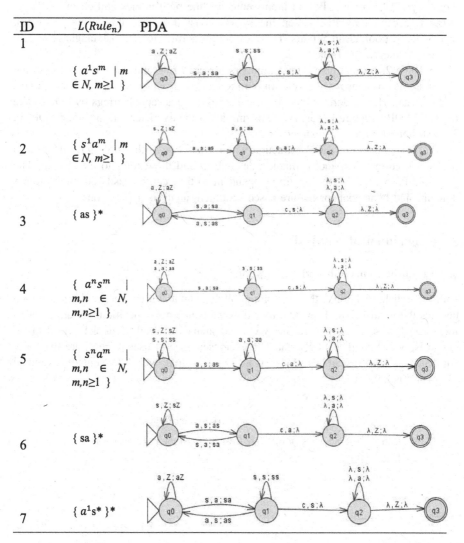

ID	$L(Rule_n)$	PDA
1	$\{a^1s^m \mid m \in N, m \geq 1\}$	
2	$\{s^1a^m \mid m \in N, m \geq 1\}$	
3	$\{as\}*$	
4	$\{a^ns^m \mid m,n \in N, m,n \geq 1\}$	
5	$\{s^na^m \mid m,n \in N, m,n \geq 1\}$	
6	$\{sa\}*$	
7	$\{a^1s*\}*$	

2.3 Push Down Automata

Context-free grammar which is frequently used in natural language processing problems can be transformed to an equivalent PDA [18]. If PDA stops at a final state and the stack is empty, the input string w which is produced by a context free grammar is recognized by PDA. For every context free grammar G, there is a PDA M recognizing L(G). That is, L(M) = L(G) [18]. A context free language recognized by PDA consists of input sets for which some sequence of moves causes the pushdown automaton to empty its stack and reach a final state [19].

After user reviews are scanned to constitute f-o sequences; the sequences which are obtained from reviews are tested if they will be accepted by PDA or not. Consequently the valid < feature, opinion > pairs determined due to the perceived f-o pairs such as "*fofo*", "*foo*", "*ofof*" etc. For an input sentence "the pool is nice and clean", if "*foo*" string is accepted by PDA, then the system must transform this string to < pool, nice > and < pool, clean > pairs. There are six more rules for Turkish language and all of them defined as in Table 2.

In Table 2, rule numbers, the grammars for each rule and the relevant PDAs are shown in the columns respectively. In our model, $\sum$ consists of f, o and c labels (inputs of automata). Here c defines the end of the string. The accepted strings by PDAs have the valid FOP structure for us. By analyzing all of the extracted strings, valid FOPs in Turkish hotel reviews are extracted.

Proposed system for FOP extraction within the scope of the study has offered as a flexible solution to expand available rules and to add these rules to the system. The proposed PDA models are domain independent so they can be easily adapted to other domains and beneficial for feature based sentiment analysis applications.

3 Experimental Analysis

3.1 Dataset Collection and Description

Unlike English, in Turkish there are not publicly available user reviews datasets which denotes the valid FOPs. Therefore user reviews concerning specific hotel are crawled form a well-known review website www.otelpuan.com. Table 3 depicts the data set which is evaluated. The FOPs stated in every sentence in user reviews are manually annotated by two researches. As seen in Table 3, there are 488 manually extracted FOPs in this dataset.

Table 3. Summary of The Hotel Dataset

Entity	# of Reviews	# of Sentences	# of Manually Extracted FOPs
Hotel	1000	5364	488

3.2 Evaluation

There are not any studies on extracting FOPs in Turkish in the literature. For this reason, the performance of the proposed system is evaluated with regard to the comparison of human generated results and computer generated results through precision, recall and F-score values. Precision, recall, F-score values have been calculated according to parameters in Table 4. The precision and recall values which determine the coverage between the human-generated and automatically generated FOPs is used as performance evaluator. The computational results show that the proposed summarization method is a promising approach to create a valid list of FOPs of Turkish hotel reviews. There is not a lexicon which indicates the accurate FOPs in Turkish, so a lexicon based comparison cannot be realized.

Table 4. Confusion Matrix

Actual	Predicted	
	FOP	*Not FOP*
FOP	476 (TP)	12 (FN)
Not FOP	155 (FP)	0 (TN)

In Table 4, TP (true positive) defines the number of FOPs which are tagged both by human and system; FN (false negative) defines the number of FOPs which are tagged by only human; FP (false positive) defines the number of FOPs which are tagged by only system and TN (True Negative) defines the number of FOPs which are tagged neither human nor system.

While the human expert confirmed 488 FOPs, the proposed system agreed 476 of them. Namely, 12 of them couldn't be detected. The 155 out of 631 FOPs which are extracted by the proposed system are not agreed by human expert. Consequently, 0.754 precision, 0.975 recall and 0.85 F-score values have been attained.

4   Conclusion

The primary goal of this research is to identify feature opinion pairs to improve the results of a feature based sentiment analysis. In terms of Turkish language rules and the mode of expressions, a PDA based FOP extraction method is proposed for Turkish user reviews in Hotel domain. With the proposed approach, firstly an automata solution is designed for the task of detecting feature-opinion relevance. From this aspect, our approach is novel. Experimental results obtained from hotel reviews illustrates that, the proposed approach provides an efficient solution for discovering accurate FOPs. The realized system can be easily adapted to other domains such as cell phones or restaurant reviews.

Moreover, when it's taken into consideration of lack of applying automata solutions for Turkish Language, the research work is in qualified so that it can be pioneer in this area. Experimental results, it can't be given as a comparative study because there hasn't been any studies FOP extraction anymore. The results and data set those are prepared allow to be compared for potential approach.

The main drawback of the proposed approach is language-dependency. In order to apply this model to any other language, the syntactic rules of the given language should be determined and PDA should be reconstructed based on these rules. FOP extraction doesn't differ in either positive or negative sentences. So finding out the negations is not a main necessity for FOP extraction.

Acknowledgement. This study is supported by The Scientific and Technological Research Council of Turkey (TUBITAK) under project number 114E422.

References

1. Liu, B.: Sentiment Analysis and Opinion Mining. Morgan & Claypool Publishers, California (2012)
2. Picazo-Vela, S., Chou, S.Y., Melcher, A.J., Pearson, J.M.: Why provide an online review? An extended theory of planned behavior and the role of Big-Five personality traits. Comput. Hum. Behav. **26**, 685–696 (2010)
3. Thelwall, M., Buckley, K., Paltoglou, G.: Sentiment strength detection for the social web. J. Am. Soc. Inf. Sci. Technol. **63**, 163–173 (2012)
4. Türkmen, H.: Ilhan Omurca S., Ekinci, E: An aspect based sentiment analysis on Turkish hotel reviews. Soc. GAU J. Appl. Sci. **6**, 9–15 (2016)
5. Hu, M., Liu, B.: Mining opinion features in customer reviews. In: 19th National Conference on Artificial Intelligence, USA, pp. 755–760 (2004)
6. Hu, M., Liu, B.: Mining and summarizing customer reviews. In: International Conference on Knowledge Discovery and Data Mining, USA, pp. 168–177 (2004)
7. Chan, K.T., King, I.: Let's tango – finding the right couple for feature-opinion association in sentiment analysis. In: Theeramunkong, T., Kijsirikul, B., Cercone, N., Ho, T.-B. (eds.) PAKDD 2009. LNCS, vol. 5476, pp. 741–748. Springer, Heidelberg (2009)
8. Yin, P., Wang, H., Guo, K.: Feature–opinion pair identification of product reviews in Chinese: a domain ontology modeling method. New Rev. Hypermedia M. **19**, 3–24 (2013)
9. Kamal, A.: Subjectivity classification using machine learning techniques for mining feature-opinion pairs from web opinion sources. IJCSI **10**, 191–200 (2013)
10. Zhou, E., Luo, X., Qin, Z.: Incorporating language patterns and domain knowledge into feature-opinion extraction. In: Sojka, P., Horák, A., Kopeček, I., Pala, K. (eds.) TSD 2014. LNCS, vol. 8655, pp. 209–216. Springer, Heidelberg (2014)
11. Quan, C., Ren, F.: Unsupervised product feature extraction for feature-oriented opinion determination. Inf. Sci. **272**, 16–28 (2014)
12. Lakkaraju, H., Socher, R., Manning, C.: Aspect specific sentiment analysis using hierarchical deep learning. In: NIPS Workshop on Deep Learning and Representation Learning, Canada, pp. 1–9 (2014)
13. Fang, L., Liu, B., Huang, M.-L.: Leveraging large data with weak supervision for joint feature and opinion word extraction. J. Comput. Sci. Technol. **30**, 903–916 (2015)
14. Che, W., Zhao, Y., Guo, H., Su, Z., Liu, T.: Sentence compression for aspect-based sentiment analysis. IEEE Audio Speech **23**, 2111–2124 (2015)
15. Qiu, G., Liu, B., Bu, J., Chen, C.: Opinion word expansion and target extraction through double propagation. Comput. Linguist. **37**, 9–27 (2011)
16. Eryiğit, G., Nivre, J., Oflazer, K.: Dependency parsing of Turkish. Comput. Linguist. **34**, 357–389 (2008)

17. Popescu, A.M., Etzioni, O.: Extracting product features and opinions from reviews. In: Kao, A., Poteet, S.R. (eds.) Natural Language Processing and Text Mining, pp. 9–28. Springer, London (2007)

18. Schützenberger, M.P.: On context-free languages and push-down automata. Inf. Contr. **6**, 246–264 (1963)

19. Antunes, C., Oliveira, A.L.: Mining patterns using relaxations of user defined constraints. In: Proceedings of the Workshop on Knowledge Discovery in Inductive Databases (2004)

20. Fokkink, W., Grune, D., Hond, B., Rutgers, P.: Detecting useless transitions in pushdown automata. arXiv preprint arXiv:1306.1947 (2013)

In Search of Credible News

Momchil Hardalov[1]([✉]), Ivan Koychev[1], and Preslav Nakov[2]

[1] FMI, Sofia University "St. Kliment Ohridski", Sofia, Bulgaria
momchil.hardalov@gmail.com, koychev@fmi.uni-sofia.bg
[2] Qatar Computing Research Institute, HBKU, Doha, Qatar
pnakov@qf.org.qa

Abstract. We study the problem of finding fake online news. This is an important problem as news of questionable credibility have recently been proliferating in social media at an alarming scale. As this is an understudied problem, especially for languages other than English, we first collect and release to the research community three new balanced credible vs. fake news datasets derived from four online sources. We then propose a language-independent approach for automatically distinguishing credible from fake news, based on a rich feature set. In particular, we use linguistic (n-gram), credibility-related (capitalization, punctuation, pronoun use, sentiment polarity), and semantic (embeddings and DBPedia data) features. Our experiments on three different testsets show that our model can distinguish credible from fake news with very high accuracy.

Keywords: Credibility · Veracity · Fact checking · Humor detection

1 Introduction

Internet and the proliferation of smart mobile devices have changed the way information spreads, e.g., social media, blogs, and micro-blogging services such as Twitter, Facebook and Google+ have become some of the main sources of information for millions of users on a daily basis. On the positive side, this has democratized and accelerated content creation and sharing. On the negative side, it has made people vulnerable to manipulation, as the information in social media is typically not monitored or moderated in any way. Thus, it has become increasingly harder to distinguish real news from misinformation, rumors, unverified, manipulative, and even fake content. Not only are online blogs nowadays flooded by biased comments and fake content, but also online news media in turn are filled with unreliable and unverified content, e.g., due to the willingness of journalists to be the first to write about a hot topic, often by-passing the verification of their information sources; there are also some online information sources created with the sole purpose of spreading manipulative and biased information. Finally, the problem extends beyond the cyberspace, as in some cases, fake news from online sources have crept into mainstream media.

© Springer International Publishing Switzerland 2016
C. Dichev and G. Agre (Eds.): AIMSA 2016, LNAI 9883, pp. 172–180, 2016.
DOI: 10.1007/978-3-319-44748-3_17

Journalists, regular online users, and researchers are well aware of the issue, and topics such as information credibility, veracity, and fact checking are becoming increasingly important research directions [3,4,19]. For example, there was a recent 2016 special issue of the ACM Transactions on Information Systems journal on Trust and Veracity of Information in Social Media [14], and there is an upcoming SemEval-2017 shared task on rumor detection.

As English is the primary language of the Web, most research on information credibility and veracity has focused on English, while other languages have been largely neglected. To bridge this gap, below we present experiments in distinguishing real from fake news in Bulgarian; yet, our approach is in principle language-independent. In particular, we distinguish between real news vs. fake news that in some cases are designed to sound funny (while still resembling real ones); thus, our task can be also seen as humor detection [10,15].

As there was no publicly available dataset that we could use, we had to create one ourselves. We collected two types of news: credible, coming from trusted online sources, and fake news, written with the intention to amuse, or sometimes confuse, the reader who is not knowledgeable enough about the subject. We then built a model to distinguish between the two, which achieved very high accuracy.

The remainder of this paper is organized as follows: Sect. 2 presents some related work. Section 3 introduces our method for distinguishing credible from fake news. Section 4 presents our data, feature selection, the experiments, and the results. Finally, Sect. 5 concludes and suggests directions for future work.

2   Related Work

Information credibility in social media is studied by Castillo et al. [3], who formulate it as a problem of finding false information about a newsworthy event. They focus on tweets using variety of features including user reputation, author writing style, and various time-based features.

Zubiaga et al. [18] studied how people handle rumors in social media. They found that users with higher reputation are more trusted, and thus can spread rumors among other users without raising suspicions about the credibility of the news or of its source.

Online personal blogs are another popular way to spread information by presenting personal opinions, even though researchers disagree about how much people trust such blogs. Johnson et al. [5] studied how blog users act in the time of newsworthy event, e.g., such as the crisis in Iraq, and how biased users try to influence other people.

It is not only social media that can spread information of questionable quality. The credibility of the information published on online news portals has also been questioned by a number of researchers [1,7]. As timing is a crucial factor when it comes to publishing breaking news, it is simply not possible to double-check the facts and the sources, as is usually standard in respectable printed newspapers and magazines. This is one of the biggest concerns about online news media that journalists have [2].

The interested reader can see [16] for a review of various methods for detecting fake news, where different approaches are compared based on linguistic analysis, discourse, linked data, and social network features.

Finally, we should also mention work on humor detection. Yang et al. [15] identify semantic structures behind humor, and then design sets of features for each structure; they further develop anchors that enable humor in a sentence. However, they mix different genres such as news, community question answers, and proverbs, as well as the One-Liner dataset [10]. In contrast, we focus on news both for positive and for negative examples, and we do not assume that the reason for a news being not credible is the humor it contains.

3   Method

We propose a language-independent approach for automatically distinguishing credible from fake news, based on a rich feature set. In particular, we use *linguistic* (*n*-gram), *credibility* (capitalization, punctuation, pronoun use, sentiment polarity), and *semantic* (embeddings and DBPedia data) features.

3.1   Features

Linguistic (*n*-gram) Features. Before generating these features, we first perform initial pre-processing: tokenization and stop word removal. We define stop words as the most common, functional words in a language (e.g., conjunctions, prepositions, interjections, etc.); while they fit well for problems such as author profiling, they turn out not to be particularly useful for distinguishing credible from fake news. Eventually, we experimented with the following linguistic features:

- *n*-**grams:** presence of individual uni-grams and bi-grams. The rationale is that some *n*-grams are more typical of credible vs. fake news, and vice versa;
- **tf-idf:** the same *n*-grams, but weighted using tf-idf;
- **vocabulary richness:** the number of unique word types used in the article, possibly normalized by the number of word tokens.

Credibility Features. We also used the following credibility features, which were previously proposed in the literature [3]:

1. Length of the article (number of tokens);
2. Fraction of words that only contain uppercase letters;
3. Fraction of words that start with an uppercase letter;
4. Fraction of words that contain at least one uppercase letter;
5. Fraction of words that only contain lowercase letters;
6. Fraction of plural pronouns;
7. Fraction of singular pronouns;
8. Fraction of first person pronouns;

9. Fraction of second person pronouns;
10. Fraction of third person pronouns;
11. Number of URLs;
12. Number of occurrences of an exclamation mark;
13. Number of occurrences of a question mark;
14. Number of occurrences of a hashtag;
15. Number of occurrences of a single quote;
16. Number of occurrences of a double quote.

We further added some sentiment-polarity features from lexicons generated from Bulgarian movie reviews [6] (5,016 positive, and 2,415 negative words), which we further expanded with some more words. Based on these lexicons, we calculated the following features:

17. Proportion of positive words;
18. Proportion of negative words;
19. Sum of the sentiment scores for the positive words;
20. Sum of the sentiment scores for the negative words.

Note that we eventually ended up using only a subset of the above features, as we performed feature selection as described in Sect. 4.2 below.

Semantic (Embedding and DBPedia) Features. Finally, we use embedding vectors to model the semantics of the documents. We wanted to model implicitly some general world knowledge, and thus we trained word2vec vectors on the text of the long abstracts from the Bulgarian DBPedia.[1] Then, we built vectors for a document as the average of the word2vec vectors of the non-stop word tokens it is composed of.

3.2 Classification

As we have a rich set of partially overlapping features, we used logistic regression for classification with L-BFGS [9] optimizer and elastic net regularization [17], which combines L1 and L2 regularization. This classification setup converges very fast, fits well in huge feature spaces, is robust to over-fitting, and handles overlapping features well. We fine-tuned the hyper-parameters of our classifier (maximum number of iterations, elastic net parameters, and regularization parameters) on the training dataset. We further applied feature selection as described below.

4   Experiments and Evaluation

4.1   Data

As there was no pre-existing suitable dataset for Bulgarian, we had to create one of our own. For this purpose, we collected a diverse dataset with enough

---

[1] http://wiki.dbpedia.org/.

samples in each category. We further wanted to make sure that our dataset will be good for modeling credible vs. fake news, i.e., that will not degenerate into related tasks such as topic detection (which might happen if the credible and the fake news are about different topics), authorship attribution (which could be the case if the fake news are written by just 1–2 authors) or source prediction (which can occur if all credible/fake news come from just one source). Thus, we used four Bulgarian news sources (from which we generated one training and three separate balanced testing datasets):

1. We retrieved most of our credible news from **Dnevnik**,[2] a respected Bulgarian newspaper; we focused mostly on politics. This dataset was previously used in research on finding opinion manipulation trolls [11–13], but its news content fits well for our task too (*5,896 credible news*);
2. As our main online source of fake news, we used a website with funny news called **Ne!Novinite**.[3] We crawled topics such as politics, sports, culture, world news, horoscopes, interviews, and user-written articles (*6,382 fake news*);
3. As an additional source of fake news, we used articles from the **Bazikileaks**[4] blog. These documents are written in the form of blog-posts and the content may be classified as "fictitious", which is another subcategory of fake news. The domain is politics (*656 fake news*);
4. And finally, we retrieved news from the **bTV Lifestyle section**,[5] which contains both credible (in the *bTV* subsection) and fake news (in the *bTV Duplex* subsection). In both subsections, the articles are about popular people and events (*69 credible and 68 fake news*);

We used the documents from Dnevnik and Ne!Novinite for training and testing: 70 % for training and 30 % for testing. We further had two additional test sets: one of bTV vs. bTV Duplex, and one on Dnevnik vs. Bazikileaks. All test datasets are near-perfectly balanced.

Finally, as we have already mentioned above, we used the long abstracts in the Bulgarian DbPedia to train word2vec vectors, which we then used to build document vectors, which we used as features for classification. (*171,444 credible samples*).

4.2   Feature Selection

We performed feature selection on the credibility features. For this purpose, we first used Learning Vector Quantization (LVQ) [8] to obtain a ranking of the features from Sect. 3.1 by their importance on the training dataset; the results are shown in Table 1. See also Fig. 1 for a comparison of the distribution of some of the credibility features in credible. vs. funny news.

---

[2] http://www.dnevnik.bg/.
[3] http://www.nenovinite.com/.
[4] https://neverojatno.wordpress.com/.
[5] http://www.btv.bg/lifestyle/all/.

Table 1. Features ranked by the LVQ importance metric.

Features	Importance
doubleQuotes 16	0.7911
upperCaseCount 4	0.7748
lowerUpperCase 5	0.7717
firstUpperCase 3	0.7708
pluralPronouns 6	0.6558
firstPersonPronouns 8	0.6346
allUpperCaseCount 2	0.6282
negativeWords 18	0.5944
positiveWords 17	0.5834
tokensCount 1	0.5779
singularPronouns 7	0.5286
thirdPersonPronouns 10	0.5273
negativeWordsScore 20	0.5206
hashtags 14	0.4998
urls 11	0.4987
positiveWordsScore 17	0.4910
singleQuotes 15	0.4884
secondPersonPronouns 9	0.4408
questionMarks 13	0.4407
exclMarks 12	0.3160

Then, we experimented with various feature combinations of the top-ranked features, and we selected the combination that worked best on cross-validation on the training dataset (compare to Table 1):

- Fraction of negative words in the text (negativeWords);
- Fraction of words that contain uppercase letters only (allUpperCaseCount);
- Fraction of words that start with an uppercase letter (firstUpperCase);
- Fraction of words that only contain lowercase letters (lowerUpperCase);
- Fraction of plural pronouns in the text (pluralPronouns);
- Number of occurrences of exclamation marks (exclMarks);
- Number of occurrences of double quotes (doubleQuotes).

4.3   Results

Table 2 shows the results when using all feature groups and when turning off some of them. We can see that the best results are achieved when experimenting with

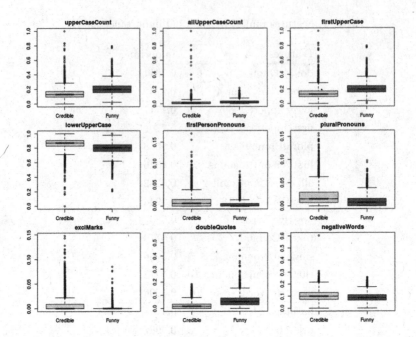

Fig. 1. Boxplots presenting the distributions of some credibility features in credible vs. funny news.

Table 2. Accuracy for different feature group combinations.

Feature groups	Dnevnik Ne!Novinite	bTV vs. bTV duplex	Dnevnik vs. Bazikileaks
Credibility + Linguistic + Semantic	**99.36**	62.04	**85.53**
Credibility + Semantic	92.67	**75.91**	82.99
Linguistic + Credibility	96.02	59.12	*61.94*
Semantic	98.95	61.31	71.01
Linguistic	95.71	*56.93*	73.25
Credibility	*83.25*	62.04	79.85
Baseline (majority class)	52.60	50.36	50.86

"Credibility + Semantic" and "Credibility + Linguistic + Semantic" feature combinations, and the results are worse when only using credibility and linguistic features.

Analyzing the results on the Dnevnik vs. the Ne!Novinite testset (first column), we can see that the linguistic features are more important than the credibility ones. Yet, the semantic features are even more important. When we combine all the feature groups, we achieve 99.36 % accuracy, but this is only marginally better than using the semantic features alone. Note, however, that using semantic features only does not perform so well on the other two test datasets, especially on the last one.

The linguistic features work relatively well on two of the test datasets, but not on bTV, where the combination of "Credibility + Semantic" is the best-performing one.

Naturally, the best results are on the Dnevnik vs. NE!Novinite, where the classifier achieves near perfect accuracy (note that this is despite the different class distribution on training vs. testing). The hardest testing dataset is bTV, where both the positive and the negative class are from sources different from those used in the training dataset; yet, we achieve up to 75.91 % accuracy, which is well above the majority class baseline of 50.36. The Dnevnik vs. Bazikileaks dataset falls somewhere in between, with up to 85.53 % accuracy; this is to be expected as the positive examples come from the same source as for the training dataset (even though the negative class is different).

Overall, on all three datasets, we achieved accuracy of 75-99 %, which is well above the majority class baseline. The strong relative performance on the three different test datasets that come from different sources suggests that our model really learns to distinguish credible vs. fake news rather than learning to classify topics, sources, or author style.

5   Conclusion and Future Work

We have presented a feature-rich language-independent approach for distinguishing credible from fake news. In particular, we used linguistic (n-gram), credibility-related (capitalization, punctuation, pronoun use, sentiment polarity, etc., with feature selection), and semantic (embeddings and DBPedia data) features. Our experiments on three different testsets, derived from four different sources, have shown that our model can distinguish credible from fake news with very high accuracy, well above a majority-class baseline.

In future work, we plan to experiment with more features, e.g., based on linked data [16], or on discourse analysis [16]. Looking at features used for related tasks such as humor- [15] and rumor-related [18] is another promising direction for future work. We also want to apply deep learning, which can eliminate the need for feature engineering altogether.

Last but not least, we would like to note that we have made our source code and datasets publicly available for research purposes at the following URL:

https://github.com/mhardalov/news-credibility

Acknowledgments. This research was performed by Momchil Hardalov, a student in Computer Science in the Sofia University "St Kliment Ohridski", as part of his M.Sc. thesis. It is also part of the Interactive sYstems for Answer Search (Iyas) project, which is developed by the Arabic Language Technologies (ALT) group at the Qatar Computing Research Institute (QCRI), HBKU, part of Qatar Foundation in collaboration with MIT-CSAIL.

References

1. Brill, A.M.: Online journalists embrace new marketing function. Newsp. Res. J. **22**(2), 28 (2001)
2. Cassidy, W.P.: Online news credibility: an examination of the perceptions of newspaper journalists. J. Comput.-Mediat. Commun. **12**(2), 478–498 (2007)
3. Castillo, C., Mendoza, M., Poblete, B.: Predicting information credibility in time-sensitive social media. Internet Res. **23**(5), 560–588 (2013)
4. Graves, L.: Deciding what's true: fact-checking journalism and the new ecology of news. Ph.D. thesis, Columbia University (2013)
5. Johnson, T.J., Kaye, B.K., Bichard, S.L., Wong, W.J.: Every blog has its day: politically-interested internet users perceptions of blog credibility. J. Comput.-Mediat. Commun. **13**(1), 100–122 (2007)
6. Kapukaranov, B., Preslav, N.: Fine-grained sentiment analysis for movie reviews in Bulgarian. In: Proceedings of Recent Advances in Natural Language Processing, RANLP 2015, Hissar, Bulgaria, pp. 266–274 (2015)
7. Ketterer, S.: Teaching students how to evaluate and use online resources. Journal. Mass Commun. Educ. **52**(4), 4 (1998)
8. Kohonen, T.: Improved versions of learning vector quantization. In: IJCNN International Joint Conference on Neural Networks, pp. 545–550 (1990)
9. Liu, D.C., Nocedal, J.: On the limited memory BFGS method for large scale optimization. Math. Program. **45**(1–3), 503–528 (1989)
10. Mihalcea, R., Strapparava, C.: Making computers laugh: investigations in automatic humor recognition. In: Proceedings of Human Language Technology Conference and Conference on Empirical Methods in Natural Language Processing, HLT-EMNLP 2005, Vancouver, British Columbia, Canada, pp. 531–538 (2005)
11. Mihaylov, T., Georgiev, G., Nakov, P.: Finding opinion manipulation trolls in news community forums. In: Proceedings of 19th Conference on Computational Natural Language Learning, CoNLL 2015, Beijing, China, pp. 310–314 (2015)
12. Mihaylov, T., Koychev, I., Georgiev, G., Nakov, P.: Exposing paid opinion manipulation trolls. In: Proceedings of International Conference Recent Advances in Natural Language Processing, RANLP 2015, Hissar, Bulgaria, pp. 443–450 (2015)
13. Mihaylov, T., Nakov, P.: Hunting for troll comments in news community forums. In: Proceedings of 54th Annual Meeting of the Association for Computational Linguistics, ACL 2016, Berlin, Germany (2016)
14. Papadopoulos, S., Bontcheva, K., Jaho, E., Lupu, M., Castillo, C.: Overview of the special issue on trust, veracity of information in social media. ACM Trans. Inf. Syst. **34**(3), 14:1–14:5 (2016)
15. Yang, D., Lavie, A., Dyer, C., Hovy, E.: Humor recognition and humor anchor extraction. In: Proceedings of 2015 Conference on Empirical Methods in Natural Language Processing, EMNLP 2015, Lisbon, Portugal, pp. 2367–2376 (2015)
16. Zaharia, M., Chowdhury, M., Franklin, M.J., Shenker, S., Stoica, I.: Spark: cluster computing with working sets. In: Proceedings of 2nd USENIX Conference on Hot Topics in Cloud Computing, HotCloud 2010, Boston, MA, p. 10 (2010)
17. Zou, H., Hastie, T.: Regularization and variable selection via the elastic net. J. Roy. Stat. Soc.: Ser. B (Stat. Methodol.) **67**(2), 301–320 (2005)
18. Zubiaga, A., Hoi, G.W.S., Liakata, M., Procter, R., Tolmie, P.: Analysing how people orient to and spread rumours in social media by looking at conversational threads (2015). arXiv preprint arXiv:1511.07487
19. Zubiaga, A., Ji, H.: Tweet, but verify: epistemic study of information verification on Twitter. Soc. Netw. Anal. Min. **4**(1), 1–12 (2014)

Image Processing

Smooth Stroke Width Transform
for Text Detection

Il-Seok Oh[1(✉)] and Jin-Seon Lee[2]

[1] Division of Computer Science and Engineering,
Chonbuk National University, Jeonju, Korea
isoh@jbnu.ac.kr
[2] Department of Information Security, Woosuk University, Jeonbuk, Korea
jslee@woosuk.ac.kr

Abstract. The stroke width transform (SWT) is a generic operation for the task of detecting texts from natural images because the characters intrinsically have the elongated shape of nearly uniform width. The edge pairing technique was recently developed by Epshtein et al. and is popularly used due to its simplicity and effectiveness. However since the natural images are noisy and sensitive to variations, high degree of artifacts arises and it hinders subsequent processing of the text detection. This paper reformulates the SWT problem in a new way that searches for an optimal solution in 3-D space. We present an effective search algorithm called the aggregation approach, borrowed from the depth image reconstruction domain. The experiments showed that the algorithm produced a smooth SWT map which is better for subsequent processes.

Keywords: Natural scene recognition · Text detection · Stroke width transform · Regularization

1 Introduction

The intelligent machine should have the ability to recognize the texts in natural scene. The first step of the text recognition pipeline is to segment a natural scene image into a set of primitive regions [1, 2]. The text areas reveal a high contrast against background areas and the color or shade within text strokes is much more uniform compared to other objects in the images. The strokes are elongated and edges on stroke contours form anti-parallelism. Use of these characteristics can help improve a lot the segmentation quality.

Our review illustrates that two approaches are widely used by the region-based text detection systems. The first one uses MSER (maximally stable extremal regions) [1, 2]. The second one uses the SWT (stroke width transform) proposed in [3] that exploits the second characteristic, stroke's elongatedness and anti-parallelism. This paper belongs to the second approach, and from now on our discussion is concentrated on the stroke width transform.

Because the characters intrinsically have the elongated shape of nearly uniform width, the stroke width transform is a generic operation for the task of detecting texts from natural images. The idea of edge pairing for estimating stroke width was recently

© Springer International Publishing Switzerland 2016
C. Dichev and G. Agre (Eds.): AIMSA 2016, LNAI 9883, pp. 183–191, 2016.
DOI: 10.1007/978-3-319-44748-3_18

proposed by Epshtein et al. [3] and it is popularly used due to its simplicity and effectiveness. Section 2 reviews research efforts for SWT accomplished during past five years.

The quality of a SWT map is critical to performance of the text detection system. Since the natural scenes are noisy and sensitive to various variations such as fonts, background, and illumination, high degree of artifacts usually appears in SWT map and it hinders subsequent processing of the text detection. Figure 1(a) shows a SWT map produced by Epshtein's algorithm. There have recently been many research efforts to reduce the artifacts, however problematic issues still remain.

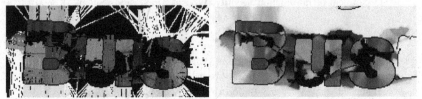

(a) SWT map produced by Epshtein's algorithm   (b) SSWT map produced by the proposed algorithm

Fig. 1. Comparing SWT and SSWT maps

This paper notices the importance of smoothness of SWT map and proposes a new method to generate a smooth SWT (SSWT) map. Figure 1(b) shows a SSWT map produced by the algorithm proposed in this paper. The smoothness is critical for next stage of identifying connected components since the process relies on the similarity of stroke width of neighboring pixels. The smoothness has also been regarded as an important factor by other classical computer vision problems such as estimating the optical flow from dynamic images and estimating the depth map from stereo images [4, 5]. An effective way of obtaining a smooth map is to use a regularization term which enforces the smoothness. To the best of our knowledge, no consideration of the smoothness factor was attempted yet explicitly for the stroke width transform.

Section 2 reviews the related works. The original SWT algorithm is explained in Sect. 3. Section 4 describes the new algorithm for accomplishing the smoothness. Section 5 presents experimental results, and Sect. 6 concludes the paper.

2 Related Works

In 2010, Epshtein et al. proposed an interesting idea called SWT [3]. An edge pixel shoots a ray toward its gradient direction to find another edge pixel at the opposite side of stroke. The distance between two paired pixels is recorded on every pixel on the ray as stroke width information. A detail description of the method is presented in Sect. 3. The Epshtein's method has been widely accepted and improved. While Epshtein used Canny edge detector, Zhang and Kasturi applied the zero crossing edge detector with the aim of obtaining closed contours [6]. Since their method discarded non-closed contours, false negative rate might be high in low contrast images. Yi and Tian used a rather stronger constraint based on both the gradient direction and gradient magnitude information and then applied morphological operation to fill the missing parts of the strokes [7]. Mosleh et al. used the same operation as Epshtein, but to alleviate the problematic situations caused by Canny edges, they used Bandlet-based edge detector that is more suitable to text areas [8]. Meng et al. proposed two pass method in which the first pass applied Epshtein algorithm and the second pass calculated energy of edge segments, and then combined two kinds of information to get final stroke width [9]. Huang et al. extended SWT to SFT (stroke feature transform) which considers both the edge information and color uniformity and generates stroke width map and stroke color map [10]. Use of dual information improved the accuracy of text detection. Karthi-keyan's technique focused the SWT operation to the salient areas of the input image [11]. Liu used the dominant stroke width to obtain superpixel image and in turn refined SWT map using superpixel image [12]. Dong attempted to improve the accuracy by localizing SWT operation to areas obtained by adaptive binarization stage [13]. Some papers applied Epshtein's algorithm as it is to implement the text detection in a series of processing pipelines [14, 15].

3   Stroke Width Transform

This section describes the original SWT algorithm in [3]. In addition we discuss the limitation of conventional smoothing operators. The algorithm first detects the edges and computes the gradient direction from a gray scale image. Each edge pixel shoots a ray along its gradient direction. If the ray meets an edge pixel with opposing gradient direction, it regards the distance between two paired edge pixels as the stroke width and records the distance at every pixel on the ray. The opposing test is $\overline{g(p)} - \overline{g(q)} \leq \cos(\Delta)$ where p and q are two paired pixels and $\overline{g(.)}$ indicates a gradient vector normalized to unit length. The parameter Δ is a tolerance factor, and $30°$ is used in [3]. For a pixel on which multiple rays pass, the minimum width is chosen. The process runs twice, one for the gradient direction and the other for the reverse gradient direction in order to cope with both the bright texts and dark texts.

Then the algorithm performs a post-processing which replaces the pixels lying on each of the valid rays with the median value of the pixels on the ray. If the median is larger than a pixel value, no replacement is done at that pixel. Figure 2 illustrates the effects of Δ and the post-processing. When we used $\Delta = 30°$ as suggested in [3] and did not apply the post-processing, we obtained Fig. 2(a). The dark pixels on 'A' and 'N' is the ones on which valid ray never passed and no width information was recorded. If we increase Δ to $45°$, it is evident in Fig. 2(c) that such holes decrease much. On the very bright pixels on two corners on left stroke of 'D', only long vertical rays were valid both in Fig. 2(a) and (c). It amounts to measuring character height rather than stroke width. The post-processing plays a very important role in diminishing the artifacts. Figure 2(b) and (d) depict the effect of the post-processing. However we observe that many holes and excavations still exist in the post-processed maps.

There are many smoothing operations including edge-preserving operators such as bilateral filter. We may think that a SWT map can be smoothed using one of those filters. However the operators cannot be directly applied since a SWT map contains

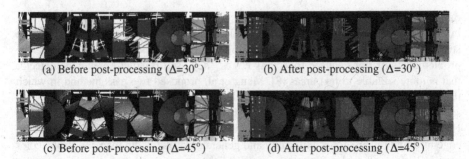

(a) Before post-processing ($\Delta=30^{\circ}$)   (b) After post-processing ($\Delta=30^{\circ}$)

(c) Before post-processing ($\Delta=45^{\circ}$)   (d) After post-processing ($\Delta=45^{\circ}$)

Fig. 2. Effects of Δ and post-processing

three kinds of pixels; void, edge, and normal pixels. This fact is the motivation for developing the new algorithm in Sect. 4.

4 Smooth Stroke Width Transform

This section explains how to construct 3-D search space and how to formulate the optimization problem. The simple but effective method for searching near-optimal solution is presented.

4.1 Formulation of Optimization Problem

We extend processing in 2-D space of the original algorithm into searching for an optimal solution in 3-D space. Equation (1) defines the 3-D search space where (x,y) represents the pixel coordinate and θ represents the direction of a ray passing through (x,y). The $i(x, y, \theta)$ is the Euclidean distance between two paired edge pixels p and q. Figure 3 illustrates the definition. We will soon explain how to compute Eq. (1).

$$i(x, y, \theta) = dist(p, q) \tag{1}$$

Equation (2) formulates the optimization problem. It searches for optimal ray direction $\hat{\theta}$ for every pixel in the map.

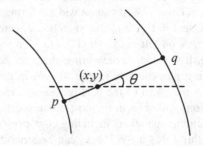

Fig. 3. Parameterization of 3-D search space

$$\left.\begin{array}{l} w(x,y) = i(x,y,\hat{\theta}(x,y)) \\ \hat{\theta}(x,y) = \underset{\theta}{\operatorname{argmin}}\, E(x,y,\theta(x,y)) \end{array}\right\} \tag{2}$$

The cost function E can be defined by Eqs. (3)$\sim$(5). Note that θ in these equations is a map with parameters (x, y).

$$E(\theta) = E_{data}(\theta) + \lambda E_{smooth}(\theta) \tag{3}$$

$$E_{data}(\theta) = \sum_{(x,y)} i(x,y,\theta(x,y)) \tag{4}$$

$$E_{smooth}(\theta) = \sum_{(x,y)} \rho(|\theta(x+1,y) - \theta(x,y)|) + \rho(|\theta(x,y+1) - \theta(x,y)|) \tag{5}$$

The E_{data} term prefers the ray with minimum distance since the minimum distance is probable to be close to the true stroke width. The second term E_{smooth} enforces the smoothness of the map $\theta(x, y)$ by preferring a low difference between neighboring pixels. The ρ is a non-decreasing function. The E_{smooth} is a sort of regularization term that enforces the smoothness of resultant maps. The λ is a weighting factor. Setting parameter λ to be zero leads to ignoring the smoothness and Eq. (2) becomes the original algorithm.

The algorithm decides the stroke width locally by choosing θ with minimum $i(x, y, \theta)$ and the procedure can be defined by Eq. (6). In other words, original SWT algorithm can be seen as a special case of SSWT proposed in this section. Note that a pixel (x, y) makes its decision independently of neighboring pixels in Eq. (6).

$$w(x,y) = \min_{\theta} i(x,y,\theta) \tag{6}$$

The limitation of the algorithm results from the fact that each pixel makes a local decision. The post-processing illustrated by Fig. 2 reduces greatly the artifacts, however it does not have any principled way of making a smooth map. On the contrary, in Eq. (2), a pixel makes its decision considering neighbor pixels. This principled technique is expected to produce a smooth SWT.

The formulation in this section borrows an idea from stereo vision community. In the stereo vision, $i(x, y, d)$ represents a difference between two image patches centered at (x, y) where d is a parameter representing the displacement between left and right images [5, 16].

4.2   Solution of Optimization Problem and Implementation Details

The parameter θ in Eq. (1) ranges $[0°,180°]$ and it is quantized into n levels. In our experiment, 16 is used for n. To compute $i(x, y, \theta)$, a ray passing through (x,y) with angle θ is shot as shown in Fig. 3. The travel distance of the ray is limited to a preset threshold T. If the ray never meets any of two edge points p and q, the ray is discarded and $i(x, y, \theta)$ is set to Null. Otherwise it checks the gradients of p and q. If they are opposing, $i(x, y, \theta)$ is set to have $dist(p, q)$. Otherwise $i(x, y, \theta)$ is set to Null. The opposing test is the same as the original algorithm.

Now the 3-D search space defined by Eq. (1) is available, we are ready to solve the optimization problem in Eq. (2). However since the search space is huge, i.e., having n^{MN} candidate solutions for $M * N$ image, an efficient algorithm should be devised to find out a global optimal or near-optimal solution. We notice that excellent strategies are already available in the stereo vision community. Scharstein and Szeliski explained and compared various approaches such as MRF, graph cut, and dynamic programming in their survey paper [5]. Another approach called aggregation approach is introduced in [16]. The paper compared various aggregation methods with respect to their accuracy. One of conclusions of Gong's benchmarking is that though the aggregation approach uses a simple scheme just applying a smoothing operation, its performance is comparable to the global optimum searching algorithms surveyed in [5]. Based on that conclusion, this paper adopts the aggregation approach in order to solve Eq. (2).

Equation (7) explains the aggregation approach. The procedure is very simple. It first applies a smoothing box filter of size $(2r + 1) * (2r + 1)$ to the map i. The smoothing is done separately on each of n planes. Note that θ in Eq. (1) was quantized into n levels. After smoothing, the procedure just chooses the ray with minimum value from the smoothed map i'.

$$\left.\begin{array}{l} \hat{\theta}(x, y) = \underset{\theta}{\operatorname{argmin}}\, i'(x, y, \theta(x, y)) \\[2mm] where\; i'(x, y, \theta) = \dfrac{1}{(2r+1)(2r+1)} \sum_{i=-r}^{r} \sum_{j=-r}^{r} i(x+i, y+j, \theta) \end{array}\right\} \qquad (7)$$

Some issues should be resolved in applying the aggregation process of Eq. (7) to the problem of Eq. (2). Firstly the map i in Eq. (1) has Null value for some points in 3-D space due to failure in forming an edge pair with opposing gradients. This issue is easily resolved by excluding points with Null value from the smoothing operation.

The second issue is related to the smoothness term E_{smooth} in Eq. (5). The term enforces the smoothness of third dimension represented by parameter θ. However a smooth θ does not necessarily lead to a smooth w. Since the stroke width map w is the final output of our algorithm, we may not obtain a satisfactory result. We resolve this problem by replacing $(2r + 1) * (2r + 1)$ 2-D box filter by $(2r + 1) * (2r + 1) * (2s + 1)$ 3-D box filter. Equation (8) explains the process.

$$i'(x, y, \theta) = \frac{1}{(2r+1)(2r+1)(2s+1)} \sum_{i=-r}^{r} \sum_{j=-r}^{r} \sum_{k=-s}^{s} i(x+i, y+j, \theta+s) \qquad (8)$$

5   Experiments

We used ICDAR2013 database for the purpose of performance evaluation [17]. The Δ and n were set to be 45° and 8, respectively. Before applying Canny edge detector, gray scale input image was smoothed with 3 * 3 Gaussian. Canny edge detector used 30 and 50 for two thresholds. We used 1 for the box filter sizes r and s in Eq. (8).

Figure 4 illustrates the effects of aggregation operation by showing width maps without and with the aggregation (3-D smoothing in Eq. (8)). The map with the aggregation in Fig. 4(b) is smoother than the one in Fig. 4(a) not using the aggregation.

(a) Without aggregation   (b) With aggregation

Fig. 4. Effects of aggregation operation

Figure 5 presents a whole view of SWT and SSWT maps. The SWT maps are susceptible to severely broken strokes around false edges such as the word "Bus Times". Many holes and excavations (indicated by dark pixels) appear in SWT maps. Between the words "National" and "Times" are many lines in SWT map, caused by incidental edge pairings. The artifacts are observed at many places. These scattered lines become regions in SSWT map due to aggregation operation. These phantom lines and areas can be easily removed by appropriate rules of the subsequent processes.

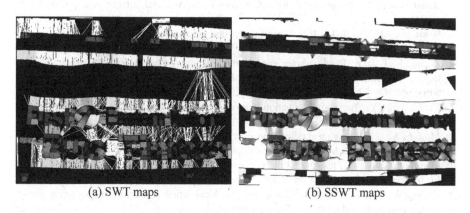

(a) SWT maps   (b) SSWT maps

Fig. 5. Comparisons of SWT and SSWT

The richer information such as 3-D map i and optimized ray direction map $\hat{\theta}$ generated by SSWT as byproduct is beneficial to the subsequent text detection stage. For example, true text areas are much denser (i.e., less pixels with Null value) in 3-D map i and have lower variance in the map $\hat{\theta}$ than other areas. Combining information from those maps with information from original color image and SSWT map must increase accuracy of the text detection stage.

6 Conclusions

This paper proposed a new method for producing a smooth stroke width transform. The basic idea lies in reformulating the SWT as an optimization problem with 3-D search space. We argued that original SWT is a special case of the proposed SSWT formulation. The algorithm for obtaining the optimal solution benefited from the classical depth map reconstruction algorithm. For a simple implementation and good performance, the aggregation approach was adopted. The experiments on real scene images showed that the smooth stroke width maps could be obtained.

References

1. Koo, H.I., Kim, D.H.: Scene text detection via connected component clustering and non-text filtering. IEEE Trans. Image Process. **22**(6), 2296–2305 (2013)
2. Yin, X.-C., Yin, X., Huang, K., Hao, H.-W.: Robust text detection in natural scene images. IEEE Trans. Pattern Recogn. Mach. Intell. **36**(5), 970–983 (2014)
3. Epshtein, B., Ofek, E., Wexler, Y., Detecting text in natural scenes with stroke width transform. In: IEEE Conference on Computer Vision and Pattern Recognition, pp. 2963–2970 (2010)
4. Lucas, B.D., Kanade, T.: An iterative image registration technique with an application to stereo vision. In: International Joint Conference on Artificial Intelligence, pp. 674–679 (1981)
5. Scharstein, D., Szeliski, R.: A taxonomy and evaluation of dense two-frame stereo correspondence algorithms. Int. J. Comput. Vis. **47**(1), 7–42 (2002)
6. Zhang, J., Kasturi, R.: Character energy and link energy-based text extraction in scene images. In: Kimmel, R., Klette, R., Sugimoto, A. (eds.) ACCV 2010, Part II. LNCS, vol. 6493, pp. 308–320. Springer, Heidelberg (2011)
7. Yi, C., Tian, Y.: Text string detection from natural scenes by structure-based partition and grouping. IEEE Trans. Image Process. **20**(9), 2594–2605 (2011)
8. Mosleh, A., et al.: Image text detection using a Bandlet-based edge detector and stroke width transform. In: British Machine Vision Conference (2012)
9. Meng, Q., Song, Y., Zhang, Y., Liu, Y.: Text detection in natural scene with edge analysis. In: International Conference on Image Processing, pp. 4151–4155 (2013)
10. Huang, W., Lin, Z., Yang, J., Wang, J.: Text localization in natural images using stroke feature transform and text covariance descriptors. In: International Conference on Computer Vision, pp. 1241–1248 (2013)
11. Karthikeyan, S.: Jagadeesh, V., Manjunath, B.S.: Learning bottom-up text attention maps for text detection using stroke width transform. In: International Conference on Image Processing, pp. 3312–3316 (2013)

12. Liu, S., Zhou, Y., Zhang, Y., Wang, Y., Lin, W.: Text detection in natural scene images with stroke width clustering and superpixel. In: Ooi, W.T., Snoek, C.G., Tan, H.K., Ho, C.-K., Huet, B., Ngo, C.-W. (eds.) PCM 2014. LNCS, vol. 8879, pp. 123–132. Springer, Heidelberg (2014)

13. Dong, W., Lian, Z., Tang, Y., Xiao, J.: Text detection in natural images using localized stroke width transform. In: He, X., Luo, S., Tao, D., Xu, C., Yang, J., Hasan, M.A. (eds.) MMM 2015, Part I. LNCS, vol. 8935, pp. 49–58. Springer, Heidelberg (2015)

14. Yao, C., Bai, X., Liu, W., Ma, Y., Tu, Z.: Detecting texts of arbitrary orientations in natural images. In: IEEE Conference on Computer Vision and Pattern Recognition, pp. 1083–1090 (2012)

15. Bhavadharani, R., Thilagavathy, A.: An efficient gaze-text-detection from images using stroke width transform. Int. J. Adv. Eng. Technol. Manag. Appl. Sci. 1(6), 1–8 (2014)

16. Gong, M., Yang, R., Wang, L., Gong, M.: A performance study on different cost aggregation approaches used in real-time stereo matching. Int. J. Comput. Vis. 75(2), 283–296 (2007)

17. Karatzas, D., et al.: ICDAR 2013 robust reading competition. In: International Conference on Document Analysis and Recognition, pp. 1484–1493 (2013)

Hearthstone Helper - Using Optical Character Recognition Techniques for Cards Detection

Costin-Gabriel Chiru[(⊠)] and Florin Oprea

University Politehnica Bucharest,
Splaiul Independenţei 313, 69121 Bucharest, Romania
costin.chiru@cs.pub.ro, oprea_florin27_04@yahoo.com

Abstract. In this paper we address the problem of capturing, processing and analyzing images from the video stream of the Hearthstone game in order to obtain relevant information on the conduct of parties in this game. Since the information needs to be presented to the user in real-time, we needed to find the most suitable methods of extracting this information. Therefore, techniques such as background subtraction, histograms comparisons, key points matching, optical character recognition were investigated. Driven by the required processing speed, we ended up using optical character recognition on limited areas of interest from the captured image. After developing the application, we tested it in real-world context, while real games were played and presented the obtained results. In the end, we also provided two examples where the application would prove useful for better decision making during the game.

Keywords: Optical character recognition · Hearthstone game · Cards logger · Image recognition · Histogram · Key point matching

1 Introduction

In this paper we address the problem of capturing, processing and analyzing images from the video stream of the Hearthstone game in order to obtain relevant information on the conduct of parties in this game.

The purpose of this application is to monitor real-time Hearthstone matches and determine various information that can improve the players gaming experience by exploiting the advantages that the computers have over humans. For ethical reasons, the information that is automatically extracted from such a game is limited to what the players can also obtain. Still, the gain provided by this application is derived from the information storage and processing power that a computer possesses, which is usually larger than what a beginner or normal player could memorize or easily deduce.

Hearthstone (Heroes of Warcraft) is a free-to-play online game developed by Blizzard Entertainment which was released in early 2014. It is a turn-based game involving two players that participate in 1vs1 match, each of them having his/her own deck of cards consisting of thirty play cards. The two players start with 30 life points each and the purpose of the game is to play their cards so that to destroy the opponent before they are themselves defeated. The first one to do so is declared the winner of the game.

© Springer International Publishing Switzerland 2016
C. Dichev and G. Agre (Eds.): AIMSA 2016, LNAI 9883, pp. 192–201, 2016.
DOI: 10.1007/978-3-319-44748-3_19

Currently, the game features between 814 and 1003 different cards, that are used to cast spells, summon minions to help in the battle or equip weapons. The cards are classified in 5 categories based on their rarity of appearance and usefulness (free, common, rare, epic and legendary), as well as into 10 different classes according to their availability to all the heroes, or to a specific one. From these cards, the player has to chose 30 to represent his/her deck of cards.

By monitoring the game, the application is able to memorize the cards that have been played by both the player and his/her opponent, which leads to several benefits: the player knows what remaining cards he/she has and can compute the likelihood of extracting a particular card next time; he/she knows which cards will no longer be available in the opponent deck (the decks may contain at most two cards of the same type, or only one if it is a legendary card); the player can predict the most probable deck of the opponent based on the cards played so far, previous history, opponent's rating, game style and current meta-game.

After we have presented the main elements of the game and what we have tried to achieve with this application, we continue the presentation by highlighting the theoretical background that supports our work. Afterwards, we will provide the details of the developed application. In section four, we will present some of the results that were obtained, along with some use cases where the application proves its value. The paper ends with our final conclusions regarding the work presented in this paper.

2 Theoretical Background

In order to determine real time in-game events, methods are needed both for image capturing and for cards recognition that are as fast, efficient and especially accurate as possible.

In terms of image capturing, this research involves acquiring a sufficient number of game snapshots per seconds so that changes in the game status to be detected. However, processing this large quantity of data needs to be done extremely fast in order not to lose information. Investigating how this problem could be solved, we found that there is an API provided by Windows that allows direct access to the opened windows from the operating system. As the information is already in the memory, copying it is fast and thus the application can perform readings of up to 15 frames per second without overwhelming the CPU [1].

For cards recognition, it is first needed to know when a new card appeared in the picture, so that not to continually process the video stream, but only at certain moments of time. Thus, we investigated "Background Subtraction" [2], a method that is commonly used for the detection of moving objects in video streams obtained from static cameras (such as surveillance cameras).

In terms of recognizing different pictures based on the elements from them, Simek [3] analyzed three such possible algorithms for determining the cards: histograms comparison, key points matching and using key points in combination with decision trees. An image histogram is "a graphical representation of the tonal distribution in the recorded image" [4]. It calculates the total number of pixels for which a given color component belongs to a given range. The idea behind key points matching is choosing specific points of interest from each image and then finding the association between the sets of points found in different images. Among the well-known key points detection algorithms are SURF (Speeded up robust features) [5] and SIFT (Scale-invariant feature transform) [6]. They solve some of the drawbacks of histograms comparisons, as they are invariant to rotation, scaling or brightness changing. The results of such algorithms are represented by sets of key points (see Fig. 1). Comparing the key points from the reference image with the ones from the preprocessed image is done using matching algorithms, such as Flann (Fast Library for approximate nearest neighbor) [7].

Fig. 1. Example of running the SURF detector and the FLANN matcher on our images.

Since the running time is one of the problems of the key points matching algorithms, improvements have been searched to accelerate the execution. One of them is represented by a combination of key points matching with decision tree classification. This is a way to use some of the extracted key points in decision trees to help classify an image. The calculation of the considered key points is faster than for the SIFT algorithm, and using the decision trees eliminates the need for associating the analyzed image with each image from the data set, leading to a faster execution time [8].

Other possible solutions to this problem are the use of perceptual hashes [9] or of the fast key point recognition using random ferns (non-hierarchical structures consisting of multiple sets of tests) for image classification [10].

An alternative option for images classification was exploiting the textual information that was present on each playing card and use an optical character recognition (OCR) system for this task. Usual OCR systems are based either on pattern matching [11] or on feature extraction [12]. Out of these two, the pattern matching method is simpler and more frequently used and focuses on comparing images with sets of binary matrices representing the character patterns. If an area from the image corresponds to one of the known patterns, then that area is assigned to the corresponding character. Features extraction is an advanced method of OCR where the algorithms seek for specific elements for the characters such as closed loops, lines direction, lines intersection, empty spaces etc. The advantage of this method is the independence (to some extent) on the font the text is written with, as it doesn't dependent on the sets of available patterns for different characters.

3   Implementation Details

As already presented in the previous section, first of all we used the Windows API to extract the snapshots at different moments of time.

The next step was to recognize the information that was present in the image. Considering the fact that the windows might have high resolutions (over 1024 × 768 pixels), this would involve a large processing time, making the use of application in a real time manner impossible. This could be reduced by scaling, but this way the items of interest from the images would also lose their details, which would have increased the margin of error when analyzing the images. Thus, we were able to gather information regarding to the current game context by analyzing specific regions in the images to determine the start and end of a game. Using this information we would know when to search for cards within the image by defining specific regions corresponding to the cards played by the opponent or the cards drawn from the user's deck. This way, an image of over a million pixels was reduced to the two areas of interest, thus reducing the magnitude of the data needed to be processed. Each of the two areas had a number of pixels of about 10 times smaller than the original size of the window, so the reduction was considerable.

Having defined the areas of interest, we needed to know the moments of time when they contained images of cards, in order to minimize the time needed for image processing and thus to minimize the degree of CPU's usage. Thus, we used "Background Subtraction" [2], and restricted the recognition analysis only to the frames with the foreground mask sufficiently highlighted.

Next, we investigated the methods for image recognition. In our case, the histograms didn't lead to good results due to the fact that all the images were created from the same template, almost 2/3 of the entire image being similar to several different cards. Thus, we went on and investigated SURF. Although the extraction of the key points from the cards from the database can be done offline and stored in memory for quick access, it is still needed to do this processing for the images that need to be recognized, and then to try to match the obtained key points against the stored ones. This process may result, depending on the image resolution, to a runtime of the order of tenths of a second for each card. Given the number of existing cards in the game, the cards identification might reach a processing time of the order of minutes, which makes this method impossible to be used for real-time recognition. For example, the analysis and detection of the opponent hero performed using SURF and Flann took approximately 3 s, given the fact that the number of heroes is much lower than the number of cards. Therefore, we concluded that this method is impractical for real-time card recognition. We also investigated the other methods for image matching, but all these methods had issues when a large number of options were available (such as it was the case of the Hearthstone cards), and therefore we finally abandoned this direction.

Since this class of techniques was abandoned, the next investigated method was OCR. To extract the characters from the images, we used Tesseract [13], one of the most accurate free recognition engine library. The decision to use such a system was influenced by the fact that the analyzed images had the same template. Therefore, extracting the text areas that help determining the playing card is easily accomplished by specifying the areas of interest: the card name and its attributes (cost, attack and life points), as this information proved to be extremely useful to exclude or filter out other cards (see Fig. 2). In order to obtain the best results, the original image had to be pre-processed before extracting the right information. The image offered to the search engine was represented by a binary matrix, where the pixels corresponding to the text were white, while the background was black. Therefore, we binarized the areas of interest and thus the OCR algorithms had better results and smaller execution times.

Fig. 2. Card with the highlighted areas of interest (left) and its pre-processing for OCR (right).

As shown in Fig. 2, the way in which the name of the card is represented is unconventional to a recognition engine, among its characteristics that might rise different problems being the fact that the used font is not a usual one and that the text line is not straight, but wavy. These factors lead to an increase in the difficulty of correctly recognizing all the text characters and thus post-processing the results was needed, such as computing the Levenshtein distance between the obtained text and the cards name and using the attribute values returned by the OCR to filter out cards, as their recognition was more accurate, given that they consisted only of numbers (which were larger and more detailed). The pre-processed images and post-processed results were achieved very quickly, with negligible execution times compared to the duration of the OCR analysis itself - which took between 0.1 and 1.5 s to finish (depending on the image and the CPU load).

The entire process, from capturing the window image to the card classification, took frequently less than two seconds (often around one second), this speed being superior to the one obtained using the key points matching method previously tested. These values allowed the image classification to be done fast enough so it would not require parallel processing (or image stacks) for ensuring that the application was not missing any cards appearing on the screen.

Since multiple steps were involved in the development of the tool, the application had to also consider a sequence of actions. Thus the application logic can be summed up by the following algorithm, where the recognition actions take place only after testing if new elements have been detected in the areas of interest:

1. Read the data and initialize the variables
2. **While** the application is running
2.1 **While** not having a game in progress
        Test if a new game has started
2.2 **While** the heroes were not detected
        Detect heroes
2.3 **While** the game is not finished
        Test for card appearance in the predefined
            zones (corresponding to user card drawn
            and opponent played card)
        Detect card attributes (using OCR)
        Filter cards based on owner hero and
            corresponding attributes
        Detect card name (OCR)
        Use edit distance to filter database (if none
            exist, remove filter and check versus
            all cards - there are scenarios where
            cards attributes might be changed)
        Verify if the game is over

4   Evaluation/Results/Examples

In this section, we will present the application functionalities and some examples from the phase of cards identification. We will also show the execution time of the interesting process and some simple use cases where this application proves to be useful.

4.1   Graphical Interface Functionality

The purpose of the GUI is to inform the player about relevant changes in the application (such as the status of the application, the cards that have been played, or the remaining ones). The monitoring and analysis modules of the application send signals to the interface in order to be displayed to the user. An example can be observed in Fig. 3, where it can be visualized the remaining cards, depending on the chosen decks of cards.

This information is updated in real-time, as the cards are played. The remaining cards are highlighted in green, while in red are presented the cards of a certain type that have been fully "consumed". Pink is used for card types that contain two cards, out of which only one has been played.

4.2   Results Obtained Using the OCR and Classification Based on Them

For an easier understanding of the cards identification methodology, we provide two examples below. The situations described in these examples are amongst the rarest and

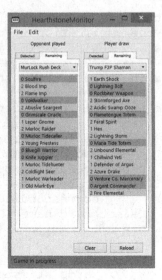

Fig. 3. Deck of cards selection and cards visualization: the initial cards (at the beginning of the game - left) and the played and remaining ones (during the game - right).

the most difficult ones, where the number and position of the words returned by OCR are not enough to determine the analyzed card.

For the first test, we have considered the card "**Stormwind Knight**". The OCR produced the following results, considering different variations of the input image such as slight text rotation to better recognize words, or image blurring to account for possible letter segmentation: *Smrwlncl Knight r, Smrmwmd Knisetu x, Snow mwind wiielu r, Smrmwind Knight, Smrmwind Knisznr, Scormwind twighr*. The four cards containing the word *knight* (that was found in the cards' names database and was correctly identified by the OCR), along with their recognition scores were: "Silver Hand Knight": 2, "Stormwind Knight": 10, "Blood Knight": 10 and "The Black Knight": 2. These scores were computed as follows: if a card name in the database contained a detected word (from each of the 6 texts obtained from OCR), it would gain 1 point; if the word would also be placed in the correct position in the detected title, the card would gain 5 points. As two different cards have been found having the same scores ("Stormwind Knight": and "Blood Knight"), in order to discriminate between them, we use the Levenshtein distance. We also add for analysis, besides the above cards, the ones having their attribute values equal to the ones identified by the OCR (cost - 4; attack - 2 and life points - 5). The smallest Levenshtein distance corresponds to the "Stormwind Knight" card and therefore the card is correctly identified.

If the results of the character recognition algorithms do not contain any words from the database, then the Levenshtein distances will be calculated only for the cards having attributes similar to those previously determined using the OCR techniques. For example, for the card "**Kor'kron Elite**", the OCR results are: *Korihron Elne, Korihron E i ire, Korikron Elire, Korhron Eme, Korhron Elire* and *Korileron Ellte* and they don't contain any word from the cards' names database. Thus, the application will consider the cards that have attributes equal to the values obtained from the OCR (cost

- 4; attack - 4 and life points - 3): "Kor'kron Elite", "Houndmaster" and "Spellbreaker". Considering the smallest Levenshtein distance, the card is correctly identified.

4.3   Runtime and Errors

The presented results have been obtained using an Intel(R) Core(TM) i7-4500U CPU @ 2.40 GHz.

The detection of a new game and of the heroes of the two players is not raising problems, as they don't need a very high response speed, since the time interval between these events and the actual beginning of the game is large enough. However, the running time of this processing is still small, with an average of 0.104 s for the detection of a new game and 0.346 for the heroes' detection.

The part where speed is needed is during the playing of the game, for the cards' detection. Thus, we present the average processing times, in seconds, obtained for the analyzed cards in Table 1. These values show that the execution time is small enough (relative to the length of cards display) to capture most of the cards, without the need to create more processing threads.

Table 1. Average time, in seconds, required for image recognition

Type of cards	OCR attributes	OCR names	Total
Cards extracted by the user	0.0146	0.7305	1.0159
Cards played by the opponent	0.0086	0.8234	0.9361

The very short time (of the order of hundredths of a second) needed to determine the cards' attributes proves the usefulness of determining them, given the fact that they are used to filter hundreds of cards.

The response speed is very important in such an application, and the chosen methods proved to be good enough for our purpose. However, another even more important factor is the precision with which the cards are identified. After testing the accuracy of cards' recognition, we obtained the results presented in Table 2. The system average accuracy was 99.1 %. The errors were caused by partial occlusion of the card that had to be identified. If the pointer is placed for a longer period of time over some elements of the game, Hearthstone will show the details of that object, which might cover certain areas of the cards, such as their attributes or name. Thus, for obtaining best results, the user should avoid occluding the cards or, when a card is (partially) occluded or incorrectly detected, to manually correct the error. For this task, functionality has been added to permit the correction of errors (such as undetected or wrongfully detected cards) directly by the user.

Table 2. Cards recognition results.

Type of cards	Identified cards	Errors
Cards extracted by the user	209	1
Cards played by the opponent	227	3
Total	436	4

5 Conclusions

In this paper, we have presented an application whose purpose was to monitor in-game events that occur in the Hearthstone game. The main requirement of the application was to be quick enough to provide answers in real time.

For the recognition of the analyzed images, we have tried two categories of methods: one based on the picture recognition, involving either computing and comparing histograms or using algorithms for key point matching and another one based on OCR. The methods from the first category required a lot of time, making them impossible to be used for real-time results. For the second method, there were considered only the important characters from the image. Thus, image comparison was reduced to comparing the texts obtained from an OCR. OCR techniques have been shown to have a running time that is small enough to allow the images identification and the save of the parties' evolution to be done in real time with an accuracy of 99.1 %.

By using the computer's memory and computational power, the application is able to provide information that otherwise players could not memorize without a significant effort. Having these pieces of data available, the system helps players in making the right decisions or in taking calculated risks, thus improving their performance and gaming experience.

Acknowledgement. The research presented in this paper was partially supported by the UPB-EX internal research grant provided by the University Politehnica of Bucharest.

References

1. Capturing an Image (Windows Dev Center). http://msdn.microsoft.com/en-us/library/windows/desktop/dd183402(v=vs.85).aspx
2. Stauffer, C., Grimson, W.E.L.: Adaptive background mixture models for real-time tracking. In: Proceedings of Computer Vision and Pattern Recognition, vol. 2 (1999)
3. Simek, K.: Image comparison - fast algorithm. http://stackoverflow.com/questions/843972/image-comparison-fast-algorithm
4. Sutton, E.: Zone System & Histograms. http://www.illustratedphotography.net/basic-photography/zone-system-histograms
5. Bay, H., Ess, A., Tuytelaars, T., Van Gool, L.: Speeded-up robust features (SURF). J. Comput. Vis. Image Underst. **110**(3), 346–359 (2008)
6. Lowe, D.G.: Distinctive image features from scale-invariant keypoints. Int. J. Comput. Vis. **60**(2), 91–110 (2004)
7. Muja, M., Lowe, D.G.: Fast aproximate nearest neighbours with automatic algorithm configuration. J. Pattern Recogn. Lett. **31**(8), 651–666 (2010)
8. Lepetit, V., Fua, P.: Keypoint recognition using randomized trees. IEEE Trans. Pattern Anal. Mach. Intell. **28**(9), 1465–1479 (2006)
9. Perceptual hash. http://phash.org/
10. Özuysal, M., Calonder, M., Lepetit, V., Fua, P.: Fast Keypoint recognition using random ferns. IEEE Trans. Pattern Analy. Mach. Intell. **32**, 448–461 (2010)

11. Muda, N., Ismail, N.K.N., Abu Bakar, S.A., Zain, J.M.: Optical Character Recognition by using Template Matching (Alphabet). http://www.academia.edu/714194/Optical_Character_Recognition_By_Using_Template_Matching_Alphabet_
12. Due Trier, O., Jain, A.K., Taxt, T.: Feature extraction methods for character recognition - a survey. Pattern Recogn. **29**(4), 641–662 (1996)
13. Tesseract-OCR. https://code.google.com/p/tesseract-ocr/

Reasoning and Search

Reasoning with Co-variations

Fadi Badra[1,2,3(✉)]

[1] Université Paris 13, Sorbonne Paris Cité, LIMICS, (UMR_S 1142),
93430 Paris, France
badra@univ-paris13.fr
[2] Sorbonne Universités, UPMC Univ Paris 06, UMR_S 1142, LIMICS,
75006 Paris, France
[3] INSERM, U1142, LIMICS, 75006 Paris, France

Abstract. Adaptation is what allows a system to maintain consistent behavior across variations in operating environments. In some previous work, a symbolic representation of the variations between two or more elements of a set was proposed. This article goes one step further and defines co-variations as functional dependencies between variations. This gives us a natural deduction rule on variations, which we show can be easily extended to perform similarity-based reasoning. A method is also proposed to learn co-variations from the data. In this method, co-variations correspond to object implication rules in a pattern structure.

Keywords: Co-variations · Adaptation · Analogical proportions

1 Introduction

Adaptation is what "allows a system to maintain consistent behavior across variations in operating environments"[9]. A variation represents a set of differences between two or more states of affairs. These differences can be "all or nothing", *i.e.*, express the gain, preservation, or loss of a property, (*e.g.*, "false morels are toxic whereas true morels are edible"), or express a change of degree (*e.g.*, "the movie *Mad Max* is more violent than the movie *Cinderella*"). In some previous work [3], we proposed to represent variations as an attribute of an ordered set of objects (at least two), all taken in a same set. The question addressed here is the formalization of the co-occurrences that may exist between two or more variations, and the definition of "rules of inference" that could be used to infer a variation from another. In particular, we would like to define a "modus ponens" inference rule, which may be summarized schematically (for pairs of objects) as follows:

$$\frac{f(z,t)}{\text{IF } f(x,y) \text{ THEN } g(x,y)}$$
$$\frac{}{g(z,t)}$$

Such deductive reasoning enables to determine the value of a variation from another by applying a "IF ... THEN" rule. This rule must express the fact that

© Springer International Publishing Switzerland 2016
C. Dichev and G. Agre (Eds.): AIMSA 2016, LNAI 9883, pp. 205–215, 2016.
DOI: 10.1007/978-3-319-44748-3_20

the variation g coïncides locally with the variation f on a set of pairs of objects that contains the pair (z, t).

In this article, we propose to define co-variations as functional dependencies between variations. This definition enables to define a natural deduction rule on variations, which can be easily extended to define a similarity-based reasoning. A method is proposed to learn co-variations from data. In this method, the co-variations correspond to object implication rules in a pattern structure.

The paper is organized as follows. The next section reviews the literature. Section 3 recalls some definitions about variations. In Sect. 4, a co-variation is defined as a functional dependency on variations. A natural deduction scheme is defined in Sect. 5, and extended in Sect. 6 to a similarity-based inference. Section 7 presents a method to learn co-variations from data. Section 8 concludes and gives future work.

2   Review of Literature

One motivation of this work is the role variations play in the adaptation step in case-based reasoning. In one of its formulations [10], adaptation is presented as the construction of a solution sol(tgt) of a target problem tgt by modifying the solution sol(srce) of a retrieved source problem srce. Adaptation can be decomposed in three steps:

① $(\texttt{srce}, \texttt{tgt}) \mapsto \Delta_{\text{pb}}$: the differences between the two problems srce and tgt are represented;

② $(\Delta_{\text{pb}}, \texttt{AK}) \mapsto \Delta_{\text{sol}}$: some adaptation knowledge AK is used to construct a variation Δ_{sol} between sol(srce) and the (future) sol(tgt) ;

③ $(\Delta_{\text{sol}}, \texttt{sol(srce)}) \mapsto \texttt{sol(tgt)}$: sol(srce) is modified into sol(tgt) by applying Δ_{sol}.

According to this model, performing adaptation requires to be able to infer variations Δ_{sol} between solutions from variations Δ_{pb} between problems (step ②).

This paper proposes a symbolic representation of co-variations, and as such can be contrasted with the statistical or graphical approaches often used for correlation detection (see for example [8,16,19]). A co-variation expresses a local coïncidence of values of two properties, each of which apply to an ordered set of objects. The notion of co-variation is therefore close to the notion of analogical dissimilarity, which is measured in [7] by counting the number of flips necessary to turn four objects into an analogical proportion, and in [11] by taking the cosine of two vectors in the euclidian space $\mathbb{R}^n$. When variations represent a change in degree of a property, co-variations express monotone correspondences between gradual changes, through rules of the form "the more x is A, the more y is B". Such correspondences are known in language semantics as argumentative *topoï*, and are defined as pairs of gradual predicates, along with the set of monotone correspondences between these gradations [2]. Gradual inference rules [5,14] may be seen as a numerical modeling of such correspondences using fuzzy logic techniques. These rules have been applied to similarity-based reasoning [13], and even to the modeling of adaptation [6]. However, their semantics is different from the one presented here. A gradual inference rule models uncertainty, whereas the semantics chosen here for co-variations rather models co-occurrence, in the spirit of association rules. Regarding co-variation learning, an algorithm is proposed in [20] to extract gradual inference rules using inductive logic programming techniques. A method is proposed in [17] to learn analogical proportions from formal contexts by reducing iteratively the analogical dissimilarity between pairs of objects.

3   Variations

This section recalls some definitions about variations.

A *variation* is modelled by a function $f : \mathcal{X}^n \longrightarrow \mathcal{V}$ which associates a value taken in a set $\mathcal{V}$ to the elements of the cartesian product $\mathcal{X}^n$. In the following, we'll assume that $n = 2$, so that variations are attributes of pairs of elements of $\mathcal{X}$. The set of all variations $f : \mathcal{X}^2 \longrightarrow \mathcal{V}$ defined on a set $\mathcal{X}^2$ and with values in $\mathcal{V}$ is denoted by $\mathcal{V}(\mathcal{X}^2, \mathcal{V})$.

Variations and Binary Relations. When $\mathcal{V} = \{0, 1\}$, the set $\mathcal{V}(\mathcal{X}^2, \{0, 1\})$ denotes the indicator functions of binary relations on $\mathcal{X}^2$. For example, if $\mathcal{X} = \mathbb{N}$ is the set of natural numbers, one can define the variation:

$$1_{\leq}(x, y) = \begin{cases} 1 & \text{if } x \leq y \\ 0 & \text{otherwise} \end{cases}$$

that returns 1 if x is lower or equal than y and 0 otherwise.

Variations Between Sets of Binary Attributes. A special case is when $\mathcal{X} = \mathscr{P}(\mathtt{M})$ denotes the powerset of a set $\mathtt{M}$ of binary attributes. For example, let $\mathtt{M} = \{\mathtt{a, b, c, d, e}\}$ denote a set of binary attributes, and $x = \{\mathtt{a, b, d}\}$ and $y = \{\mathtt{a, e}\}$ be two sets of attributes of $\mathtt{M}$. A natural way to represent the variations between x and y is to introduce the four sets $x \cap y$, $x \cap \overline{y}$, $\overline{x} \cap y$, and $\overline{x} \cap \overline{y}$, which together form a partition of $\mathtt{M}$. In our example, we get:

$$x \cap y = \{\mathtt{a}\} \quad x \cap \overline{y} = \{\mathtt{b, d}\} \quad \overline{x} \cap y = \{\mathtt{e}\} \quad \overline{x} \cap \overline{y} = \{\mathtt{c}\}$$

When x, y, z, and t represent sets of binary attributes, the analogical proportion $x : y :: z : t$ is defined [18] by $x \cap \overline{y} = z \cap \overline{t}$ and $\overline{x} \cap y = \overline{z} \cap t$, which means that the analogical proportion $x : y :: z : t$ holds for two pairs (x,y) and (z,t) if and only if the two pairs take the same value for the variation:

$$\upsilon(x, y) = \{ x \cap \overline{y}, \overline{x} \cap y \}$$

If the three sets x, y, and z are known, solving the analogical proportion equation enables to determine the set t. For example, if $x = \{a, b, d\}$, $y = \{a, e\}$, and $z = \{a, b, c\}$, then $\upsilon(x, y) = \{\{b, d\}, \{e\}\}$ and all sets of attributes t that verify $\upsilon(z, t) = \upsilon(x, y)$ are in analogical proportion with x, y, and z. Unfortunately, in this example, the equation has no solution (Fig. 1) since z does not contain the attribute d.

	a b c d e
x	1 1 0 1 0
y	1 0 0 0 1
z	1 1 1 1 0 0
t	1 0 1 ? 1

Fig. 1. Resolution of the equation $x:y :: z:t = 1$.

4   Co-variations

In this section, co-variations are defined as functional dependencies between two (sets of) variations. Let us first define the notion of co-variation between two single variations.

Definition 1. *Let $f, g \in \mathcal{V}(\mathcal{X}^2, \mathcal{V})$ be two variations. A variation g co-varies with a variation f on a subset R of $\mathcal{X}^2$, denoted by $f \overset{R}{\curvearrowright} g$, iff for all (x, y) and (z, t) of R:*

$$f(x, y) = f(z, t) \Rightarrow g(x, y) = g(z, t)$$

This definition expresses the fact that whenever two elements of the subset R of $\mathcal{X}^2$ share a same value for the variation f, they must also share a same value for g. If $R = \mathcal{X}^2$, the set R can be omited, and we will write $f \curvearrowright g$ for $f \overset{\mathcal{X}^2}{\curvearrowright} g$.

This definition can be extended to the co-variation between two sets of variations.

Definition 2. *Let $F \subseteq \mathcal{V}(\mathcal{X}^2, \mathcal{V})$ and $G \subseteq \mathcal{V}(\mathcal{X}^2, \mathcal{V})$ be two sets of variations. The set G co-varies with the set F on a subset R of $\mathcal{X}^2$, denoted by $F \overset{R}{\curvearrowright} G$, iff for all (x, y) and (z, t) of R :*

$$\forall f \in F, f(x, y) = f(z, t) \Rightarrow \forall g \in G, g(x, y) = g(z, t)$$

This definition expresses the fact that each subset of R that shares the same value for all variations of F must also share a same value for all variations of G. Similarly, we will write $F \curvearrowright G$ for $F \overset{\mathcal{X}^2}{\curvearrowright} G$.

Co-Variations and Analogical Proportions. Let us assume that $\mathcal{X} = \mathscr{P}(\mathtt{M})$ denotes the powerset of a set $\mathtt{M}$ of binary attributes. Let us consider the variations $\upsilon_\mathtt{T}$ defined for $\mathtt{T} \subseteq \mathtt{M}$ by: $\upsilon_\mathtt{T}(x, y) = \{x \cap \bar{y} \cap \mathtt{T}, \bar{x} \cap y \cap \mathtt{T}\}$. The variations $\upsilon_\mathtt{T}$ represent the differences that exist between two sets of attributes, but on a subset $\mathtt{T} \subseteq \mathtt{M}$ only. They take the same value for all pairs (x, y) that are in analogical proportion on the subset $\mathtt{T}$. One can write the co-variations $\upsilon_\mathtt{S} \overset{R}{\curvearrowright} \upsilon_\mathtt{T}$. These co-variations express the fact that every subset of R that is in analogical proportion on $\mathtt{S}$ (*i.e.*, for which the variation $\upsilon_\mathtt{S}$ takes the same value) is also in analogical proportion on $\mathtt{T}$. More formally:

Definition 3. *Let $R \subseteq \mathcal{X}^2$ and $T \subseteq M$. The set R is in analogical proportion on T iff:*

$$\exists v \in \mathscr{P}(M) \times \mathscr{P}(M) \mid \forall (x,y) \in R, \, \upsilon_T(x,y) = v$$

A subset R of $\mathcal{X}^2$ is in analogical proportion on a set of attributes T if the variation υ_T takes the same value on R. Here, $\mathscr{P}(M)$ denotes the powerset of M.

Proposition 1. *The co-variation $\upsilon_S \overset{R}{\curvearrowright} \upsilon_T$ holds iff every subset of R which is in analogical proportion on S is also in analogical proportion on T.*

Proof. $\Rightarrow$: Assume that $\upsilon_S \overset{R}{\curvearrowright} \upsilon_T$ and that $A \subseteq R$ is in analogical proportion on S. Then, (Definition 3) there exists a v such that $\forall (x,y) \in R, \, \upsilon_S(x,y) = v$. Assume that there exists $(x,y), (z,t) \in R$ such that $\upsilon_T(x,y) \neq \upsilon_T(z,t)$. We have $\upsilon_S(x,y) = \upsilon_S(z,t) = v$ so by Definition 1, $\upsilon_T(x,y) = \upsilon_T(z,t)$. Contradiction.

$\Leftarrow$: Let $(x,y), (z,t) \in R$ be such that $\upsilon_S(x,y) = \upsilon_S(z,t)$. The set $\{(x,y), (z,t)\} \subseteq R$ is in analogical proportion on S and therefore on T. As a result, (Definition 3) there exists a v such that $\upsilon_T(x,y) = \upsilon_T(z,t) = v$.   □

5   A Rule-Based Inference

In this section, co-variations are used in a rule-based inference to predict the value of a variation from the values of one or many other variations.

Definition 4. *Let $f, g \in \mathcal{V}(\mathcal{X}^2, \mathcal{V})$ be two variations. The "modus ponens" inference rule on variations is as follows:*

$$\frac{f(x,y) = f(z,t) \text{ for } (x,y), (z,t) \in R \qquad f \overset{R}{\curvearrowright} g}{g(x,y) = g(z,t)} \tag{MP}$$

This rule states that knowing that g co-varies with f on a subset R, if we know that two pairs (x,y) and (z,t) take the same value for f, then we can infer that they also take the same value for g.

This rule can be extended to co-variations $F \overset{R}{\curvearrowright} G$ between two sets of variations F and G.

Definition 5. *Let $F \subseteq \mathcal{V}(\mathcal{X}^2, \mathcal{V})$ and $G \subseteq \mathcal{V}(\mathcal{X}^2, \mathcal{V})$ be two sets of variations, $f \in F$ and $g \in G$. The "modus ponens" inference rule on sets of variations is as follows:*

$$\frac{\forall f \in F, \, f(x,y) = f(z,t) \text{ for } (x,y), (z,t) \in R \qquad F \overset{R}{\curvearrowright} G}{\forall g \in G, \, g(x,y) = g(z,t)}$$

Example #1: The A Fortiori *Inference.* When the two variations f and g are indicator functions of partial orders, the modus ponens inference corresponds to an *a fortiori* inference [1]. This type of inference exploits the monotony of two partial orders to estimate the value of an attribute. The authors of [12] give the following example. If we know that whiskey is stronger than beer, and that buying beer is illegal under the age of 18, then we can plausibly derive that buying whiskey is illegal under the age of 18. Let us call $\mathcal{A}$ the set of alcohols, and $\leq_{\text{degree}}$ and $\leq_{\text{legal\_age}}$ the partial orders on $\mathcal{A}$ that order the alcohols respectively on their degree and minimum legal age. The example can be interpreted as the co-variation $\mathbf{1}_{\leq \text{degree}} \curvearrowright \mathbf{1}_{\leq \text{legal\_age}}$ between the two variations $\mathbf{1}_{\leq \text{degree}}$ and $\mathbf{1}_{\leq \text{legal\_age}}$ of $\mathscr{V}(\mathcal{A}^2, \{0,1\})$ which represent respectively the indicator functions of the two relations $\leq_{\text{degree}}$ and $\leq_{\text{legal\_age}}$. The inference rule (MP) gives[1]:

$$\frac{\mathbf{1}_{\leq \text{degree}}(\texttt{beer}, \texttt{whiskey}) = 1 \qquad \mathbf{1}_{\leq \text{degree}} \curvearrowright \mathbf{1}_{\leq \text{legal\_age}}}{\mathbf{1}_{\leq \text{legal\_age}}(\texttt{beer}, \texttt{whiskey}) = 1}$$

and we conclude that the minimum legal age to drink whiskey is at least equal to the minimum legal age to·drink beer.

Example #2: Analogical Proportions. Let us assume that $\texttt{M} = \{\texttt{a}, \texttt{b}, \texttt{c}, \texttt{d}, \texttt{e}\}$ is a set of binary attributes. For $\texttt{T} \subseteq \texttt{M}$, the variations υ_T are the ones defined in Sect. 4. Assume that we know the rule $\upsilon_{\{\texttt{a},\texttt{b}\}} \overset{R}{\curvearrowright} \upsilon_{\{\texttt{e}\}}$ on the set $R = \{(x, y), (z, t)\}$. This rule says that every pair of R that is in analogical proportion on $\{\texttt{a}, \texttt{b}\}$ is also in analogical proportion on $\{\texttt{e}\}$. If we have, for example, $x = \{\texttt{a}, \texttt{b}, \texttt{d}\}$, $y = \{\texttt{a}, \texttt{e}\}$, and $z = \{\texttt{a}, \texttt{b}, \texttt{c}\}$, and we know that t contains $\texttt{a}$ and not $\texttt{b}$, then the inference rule (MP) gives:

$$\frac{\upsilon_{\{\texttt{a},\texttt{b}\}}(z, t) = \upsilon_{\{\texttt{a},\texttt{b}\}}(x, y) \qquad \upsilon_{\{\texttt{a},\texttt{b}\}} \overset{R}{\curvearrowright} \upsilon_{\{\texttt{e}\}}}{\upsilon_{\{\texttt{e}\}}(z, t) = \upsilon_{\{\texttt{e}\}}(x, y)}$$

So, we can deduce that (z, t) is in analogical proportion on $\{\texttt{e}\}$. By solving the analogical proportion equation, we deduce that t contains $\texttt{e}$.

The "modus ponens" inference is a kind of deductive reasoning, in which a co-variation $f \overset{R}{\curvearrowright} g$ learned on R can only be applied to a point (z, t) of R. What happens if $(z, t) \notin R$? Can we still apply the rule? Under what conditions?

6   A Similarity-Based Inference

This section defines a hypothetical reasoning scheme, in which co-variations may be applied to some points that are outside of their known domain R of validity.

---

[1] Here the rule $\mathbf{1}_{\leq \text{degree}} \curvearrowright \mathbf{1}_{\leq \text{legal\_age}}$ says that the two relations vary in the same direction but does not explicitly give the direction of the co-variation. So a more rigorous application of the inference rule (MP) would require to compare the pair $(\texttt{beer}, \texttt{whiskey})$ with one or many other pairs of alcohols.

The idea of the method is to generalize the inference rule (MP) to any pair $(z,t) \in \mathcal{X}^2$. This leads to the following schema:

$$f(z,t) = f(x,y) \text{ for } (x,y) \in R$$
$$\frac{f \overset{R}{\curvearrowright} g}{g(z,t) = g(x,y)}$$

When $(z,t) \in R$, this schema corresponds to previously defined inference rule (MP). But here, the inference rule may be applied to pairs $(z,t) \notin R$, on the basis of the similarity between (z,t) and some points (x,y) of R, which take the same value for f.

Example #1: The A Fortiori *Inference.* In the example of the alcohols, suppose that the rule $\mathbf{1}_{\leq \text{ degree}} \overset{R}{\curvearrowright} \mathbf{1}_{\leq \text{ legal\_age}}$ is known on the set $R = \{\texttt{beer}, \texttt{whiskey}\}^2$ and that we want to estimate the minimum legal age for cider consumption. If we know that $\texttt{cider} \leq_{\text{degree}} \texttt{beer}$, the inference rule gives:

$$\mathbf{1}_{\leq \text{ degree}}(\texttt{cider}, \texttt{beer}) = \mathbf{1}_{\leq \text{ degree}}(\texttt{beer}, \texttt{whiskey})$$
$$\frac{\mathbf{1}_{\leq \text{ degree}} \overset{R}{\curvearrowright} \mathbf{1}_{\leq \text{ legal\_age}}}{\mathbf{1}_{\leq \text{ legal\_age}}(\texttt{cider}, \texttt{beer}) = \mathbf{1}_{\leq \text{ legal\_age}}(\texttt{beer}, \texttt{whiskey})}$$

which makes it possible to formulate the hypothesis that the minimum legal age for cider consumption is lower or equal than the minimum legal age for beer consumption.

Example #2: Analogical Proportions. In the example of analogical proportions, this inference rule allows to apply the co-variation $\upsilon_{\{a,b\}} \overset{R}{\curvearrowright} \upsilon_{\{e\}}$ to the pair (z,t) even though it is known only on a subset $R = \{(x,y)\}$ that does not contain the pair (z,t). The inference rule gives:

$$\delta_{\{a,b\}}(z,t) = \delta_{\{a,b\}}(x,y)$$
$$\frac{\delta_{\{a,b\}} \overset{R}{\curvearrowright} \delta_{\{e\}}}{\delta_{\{e\}}(z,t) = \delta_{\{e\}}(x,y)}$$

which makes it possible to formulate the hypothesis that t contains e.

Co-Variations and Adaptation. The modelling of adaptation presented in Sect. 2 can be formulated as a similarity-based reasoning on variations. The set $R \subseteq \mathcal{X}^2$ represents a set of pairs of source cases. The element z represents a retrieved source case $(\texttt{srce}, \texttt{sol}(\texttt{srce}))$ and the element t represents the target case $(\texttt{tgt}, \texttt{sol}(\texttt{tgt}))$ for which we want to construct the solution part $\texttt{sol}(\texttt{tgt})$. An adaptation rule $\Delta_{\text{pb}} \overset{R}{\curvearrowright} \Delta_{\text{sol}}$ learned on R relates a set Δ_{pb} of variations between problems to a set Δ_{sol} of variations between solutions. The similarity-based reasoning that we have presented can be used to construct $\texttt{sol}(\texttt{tgt})$ from adaptation rules that are known on a set of source cases that do not contain $\texttt{tgt}$, and still use them to construct $\texttt{sol}(\texttt{tgt})$.

Such an approach requires to have acquired some co-variations, along with their domains R of validity. The next section proposes a method to learn co-variations from data.

7 Learning Co-variations

This section describes a method to learn co-variations from data. The idea of the method is to extract co-variations from a partition pattern structure, in the spirit of what is done in [4].

Pattern Structures. Let G be a set of objects, $(D, \sqcap)$ a meet-semilattice[2], and δ a mapping $\delta : G \longrightarrow D$ that associates to each element of G its "description" in D. Then, $(G, (D, \sqcap), \delta)$ is a *pattern structure* [15]. The elements of D are called patterns and are ordered by a subsumption relation $\sqsubseteq$: $c \sqsubseteq d$ iff $c \sqcap d = c$. The derivation operators $(.)^{\square}$ defined by:

$$A^{\square} = \prod_{g \in A} \delta(g) \text{ for } A \subseteq G \quad \text{and} \quad d^{\square} = \{g \in G \mid d \sqsubseteq \delta(g)\} \text{ for } d \in D$$

form a Galois connection between $(\mathscr{P}(G), \subseteq)$ and $(D, \sqsubseteq)$. For $A, B \subseteq G$, an *object implication* $A \rightarrow B$ holds if $A^{\square} \sqsubseteq B^{\square}$.

Partition Structures. A *partition structure* [4] is a pattern structure $(G, (D, \sqcap), \delta)$ in which the set of descriptions D is the set of partitions of a set $\mathcal{U}$, and the relation $\sqcap$ gives the meet of two partitions. Let $\mathscr{E}$ denote the set of equivalence relations on a set $\mathcal{U}^2$, and $\cap, \cup$ denote respectively the intersection and union operation on $\mathcal{U}^2$. It can be shown that $(\mathscr{E}, \cap, \cup)$ forms a lattice, *i.e.*, every pair of equivalence relations of $\mathscr{E}$ has an infimum and a supremum. There is a one-to-one correspondence between the set $\mathscr{E}$ of equivalence relations on $\mathcal{U}^2$ and the set of partitions of $\mathcal{U}$. A partition of $\mathcal{U}$ is a set $P \subseteq \mathscr{P}(\mathcal{U})$ such that $\bigcup_{p_i \in P} p_i = \mathcal{U}$ and $p_i \cap p_j = \emptyset$ for all $i, j, i \neq j$. For example, if $\mathcal{U} = \{1, 2, 3\}$, the partition $\{\{1, 2\}, \{3\}\}$ represents the relation $\{(1, 2), (2, 1), (1, 1), (2, 2), (3, 3)\}$. Let D be the set of partitions of $\mathcal{U}$. An intersection operator $\sqcap$ and an union operator $\sqcup$ can be defined that correspond to the $\cap$ and $\cup$ operators on equivalence relations. For example, if $\mathcal{U} = \{1, 2, 3, 4\}$, $\{\{1, 3\}, \{2, 4\}\} \sqcap \{\{1, 2, 3\}, \{4\}\} = \{\{1, 3\}, \{2\}, \{4\}\}$ since $\{(1, 3), (2, 4)\} \cap \{(1, 2), (1, 3), (2, 3)\} = \{(1, 3)\}$ (reflexivity is omitted here for the sake of readability). As its relational counterpart, the set $(D, \sqcap, \sqcup)$ forms a lattice, and $(D, \sqcap)$ is a meet-semilattice ($\sqcap$ is idempotent, associative, and commutative) so it can be used as a set of descriptions in a pattern structure.

Variation Structures. A *variation structure* is a partition pattern structure that associates to each variation a partition of $\mathcal{X}^2$ in which each class groups the pairs (x, y) that take the same value for this variation.

Definition 6. *A variation structure on a subset $R \subseteq \mathcal{X}^2$ is a partition structure $(G, (D, \sqcap), \delta)$ such that:*

---

[2] A meet-semilattice is a partially ordered set which has a greatest lower bound for any non-empty finite subset.

- $G \subseteq \mathscr{V}(\mathcal{X}^2, \mathcal{V})$;
- D is the set of partitions of R;
- $\sqcap$ gives the meet of two partitions;
- $\delta(\upsilon_i)$ is defined by the following $\equiv_{\upsilon_i}$ equivalence relation:

$$(x, y) \equiv_{\upsilon_i} (z, t) \text{ iff } \upsilon_i(x, y) = \upsilon_i(z, t)$$

In this structure, the objects are variations $\upsilon_i \in \mathscr{V}(\mathcal{X}^2, \mathcal{V})$ that can be seen as attributes of pairs of elements of $\mathcal{X}$. To each variation υ_i is associated a partition $\delta(\upsilon_i)$ of R such that two pairs (x, y) and (z, t) are in the same class if they take the same value for the variation υ_i.

Co-variations on R correspond to object implications in this structure.

Proposition 2. *Let $(G, (D, \sqcap), \delta)$ be a variation structure on R and $A, B \subseteq G$:*

$$A \overset{R}{\curvearrowright} B \text{ iff } A^\square \subseteq B^\square$$

Proof. $\Rightarrow$: Assume two pairs $(x, y), (z, t) \in \mathcal{X}^2$ are in the same equivalence class of the partition $A^\square$. Then, $\forall f \in A$, $f(x, y) = f(z, t)$ (Definition 6). By definition of the co-variation $A \curvearrowright B$ (Definition 2), we have that $\forall g \in B$, $g(x, y) = g(z, t)$, which means that $(x, y) \equiv_g (z, t)$ (Definition 6) for all $g \in B$, so the two pairs $(x, y), (z, t)$ are in the same equivalence class for the partition $B^\square$.

$\Leftarrow$: Let two pairs $(x, y), (z, t) \in \mathcal{X}^2$ be such that $\forall f \in A$, $f(x, y) = f(z, t)$. Then, $(x, y) \equiv_f (z, t)$ (Definition 6) for all $f \in A$, and as $A^\square \subseteq B^\square$, two pairs $(x, y), (z, t)$ that are in the same equivalence class for the partition $A^\square$ are also in the same class for the partition $B^\square$. Thus, $\forall g \in B$, $g(x, y) = g(z, t)$ (Definition 6). $\qquad\square$

This result is interesting because it shows that co-variations can be extracted from data by adapting existing pattern mining algorithms.

8   Conclusion and Future Work

In this article, co-variations are defined as functional dependencies between variations, which enables to come up with a natural deduction rule on variations. We showed that the natural "modus ponens" inference on variations can be easily extended to perform similarity-based reasoning on variations, and a method was proposed to learn co-variations from data.

Future work include implementing an algorithm to extract co-variations from data and testing on various data sets. In particular, we would like to apply such an algorithm to mine patient trajectories of patients in french hospitals, in order to explain how health care practices evolve over time or vary between two care units of a same hospital, or between two hospitals. Besides, an idea that this work suggests is that adaptation in case-based reasoning can be viewed as a similarity-based reasoning on variations. It would be interesting to further develop this idea and to compare a modeling of adaptation based on this principle to existing approaches.

Aknowledgements. The author wishes to thank the reviewers for their constructive remarks.

References

1. Abraham, M., Gabbay, D.M., Schild, U.: Analysis of the Talmudic Argumentum a Fortiori inference rule (Kal Vachomer) using Matrix Abduction. Stud. Log. **92**(3), 281–364 (2009)
2. Anscombre, J.C.: La Théorie Des Topoï: Sémantique Ou Rhétorique? Hermes (Wiesb). **1**(15), 185–198 (1995)
3. Badra, F.: Representing and learning variations. In: International Conference Tools Artificial Intelligence, pp. 950–957. IEEE, Vietri sul Mare (2015)
4. Baixeries, J., Kaytoue, M., Napoli, A.: Characterizing functional dependencies in formal concept analysis with pattern structures. Ann. Math. Artif. Intell. **72**(1–2), 129–149 (2014)
5. Bouchon-Meunier, B., Laurent, A., Lesot, M.J., Rifqi, M.: Strengthening fuzzy gradual rules through "all the more" clauses. In: IEEE International Conference on Fuzzy Systems, FUZZ-IEEE 2010, Barcelona, Spain, pp. 1–7 (2010)
6. Bouchon-Meunier, B., Marsala, C., Rifqi, M.: Fuzzy analogical model of adaptation for case-based reasoning. In: IFSA-EUSFLAT, pp. 1625–1630 (2009)
7. Bounhas, M., Prade, H., Richard, G.: Analogical classification: a new way to deal with examples. Front. Artif. Intell. Appl. **263**, 135–140 (2014)
8. Calders, T., Goethals, B., Jaroszewicz, S.: Mining rank-correlated sets of numerical attributes. In: Proceedings of 12th ACM SIGKDD International Conference on Knowledge Discovery and Data Mining - KDD 2006, p. 96 (2006)
9. Coutaz, J., Crowley, J.L., Dobson, S., Garlan, D.: Context is key. Commun. ACM **48**(3), 49–53 (2005)
10. D'Aquin, M., Badra, F., Lafrogne, S., Lieber, J., Napoli, A., Szathmary, L.: Case base mining for adaptation knowledge acquisition. In: IJCAI - International Joint Conference on Artificial Intelligence, Hyderabad, India, pp. 750–755 (2007)
11. Derrac, J., Schockaert, S.: Characterising semantic relatedness using interpretable directions in conceptual spaces. Front. Artif. Intell. Appl. **263**, 243–248 (2014)
12. Derrac, J., Schockaert, S.: Inducing semantic relations from conceptual spaces: a data-driven approach to plausible reasoning. Artif. Intell. **228**, 66–94 (2015)
13. Dubois, D., Hüllermeier, E., Prade, H.: Fuzzy set-based methods in instance-based reasoning. IEEE Trans. Fuzzy Syst. **10**(3), 322–332 (2002)
14. Dubois, D., Prade, H.: Gradual inference rules in approximate reasoning. Inf. Sci. (Ny) **61**(1–2), 103–122 (1992)
15. Ganter, B., Kuznetsov, S.O.: Pattern structures and their projections. In: Delugach, H.S., Stumme, G. (eds.) ICCS 2001. LNCS (LNAI), vol. 2120, pp. 129–142. Springer, Heidelberg (2001)
16. Li, Y., Zhang, M., Jiang, Y., Wu, F.: Co-variations and clustering of Chronic Disease behavioral risk factors in China: China Chronic Disease and Risk Factor Surveillance, 2007. PLoS ONE **7**(3), e33881 (2012)
17. Miclet, L., Prade, H., Guennec, D.: Looking for analogical proportions in a formal concept analysis setting. In: Concept Lattices Applications, vol. 959, pp. 295–307. CEUR-WS (2011)

18. Prade, H., Richard, G.: Reasoning with Logical Proportions. In: Proceedings of 12th International Conference on Principles of Knowledge Representation and Reasoning, Toronto, Canada, pp. 545–555 (2010)

19. Sedki, K., Beaufort, L.B.D.: Cognitive maps and Bayesian networks for knowledge representation and reasoning. In: IEEE 24th International Conference on Tools with Artificial Intelligence, ICTAI 2012, Athens, Greece, pp. 1035–1040 (2012)

20. Serrurier, M., Dubois, D., Prade, H., Sudkamp, T.: Learning fuzzy rules with their implication operators. Data Knowl. Eng. 60(1), 71–89 (2007)

Influencing the Beliefs of a Dialogue Partner

Mare Koit[✉]

University of Tartu, J. Liivi 2, 50409 Tartu, Estonia
mare.koit@ut.ee
http://www.cl.ut.ee

Abstract. A model of argumentation dialogue which includes reasoning is introduced in the paper. The communicative goal of the initiator is to convince the partner to do an action. The choice of an argument depends, on the one hand, on the needed resources and the beliefs about the positive and negative aspects of doing the action, and on the other hand, on the result of reasoning based on these beliefs. The initiator of dialogue is using a partner model – the hypothetical beliefs about the partner who at the same time operates with the actual beliefs. Both the participants' models are changing during a dialogue as influenced by the partners' arguments. Two implementations have been created. In one implementation, the computer initiates a dialogue and attempts to influence the user to make a decision about doing an action. In the other implementation, the roles of the computer and the user are reversed. Interaction is text-based, participants are using ready-made sentences in natural language which are classified semantically. The paper studies how the participants are updating their beliefs in dialogue. The study is based on the interactions with our dialogue systems.

Keywords: Beliefs · Updating · Reasoning model · Argumentation dialogue

1 Introduction

A dialogue system (DS), or conversational agent, is a computer system intended to interact with a human using text, speech, graphics, gestures and other modes for communication. A dialogue manager is a component of a DS which controls the conversation. The dialogue manager reads the input modalities, updates the current state of the dialogue, decides what to do next, and generates output [1].

Four kinds of dialogue management architectures are most common: plan-based, finite-state, frame-based, and information-state [2, chap. 24]. One of the earliest models of conversational agent is based on the use of artificial intelligence planning techniques. Using plans to generate and interpret sentences require the models of beliefs, desires, and intentions (BDI) [3,4]. Plan-based approaches, though complex and difficult to embed in practical dialogue systems, are seen as more amenable to flexible dialogue behavior [5].

The simplest dialogue manager architecture, used in many practical implementations, is a finite-state manager. Frame-based dialogue managers ask the

© Springer International Publishing Switzerland 2016
C. Dichev and G. Agre (Eds.): AIMSA 2016, LNAI 9883, pp. 216–225, 2016.
DOI: 10.1007/978-3-319-44748-3_21

user questions to fill slots in a frame until there is enough information to perform a data base query, and then return the result to the user. If the user answers more than one question at a time, the system has to fill in these slots and then remember not to ask the user the associated questions for the slots. In this way, the user can also guide the dialogue [6].

More advanced architecture for dialogue management which allows for sophisticated components is the information-state architecture [5,7]. We use this approach in our implementations.

'Dialogue state tracking' refers to accurately estimating the user's goal as a dialogue progresses. It is sometimes also called 'belief tracking' [8].

A conventional DS is a speech-based system that gives information to a user. In this paper, however, we will consider another kind of dialogues in natural language – negotiations that include argumentation. We suppose that there are two participants – A and B – and one of them – let it be A – initiates the dialogue making a proposal to his partner B to do (or not) an action D. If B refuses then A attempts to influence her in a dialogue, proposing several arguments for/against doing D. A's arguments are based on the partner model – his image about B's beliefs. At the same time, B can present counter arguments if her goal is opposite to A. The counter arguments indicate which beliefs of A about B are wrong and how the partner model has to be updated by A. One possible scenario for such an interaction is that A is a conversational agent (DS) and B is a human user. However, we do not exclude the other scenarios: (a) A is a user and B is a conversational agent, or (b) both A and B are artificial agents, or (c) both of them are humans. These scenarios give us an opportunity to study and to model behaviour of both participants in order to understand how their beliefs are changing during a dialogue as influenced by the partner's arguments. We have created two different experimental dialogue systems – one is playing the role of A and the other the role of B in an argumentation dialogue with a user. At the moment, the interaction is text-based. A possible future practical application, as we see, could be to train the user's argumentation skills in interaction with a DS.

The remainder of this paper is organised as follows. Section 2 describes our model of argumentation dialogue. Section 3 studies how the partner model (i.e. the agent's beliefs about the partner) is updated in a conversation and how the actual beliefs of an agent are changing due to the partner's arguments. The relationship between the partner model and the actual beliefs will be discussed in Sect. 4. Section 5 makes conclusions.

2   A Model of Argumentation Dialogue

Let us consider a dialogue in natural language between two participants A and B (human or artificial agents). Let A be the initiator of dialogue, and let his communicative goal be "B makes A decision to do an action D" or, respectively, "B makes a decision not to do D". B's communicative goal can be either the same or opposite. In interaction, A is influencing B to make the decision which coincides with his communicative goal. The following cases can occur:

1. A's goal is "B decides to do D" but B's goal is "B will not do D"
2. A's goal is "B decides not to do D" but B's goal is "B will do D"
3. A's goal is "B decides to do D", and B's goal is "B will do D"
4. A's goal is "B decides not to do D", and B's goal is "B will not do D".

A's and B's communicative goals are opposite in the cases (1) and (2). When interacting, the initiator A presents arguments in order to influence B to adopt A's goal and to abandon her own initial goal. At the same time, B can present counter arguments which should bring A to adopt B's goal and to abandon his own initial goal.

A's and B's communicative goals coincide in the cases (3) and (4). When interacting, they cooperatively look for arguments in support of doing (respectively, not doing) D and find out how to overcome possible obstacles before doing D or, respectively, to prevent possible undesirable consequences of not doing D.

Let us, for example, consider the case (1). The initiator A has a partner model – an image about B's beliefs. The partner model gives him an opportunity to suppose that B will agree to accept his communicative goal (to do the action D). When constructing his first turn, A chooses the dialogue acts (e.g. request, proposal, question, etc. depending on his image about B) and determines their verbal form (utterances). The partner B analyses A's turn and in order to make a decision – to do D or not – she triggers a reasoning procedure in her mind. In the reasoning process, B is weighing her resources for doing D, positive and negative aspects of doing D and finally, she makes a decision. Then B in her turn chooses the dialogue acts (agreement, refusal) and their verbal form in order to inform A about her decision. If B agrees to do D then the dialogue finishes (A has reached his communicative goal). If B refuses then A has to change his partner model (it did not correspond to the reality) and find out new arguments in order to convince B to make a positive decision, cf. [9].

B can add arguments to her refusal. These (counter) arguments give information about the reasoning process that brought B to the (negative) decision. A uses the arguments given by B for updating his current partner model.

2.1   Reasoning Model

Our reasoning model is introduced in [9]. In general, it follows the ideas realised in the well-known BDI model.

The reasoning model consists of two parts: (1) a model of human motivational sphere; (2) reasoning procedures. In the motivational sphere three basic factors are differentiated that regulate reasoning of a subject concerning of doing an action D. First, a subject may *wish* to do D if the pleasant aspects of D for him/her overweight the unpleasant ones; secondly, a subject may find it reasonable to do D if D is *needed* to reach some higher goal, and the useful aspects of D overweight the harmful ones; and thirdly, a subject can be in a situation where s/he *must* (is obliged) to do D – if not doing D will lead to some kind of punishment.

If the subject is reasoning about not doing D then the basic factors which trigger the reasoning are analogous: first, the subject *does not wish* to do D if the unpleasant aspects of D overweight the pleasant ones; secondly, doing D is *not needed* for him/her if the harmful aspects of D overweight the useful aspects; and thirdly, doing D is *not allowed (prohibited)* for him/her if it will cause some punishment.

We represent the model of motivational sphere of a reasoning subject by the following vector of 'weights' of beliefs (with numerical values of its components):

wD = (wD(resources), wD(pleasant), wD(unpleasant), wD(useful), wD(harmful), wD(obligatory), wD(prohibited), wD(punishment-do), wD(punishment-not)).

In the description, wD(pleasant), etc. mean the weight of pleasant, etc. aspects of D; wD(punishment-do) – the weight of punishment for doing D if it is prohibited, and wD(punishment-not) – the weight of punishment for not doing D if it is obligatory. Further, wD(resources) = 1 if subject has all the resources necessary to do D (otherwise 0); wD(obligatory) = 1 if D is obligatory for the reasoning subject (otherwise 0); wD(prohibited) = 1 if D is prohibited (otherwise 0). The values of other weights can be non-negative natural numbers.

The second part of the reasoning model consists of reasoning procedures that supposedly regulate human action-oriented reasoning. Every reasoning procedure represents steps that the subject goes through in his/her reasoning process; these consist in comparing the summarised weights of different aspects of D; and the result is the decision: to do D or not.

We use two vectors of motivational sphere. The vector **wDAB** represents A's beliefs concerning B's evaluations and it is used as a partner model. The vector **wDB** represents B's actual evaluations of D's aspects (which exact values A does not know) and it is used as the model of B herself. In the paper, we will consider the needed changes that will be made and tracked by the participants due to arguments presented in a dialogue. In the following, we suppose that the action D is fixed and we do not indicate it in the vectors.

A reasoning procedure depends on the determinant which triggers it. As an example, let us present the reasoning procedure WISH as a step-form algorithm triggered by the *wish* of the reasoning subject to do D, that is, D is more pleasant than unpleasant for the subject, cf. [9].

Presumption: w(pleasant) $\geq$ w(unpleasant).

1. Is w(resources) = 1? If not then go to 11.
2. Is w(pleasant) > w(unpleasant) + w(harmful)? If not then go to 6.
3. Is w(prohibited) = 1? If not then go to 10.
4. Is w(pleasant) > w(unpleasant) + w(harmful) + w(punishment-do)? If yes then go to 10.
5. Is w(pleasant) + w(useful) > w(unpleasant) + w(harmful) + w(punishment-do)? If yes then go to 10 else go to 11.
6. Is w(pleasant) + w(useful) $\leq$ w(unpleasant) + w(harmful)? If not then go to 9.
7. Is w(obligatory) = 1? If not then go to 11.
8. Is w(pleasant) + w(useful) + w(punishment-not) > w(unpleasant) + w(harmful)? If yes then go to 10 else go to 11.

9. Is w(prohibited) = 1? If yes then go to 5 else go to 10.
10. Decide: do D. End.
11. Decide: do not do D.

The idea is quite simple: a reasoning subject is step-by-step weighing positive and negative aspects of doing D. If the positive aspects in sum weigh more then the decision will be "do D" else "do not do D". The reasoning can also be considered as argumentation for/against doing D.

If D is more unpleasant than pleasant but more useful than harmful then the reasoning procedure NEEDED can be triggered by the reasoning subject. If D is obligatory and not doing D involves a punishment then the subject can use the reasoning procedure MUST, cf. [9].

A *communicative strategy* is an algorithm used by a participant for achieving his/her goal in the interaction. In order to achieve the goal in argumentation dialogue, a participant can present different arguments for/against D in a systematic way. For example, if the initiator A has the communicative goal "B will do D" then he can over and over again stress the pleasant aspects of D (i.e. *entice* the partner B to do D), or stress usefulness of D for B (i.e. *persuade* B), or stress punishment for not doing D if it is obligatory (*threaten* B), etc. We call *communicative tactics* these concrete ways of applying a communicative strategy [9]. The participant A, trying to direct B's reasoning to the desirable decision, proposes arguments for doing D (respectively, not doing D) while B, when opposing, proposes counter arguments. While enticing (respectively, persuading or threatening) the partner for doing D, A attempts to trigger the reasoning procedure WISH (respectively, NEEDED or MUST) in B's mind.

2.2   Argumentation-Based Dialogue

Let us make some more assumptions. First, let us consider the general scenario (b, Sect. 1) supposing that both A and B are conversational agents interacting in a natural language. (If one of them actually is a human user then some of the introduced models and formalisms are not needed.) Both A and B have access to a common set of reasoning procedures. Both A and B can use fixed sets of dialogue acts and the corresponding utterances in a natural language which are pre-classified semantically, e.g. the set Pincreasingresources for indicating that there exist resources for doing a certain action D (e.g. *The company will cover all your expenses*), Pincreasingpleasantness for stressing the pleasantness of D (e.g. *You can meet interesting people*), Pmissingresources for indicating that some resources for doing D are missing (e.g. *I don't have proper dresses*), etc. Therefore, no linguistic analysis or generation will be made during a dialogue (in our implementation). However, these restrictions will involve that the generated dialogues can be not quite coherent.

A, starting an interaction, generates a partner model $\mathbf{w}_{AB}$ (using his knowledge) and determines the communicative tactics T which he will use (e.g. enticement), i.e. he accordingly fixes a reasoning procedure R which he will try to trigger in B's mind (e.g. WISH). B has her own model $\mathbf{w}_B$ (which exact values

A does not know). She in her turn determines a reasoning procedure RB which she will use in order to make a decision about doing *D* (which can be different from R fixed by *A*) and her communicative tactics TB.

In the following dialogue example (generated with our DS), *A* is the manager of a company and *B* is working for the company but is at the same time studying at a university. *A* presents arguments for doing *D* by *B* (to travel to N. in order to conclude a contract). He succeeds to avert *B*'s counter arguments and convince *B* to accept his goal. In the example, we also annotate the arguments in **RC** (reason-claim) formalism [10] using additional notations for statements (given in parentheses).

(1) A: *The company offers you a trip to N. in order to conclude a contract.* (trip)
    You can meet interesting people. (people) **R**(people): **C**(trip)
(2) B: *I don't have proper dresses.* (dresses)
    R(**R**(dresses) : **C**(¬ trip)) : **C**(–**R**(people) : **C**(trip)) [strong rebuttal]
(3) A: *The company will pay your executive expenses.* (expenses) *The nature is very nice in N.* (nature)
    R(**R**(expenses) : **C**(¬ dresses)) : **C**(–**R**(dresses) : **C** (¬ trip)) [strong premise attack]
    & **R**(nature) : **C**(trip)
(4) B: *I can have some problems at my university.* (university)
    R(**R**(university) : –**C**(trip)) : **C**(–**R**(nature) : **C**(trip)) [weak rebuttal]
(5) A: *It's all right – your examinations period will be extended.* (extension) *You can sunbathe in N. early in spring already.* (sunbathe)
    R(**R**(extension) : **C**(¬ university)) : **C**(–**R**(university) : **C**(¬ trip)) [strong premise attack]
    & **R**(sunbathe) : **C**(trip)
(6) B: *OK, I'll do it.*
(7) A: *I am glad.*

Let us point out that the participants are explicitly presenting only reasons of arguments (when speaking in terms of **RC** formalism); the claim, or conclusion (doing resp. not doing the action) is implicit.

3   Changes of Beliefs

3.1   Incremental Update of the Partner Model

Let us consider the example dialogue in Sect. 2.2 in order to demonstrate in more details how the partner model is used in interaction. Let us suppose that a conversational agent is playing *A*'s role. The communicative goal of *A* is to reach *B*'s decision to do the action *D* = 'to travel to N. in order to conclude a contract'. *A* will implement the tactics of enticement and generates a partner model, let it be **wAB** = (w(resources)=1, w(pleasant) = 3, w(unpleasant) = 2, w(useful) = 2, w(harmful) = 1, w(obligatory) = 0, w(prohibited) = 0, w(punishment-do) = 0,

−w(punishment-not) = 0). The reasoning procedure WISH (Sect. 2.1) is applicable and yields a positive decision in this model. A tries to trigger the reasoning procedure WISH in B.

We suppose here that every statement (argument) presented in dialogue will increase (or respectively, decrease) the corresponding weight in a model of beliefs by one unit.

A starts the dialogue with a proposal. Using the tactics of enticement and attempting to trigger the reasoning procedure WISH in B he adds an argument to the proposal for increasing the pleasantness (turn 1). Therefore, he increases the initial value of the pleasantness in his partner model by 1. The current reasoning procedure WISH still gives a positive decision in the updated model. However, B's counter argument (turn 2) demonstrates that B actually has resources missing (*I don't have proper dresses*) therefore, A has to change the value of w(resources) in his partner model from 1 to 0. Now A has to find an argument indicating that the resources actually exist: he selects an utterance from the set Pincreasingresources (*The company will pay your executive expenses*) and when following the tactics of enticement in turn 3 he adds an argument for increasing the pleasantness (*You can meet interesting people*). The value of w(resources) will be 1 and the value of w(pleasant) will be increased by 1 in the updated partner model. The reasoning in the updated model gives a positive decision. Nevertheless, B has a new counter argument indicating the harmfulness of the action: *I can have some problems at my university* (turn 4). This turn needs more comments. B's statement for harmfulness increases the weight w(harmful) in the partner model by 4 and not by 1. Why? Let us consider the step-form reasoning algorithm WISH (Sect. 2.1). There are four possible paths to achieve a negative decision (do not do D): coming through the steps 1-2-3-4-5 or 1-2-6-9-5 (if D is prohibited); 1-2-6-7-8 (if D is obligatory); or 1-2-6-7 (if D is neither obligatory nor prohibited). The last path is acceptable according the current model **w**AB and the weight w(harmful) indicated by B has to be increased so much that the condition checked in step 6 will be fulfilled − (at least) by 4.

Responding to B's counter argument A decreases the value of w(harmful) by 1 using the utterance *It's all right – your examinations period will be extended* and increases the value of w(pleasant) once more using the utterance *You can sunbathe in N. early in spring already* (turn 5). The reasoning procedure WISH gives a positive decision in the updated partner model. Now it turns out that B has made this same decision (turn 6). A has achieved his communicative goal and finishes the dialogue (turn 7).

3.2   Changes of Own Beliefs

The last example (Sect. 3.1) demonstrates how A is updating the partner model **w**AB in argumentation dialogue with B. As compared with the initial model, the values of two weights have been increased: w(pleasantness) from 3 to 6 and w(harmfulness) from 1 to 4. The changes have been caused by A's arguments and B's counter arguments.

Does the final model **wAB** coincide with B's actual model **wB**, i.e. has A correctly guessed all the actual weights of B's beliefs? The answer is 'not'. Let us discuss why. Let us again consider the example dialogue (Sect. 2.2). Let us suppose that B is a conversational agent (not a human user) and that B's actual model is **wB** = (0, 3, 2, 1, 5, 1, 1, 0, 0) at the beginning of the dialogue (different from **wAB** as in Sect. 3.1). Thus, B considers D as an obligatory action therefore not doing D involves a punishment (differently from A's picture about B). In addition, let us suppose that B's communicative goal coincides with A's one: B has a wish to do D (it is more pleasant than unpleasant). She triggers a reasoning procedure WISH in her model of beliefs in order to check her resources and the other aspects of doing D and to make a decision.

In dialogue, A is updating his partner model **wAB**. At the same time, B has to update the model **wB** of herself as based on the arguments presented by A. Similarly with A who does not know the exact values of B's beliefs in **wB**, also B does not know the exact values of beliefs in the model **wAB**. Both participants can make conclusions only based on arguments presented by the partner.

Let us suppose that A is acting as considered in Sect. 3.1. A makes a proposal to B and adds an argument which increases the weight w(pleasantness) in the initial model of B herself by 1. The reasoning procedure WISH triggered by B in **wB** gives a negative decision: resources are missing (*I don't have proper dresses*, turn 2). A's next utterances (turn 3) increase the weights wB(resources) and wB(pleasant) by 1. Constructing her next turn (5) B again triggers the reasoning procedure WISH in the updated model **wB** and over again comes to a negative decision. She chooses to indicate the harmfulness (*I can have some problems at my university*). When responding (turn 6) A presents an argument which decreases the harmfulness by 1 and when enticing he adds an argument which increases the pleasantness by 1 in the model **wB**. Now B, after triggering the reasoning procedure WISH in her updated model, gets a positive decision (turn 6). Her final model will be **wB** = (1, 6, 2, 1, 4, 1, 1, 0, 0). The dialogue finishes, both A and B have achieved their common communicative goal. A has been able to convince B to make a decision to do D using the arguments by which B updated her initial model of beliefs in order to come to a positive decision. Although the models **wAB** and **wB** do not coincide at the end of the dialogue, the proportions of the weights of the positive (pleasantness, usefulness) and negative aspects of doing D (unpleasantness, harmfulness) are similar.

4   Discussion

When attempting to direct B's reasoning to the desirable decision ("do D" in the considered example), A presents several arguments stressing the positive and downgrading the negative aspects of D. The choice of A's argument is based, on one hand, on the (counter) argument presented by the partner and on the other hand, on the partner model. When choosing the next argument, A triggers a reasoning procedure in his partner model, in order to be sure that the reasoning will give a positive decision after presenting this argument. B herself can use the same or a different reasoning procedure triggering it in her own model. (In the example, both participants are using the same reasoning procedure WISH.) After the updates made both by A and B in the two models during a dialogue, the models will approach each to another but, in general, do not equalise. Altough, the results of reasoning in both models can be equal as demonstrated the example.

Therefore, A can convince B to do D even if not having a right picture of her. Our dialogue model considers only a limited kind of dialogues but although, it illustrates the situation where the dialogue participants are able to change their beliefs and bring them closer one to another by using arguments. The initiator A does not need to know whether the counter arguments of the partner B have been caused by B's opposite goal or are there simply obstacles before their common goal and can be eliminated by arguments. A's goal, on the contrary is not hidden from B. Secondly, as said in Sect. 2.1 the different communicative tactics used by A are aimed to trigger different reasoning procedures in B's mind. A can fail to trigger the pursued communicative tactics but however, he can achieve his communicative goal when having a sufficient number of statements for supporting the communicative goal.

5   Conclusion and Further Work

We are considering the dialogues where two (human or artificial) agents A and B discuss about doing an action D by one of them (B). Their initial communicative goals can conform or be opposite. They present arguments for and against of doing D, in order to achieve their goals. A's arguments are based on his partner model whilst B's arguments are based on her model of herself. Both models include the beliefs about the resources, positive and negative aspects of doing D which have numerical values (weights) in our implementation. Both models are changing during a dialogue. We study how the models are updated in a dialogue, and track the changes.

We have created two implementations – two experimental DSs which interact with a user using texts in a natural language. In one implementation, the computer is playing A's role and in the other – B's role. When attempting to direct B's reasoning to the decision "do D", A presents several arguments (statements) stressing the positive and downgrading the negative aspects of D. The choice of statements is based on the partner model. Before bringing out an argument, A triggers a reasoning procedure in his partner model, in order to be sure that the reasoning will give a positive decision. When opposing, B can use the same or a different reasoning procedure triggering it in the model of herself. After the changes have been made by the participants during a dialogue, the two models of beliefs (A's model of B and B's model of herself) will approach each to other. The results of reasoning in both models can be (or not be) equal.

Our future work includes development of the implementations. When adding text and speech processing tools to a DS we can achieve more natural interaction of a user with the system.

Acknowledgments. This study was supported by the Estonian Ministry of Education and Research (IUT20-56), and by the European Union through the European Regional Development Fund (Centre of Excellence in Estonian Studies).

References

1. Bos, J., Klein, E., Lemon, O., Oka, T.: DIPPER: description and formalisation of an information-state update dialogue system architecture. In: Proceedings of SIGDial Workshop on Discourse and Dialogue, Sapporo, pp. 115–124 (2003). http://sigdial. org/workshops/workshop.4/proceedings/11_SHORT_bos_dipper.pdf
2. Jurafsky, D., Martin, J.M.: Speech and Language Processing: An Introduction to Natural Language Processing, Computational Linguistics, and Speech Recognition. Prentice Hall, Upper Saddle River (2009)
3. Cohen, P.R., Perrault, C.R.: Elements of a plan-based theory of speech acts. Cogn. Sci. **3**, 177–212 (1979)
4. Galitsky, B.: Exhaustive simulation of consecutive mental states of human agents. Knowl.-Based Syst. **43**, 1–20 (2012). http://dx.doi.org/10.1016/j.knosys. 2012.11.001
5. Traum, D.R., Larsson, S.: The information state approach to dialogue management. In: van Kuppevelt, J., Smith, R.W. (eds.) Current and New Directions in Discourse and Dialogue. TSLT, vol. 22, pp. 325–353. Springer, Dordrecht (2003)
6. Nestorovič, T.: A frame-based dialogue management approach. In: 2nd International Conference on the Applications of Digital Information and Web Technologies, ICADIWT 2009, pp. 327–332 (2009). http://dx.doi.org/10.1109/ICADIWT. 2009.5273964
7. Young, S., Schatzmann, J., Thomson, B., Weilhammer, K., Ye, H.: The hidden information state dialogue manager: a real-world POMDP-based system. In: Proceedings of NAACL HLT, Rochester pp. 27–28 (2007)
8. Williams, J.D.: A belief tracking challenge task for spoken dialog systems. In: Proceedings of NAACL HLT 2012 Workshop on Future Directions and Needs in the Spoken Dialog Community: Tools and Data, Association for Computational Linguistics, p. 2 (2012). http://research.microsoft.com/pubs/163683/naaclhlt2012. pdf
9. Koit, M., Õim, H.: A computational model of argumentation in agreement negotiation processes. Argum. Comput. **5**(2–3), 209–236 (2014). Taylor & Francis, http://dx.doi.org/10.1080/19462166.2014.915233
10. Amgoud, L., Besnard, P., Hunter, A.: Logical representation and analysis for RC-arguments. In: Proceedings of ICTAI, p. 8 (2015). www.irit.fr/~Leila.Amgoud/ ictai2015-1.pdf

Combining Ontologies and IFML Models Regarding the GUIs of Rich Internet Applications

Naziha Laaz[✉] and Samir Mbarki[✉]

MISC Laboratory, Faculty of Science Kenitra, Ibn Tofail University, Kenitra, Morocco
laaznaziha@gmail.com, mbarkisamir@hotmail.com

Abstract. Rich Internet Applications (RIAs) is a new kind of web applications. These applications provide more effective graphical components and promote the fusion of traditional applications and client-server applications. They also furnish convivial and interactive interfaces similar to desktop applications. However RIAs designing and implementation are time and cost consuming. To meet RIAs requirements, we propose a new approach based on Model Driven Engineering methodology to generate GUIs from abstract models. The structural and dynamic aspects of GUIs are modeled to represent complete RIA interfaces. Our model driven development process is based on Ontology and IFML; The logical description of UI components is presented by the ontology domain and their interactions are captured by IFML. The proposed process takes as input abstract models. Then, we apply transformations on these models to produce a code representing Flex rich interfaces. Our approach is illustrated by an example of an e-commerce web site interface.

Keywords: Ontology Definition Metamodel (ODM) · Ontologies · Interaction Flow Modeling Language (IFML) · Platform Independent Model (PIM) · Model Driven Engineering (MDE) · Graphical User Interface (GUI) · Rich Internet Application (RIA)

1 Introduction

Last decade and half ago, systems have been equipped with sophisticated GUIs, and their complexity increases in time. Powerful interaction functionalities are implemented on top of variety of technologies and platforms whose boundaries are becoming less distinguishable: client-server applications, Web applications, rich Internet applications, mobile applications. Our proposal is focused on GUIs of RIAs; these applications have combined the richness and interactivity of desktop interfaces into the web distribution model.

However, software development needs to be more abstract with developed practices [1]. So, researches in software development have focused on abstract models of user interfaces and new modeling language standards have appeared. They have become more powerful in expressing requirements at a high abstraction level. Indeed, the Object Management Group (OMG) launched an effort known as the Model Driven Architecture [2] to align with these changes in technology, and raise the level of abstraction of

© Springer International Publishing Switzerland 2016
C. Dichev and G. Agre (Eds.): AIMSA 2016, LNAI 9883, pp. 226–236, 2016.
DOI: 10.1007/978-3-319-44748-3_22

physical systems. Consequently, many solutions have been emerged to describe and generate graphical interfaces; most of them respect MDA approach. In parallel, many ideas have been proposed which are based on ontologies by integrating it in the description and generation of graphical interfaces. In order to make these ideas functional and evolve to better ontology-driven development practices using, OMG took the initiative to define the Ontology Definition Metamodel [3] by marrying ontologies with meta-modeling. Most of works based on ODM are focused on the database and business layers [4], while some works was done to generate presentation form of GUIs. In the other hand, IFML was recently defined to describe the elements and behavior of user interfaces in order to generate code implementation of these interfaces [5].

To meet these requirements, we present a new MDE approach combining the two new standards IFML and ODM to benefit from each other in order to generate rich user interfaces of RIA platform. We used known frameworks and technologies of model-driven engineering, such as Eclipse Modeling Framework (EMF) for Meta Models, Query View Transformation (QVT) for model transformations and Acceleo for code generation. The approach allows to quickly and efficiently generate a RIA focusing on the graphical aspect of the application. It can be replicated for different target technologies and platforms.

The rest of this paper is organized as follows: Sect. 2 describes our contribution; we have two sub-sections in this part: The formal ontology of GUIs respecting the syntax OWL 2.0 and IFML. Section 3 is dedicated to the related work. In Sect. 4, we explain the process of Ontology-Driven UIs; it is divided into tree sub-parts: The definition of PIM models, PIM to PSM transformation and code generation from PSM model. Finally, the last part will present a conclusion indicating the status of the objectives and describing future work.

2   Methodology and Contribution

Nowadays, the GUIs are deployed in heterogeneous and interactive spaces: They are spread over a different kind of platforms. This allows thinking of adapting new ways of developing the presentation layer of the application. In this work, we present an approach based on MDA, which proposes a solution for the GUIs abstract representation and their automatic generation to a specific platform.

Since, our ultimate goal is to raise the abstraction level of the GUIs definition that include the structural and dynamic aspects. These two aspects cover all information related to the nature of graphical components, their properties and interactions. Instead, we have developed a process to automatically map from a high-level representation to a lower-level language. The proposed approach specifies two transformations during the development cycle, models becoming more and more concrete until obtaining the code by successive transformations, We have two PIM; two models of the most abstract level, are transformed into a PSM, then a second transformation is established to generate code from the specific model to the Flex platform. After a study on the various GUIs PIM, we detected that the difficult resides in the choice of abstract input models. The two chosen PIM represent two specifications proposed by the OMG. The first PIM, ODM, is the metamodel that defines ontologies which supports several different ontology representation systems. With this metamodel, we developed our ontology of graphical components. The approach exploits the new language IFML as a second PIM by extending the graphical part of the MetaModel to fit the RIAs' needs. We choose IFML because it allows obtaining all information of interactions between components represented in the GUI ontology.

We established an analysis to represent different aspects of GUIs with these two specifications as presented in the following sub sections.

2.1 GUIs Ontology Respecting OWL2.0 Syntax

We present a semantic approach to the problem by defining ontology for graphical user interfaces. In this section, we show how the GUI domain concepts are presented in OWL 2.0 using OWL DL. The GUI ontology is formed by three concepts: Declarations, Axioms and Assertions. The basic elements of user interfaces domain are expressed by Declarations. There are five declaration types: Classes, ObjectProperty, DataProperty, DataTypes and Individuals. The GUI contains elements divided into: containers and controls; The containers provide a space where controls can be located, and the controls are the elements that display content or accept user input in interfaces (buttons, fields, lists e.g.). These two concepts are represented as classes to group items with similar characteristics resources (buttons, Menus, Field, Window e.g.), and individuals represent instances of classes.

Object and data properties can be used to represent relationships in the domain [6]. The relationship between individuals of the two classes is represented by ObjectProperty, we use this property to link a container with controls or a container with another container. For example, we have defined an objectProperty named **"composedOf"** to say that the class Menu consists of several MenuItem classes, or, to define the relationship between the two classes List and ListItem, etc. Howerver, datatypes are sets of literals such as strings or integers. All these declarations are grouped by axioms, in order to form complex descriptions from the basic entities.

There are three kinds of axioms: Class Axioms, ObjectProperty Axioms and DataProperty Axioms. Each axiom is associated with an expression. Properties Expressions are characterized by a domain and range; a domain is represented by classes and ranges can be classes or dataTypes. However, there are other types of axioms linked to ClassExpression. We consider two classes: Control and List. Control is more general than List, which means that every time we know that an individual is a list, this individual must be a control. In OWL 2, this is done by a so-called axiom subClassOf, that has List as subClass Expression and Control as superClass Expression.

To distinguish between the different types of containers and controls, we defined DataProperties Axioms which have Boolean datatype as Range. The sets of this data Properties are represented in Table 1. For example, we chose isDefault for container when it is a default container such as homepage or welcome interface. In addition to this data properties that give semantics to widgets, there are another dataproperties that define several characteristics such as a geometry (x, y, width, height) and "hasText", "hasname"; text and name attaching to widgets.

Table 1. Nature of GUI Concepts

GUI concept	DataProperties	Description
List	isSimple	Scrolls elements vertically, on a single column.
	isCombo	list items vertically by displaying only the selected item.
Field	isStatic	the components that display text
	isEntry	the controls that allow the user to enter text
Button	isPush	graphical control element that provides the user a simple way to trigger an event
	isRadio	Represents a selection of one item from a list of items. Occur only in groups. Selecting one radio button deselects the others.
	isCheck	Possibility of multiple selections
Window	isModal	Designed to block any user interaction in all other previously active containers.
	isDefault	A home page or welcome default container.
	isXOR	the Container comprising child or containers that are displayed alternatively
	isLandmark	container is reachable from any other element of the user interface without having interactions with other containers

We can add some subPropertyOf Axioms by attaching them with dataProperty and their subProperty Expressions (subPropertyOf Axiom has "hasSize" as dataProperty and "width", "height" as subProperties). After detailed declarations and axioms, it remains to talk about the role of the assertions in the concepts definition of GUI domain. The assertions can be Class Assertions, ObjectProperty Assertions, or DataProperty Assertions. The assertions are intended to clarify how individuals relate to other individuals. In the section reserved to the ontology-driven UI development process, we will present the logic model of GUIs as result of this analysis, respecting the ODM metamodel that meets the syntax OWL2.0.

2.2 Extending Interaction Flow Modeling Language

The new OMG Interaction Flow Modeling Language standard (IFML) is defined in March 2014 [7]. IFML is a platform independent model (PIM) that can be used to express interaction design decisions independently of the implementation platform. It allows to capture the user interaction and content of the front-end (user interface) and model the control behavior of that system's user interface.

In the approach presented in this paper, we used IFML as PIM-level interaction flow modeling, it brings several benefits to user interfaces development process of web applications. It permits the formal specification of the different perspectives of the user interface such as interface composition, interaction and navigation options. This work uses one of four technical artifacts defined by the present specification [5], which is the IFML metamodel.

IFML metamodel is composed of three packages: The Core package, the Extensions package and the DataTypes package. The Core package contains the concepts that build up the interaction infrastructure of the language in terms of InteractionFlowElements, InteractionFlows, and Parameters. The Extension package extends the concepts defined by Core package by concrete concepts with more complex behaviors. While The Data-Types package contains the basic data types defining in the UML metamodel, and specializes a number of UML metaclasses as the basis for IFML metaclasses, and presumes that the IFML DomainModel is represented in UML. After studying the various packages of IFML metamodel, we noted that the part defining the graphical application components and their characteristics did not provide sufficient information. IFML is extensible, in fact, we thought to expand it to meet the needs of RIA development and implementation of overall output platforms.

We have completed the IFML metamodel by defining the GUI ontology. In addition, we modify the core package of the metamodel. So, we added Ereference to ViewComponent metaclass which has isContainment() as a method because ViewComponent can be composed of viewcomponents like a form composed of List As seen in (Fig. 1).

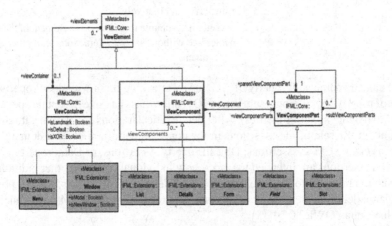

Fig. 1. IFML extension.

3   Related Work

Ontologies provide a formal representation of knowledge and the relationships between concepts [8]. Recently, a number of use cases have been proposed that employ ontologies for modeling user interfaces and their components, Examples are automatic generation of explanations for user interfaces, adaptation of user interfaces for different needs and contexts, and integration of user interface components [9].

Those use cases require a strongly formalized ontology of the domain of user interfaces and interactions. In this regard, many works have been developed. In [10], the authors discuss the differences between the UI description languages and formal ontologies and how they can benefit from each other. Their goal is to define a formal ontology of user interfaces and interactions domain. The formal ontology will not replace user interface description languages, but will be a valuable enhancement.

With the appearance of MDE, great research effort has been dedicated to this methodology marrying it with ontology. Lot of them are focused on conceptual models. The most relevant are: [11–15]. In the other hand, few UI generation approaches based on MDE have been defined in recent years. In [16], the authors describe a method for rich UI development for data-intensive Web applications based on OWL2 ontologies, which applies model-driven engineering to derive a user interface from the domain ontology, incorporating modern rich components for Web-based interfaces. The model-driven process proposed is supported by the TwoUse Toolkit [19] for OWL2 authoring and management. In our approach we also use ontologies as domain model, but we combine it with IFML model to derive a complete presentation of Web UIs.

However, [17] presents an approach to the problem of porting graphical user interfaces by aligning representations of user interfaces in different technologies to an abstract semantic web model for graphical user interfaces. Our proposal has similarities with this approach in sense that we also assemble ontology concepts to give a high abstract representation of UIs.

4   The Model Driven Development Process

We present an approach deriving User Interface (UI) of Rich Internet Applications from the combination of OWL2 ontology and IFML to automatically generate UI according to models specifications. Semantics of UI elements and their characteristics can be induced from the GUIs ontology's domain. However, IFML is responsible for capturing the interactions and actions related to concepts defined in the logical model of UIs. The model-driven process proposed was implemented using MDE tools of Eclipse Modeling Project.

This process is projected on a case study shown in (Fig. 2). It starts with abstract Models, in order to produce a flex Model as a target model. The choice of RIA as destination was not arbitrary because the design and implantation of GUI for RIAs is known for its complexity and difficulty in using existing tools. The chosen case study represents an interface containing a form for billing in e-commerce website. We will see later, how

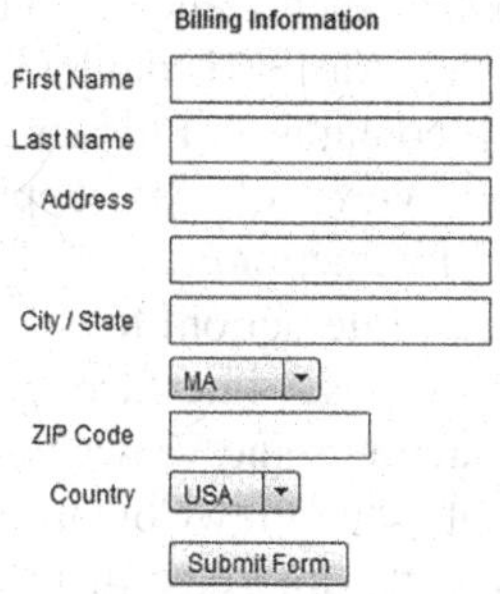

Fig. 2. Case study: billing form.

the elements composing this interface are represented with their various types of inter-action. The process is divided into three steps: The definition of PIM Models, the definition of PSM Model and M2M Transformation. Figure 3 shows the Model-driven process combining ODM and IFML to generate UIs of Rich Internet Applications.

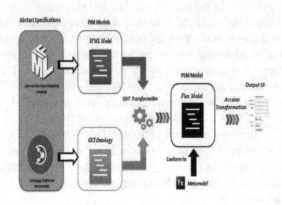

Fig. 3. Process overview.

4.1 PIM Models

In a first step, we define two PIM models: Logic model respecting the syntax of ODM metamodel, and Interaction model deriving from IFML metamodel. The two metamodels are defined with EMF (Eclipse Modeling Framework) in ecore format.

To define the logic model, we applied the analysis detailed in Sect. 2 to our case study. As Depicted in (Fig. 2). The interface comprises two containers; Window and Form which contain controls; six static fields, seven entry fields, two lists and a submit button. The representation of these elements in the logic model is defined as follows: containers and controls are represented by Individuals instanced from owl: Class distinguished by their data Properties as an example see (Fig. 4).

The second abstract model respects the syntax of IFML metamodel. We analyzed the IFML metamodel to select the required packages for an independent intermediate representation of interactions related to UIs. In our process, we will focus on the two packages: Core and extensions. These packages represent abstractly the structure of user interfaces, and dependencies between their elements in terms of interactions.

According to IFML metamodel, An IFML model is the top-level container of all the rest of the model elements. It contains an InteractionFlowModel, which contains all the elements of the user view of the application represented by the InteractionFlowModelElement. InteractionFlowModelElement has seven direct subtypes: InteractionFlowElement, InteractionFlow, Expression, The elements of an IFML model that are visible at the user interface level are called ViewElements, which are specialized in ViewContainers and ViewComponents. ViewContainers like windows, menus are containers of other ViewContainers or ViewComponents, while ViewComponents are elements of the interface that display content or accept input from the user. The extension package includes concrete examples of

Fig. 4. Defining submit button with ODM.

ViewComponents such as List, Details and Form, and ViewComponentParts such as Fields and Slots [5].

In the example shown in (Fig. 2), a view container is tagged as «window» and marked as Default. The Form is composed of OnSubmit Event. The effect is represented by a navigation flow connecting the event associated with the OnSubmit Event. The navigation flow expresses a change of state of the user interface. The occurrence of the event causes a transition from a Form (source Interaction Flow element) to other windows (targets Interaction Flow elements). We associated the form with two Expressions; an activation Expression that denotes the condition that must be satisfied by the current interaction context for the event that triggers an action to become active and Interaction Flow Expression that determines which InteractionFlow is followed after an event occurrence. These expressions are expressed by javascript Language. The model instance is an abstract form to our case study respecting the IFML metamodel syntax [5]. Thus, with this two ODM and IFML models, we can easily have a UI target model in multiple platform as desktop, web or mobile.

4.2   Model to Model Transformation (PIM 2 PSM)

Once the Meta Modeling phase established, we defined the transformation rules. For this work, we used the QVT-Operational mappings language implemented by Eclipse modeling Framework [18]. The PIM, the most abstract level model is transformed into PSM Model. Since we have defined PIM models and PSM Metamodel for RIA, we define the Model To Model transformation using the QVTo standard respecting a defined algorithm. The entry metamodels are ODM and IFML metamdels, and the target one is Flex metamodel.

The entry point of the transformation is the main method. This method makes the correspondence between all elements of the IFML and Ontology models of the input models and the elements of type Flex output model.

For instance, each NamedIndividual which is an instance of "Window" class that has a dataProperty "isDefault", it would be mapped to BorderContainer. Form is obtained from "Form" Individual, and the NamedIndividuals of "Button" and "Field" and "List" classes, are mapped as Button, Text Input and List in Flex model. The data-properties (width, height, name, size, etc.) are mapped to properties figured in the Flex metaclasses. The (Fig. 5) below shows an excerpt of the Transformation program:

```
modeltype MyOWL uses "http://owl2/1.0";
modeltype MyIFMLCORE uses "http://www.omg.org/spec/IFML/core";
modeltype MyIFMLEXT uses "http://www.omg.org/spec/IFML/ext";
modeltype MyFlex uses "http://Flex/1.0";
//INPUT AND OUTPUT MODELS
transformation Transforms(in owl : MyOWL, core : MyIFMLCORE, out flex : MyFlex);
main() {
owl.objectsOfType(Ontology) -> map OntologyToFlexModel();
}
//MAPPING DES RACINES
mapping Ontology::OntologyToFlexModel() : FlexModel
{
bordercontainers += self.declaration.entity[NamedIndividual]->map
individual2borderContainer();
}
//MAPPING NAMEDINDIVIDUALS which have class window as class expression TO Border
Containers
mapping NamedIndividual:: individual2borderContainer() : BorderContainer

when {self.classAssertion.classExpression[Class].entityURI="owl:Window"->asBag();}
{
 name:=self.value;

self.dataPropertyAssertion->forEach(element)
    {
      if element.dataPropertyExpression.entityURI.toString().indexOf("width")-
```

Fig. 5. Query view transformation code excerpt.

4.3 PSM Model and Code Generation

This step describes the gradual refinement from a higher to a lower level of abstraction [19]. By applying the transformation rules mentioned before, each element figured in ODM and IFML models will be transformed to an element of Flex metamodel. It includes the generation of the target model in compliance with Flex metamodel. This model contains all the elements, properties, interactions collected from the two PIM models. This file is used to produce the necessary code by applying M2T transformation.

5   Conclusion and Future Work

With this paper, we have given a new approach MDE assembling two important abstract specifications defined by the OMG to derive UIs Rich Internet Apps. Our approach is based on the assumption that a UI for rich internet applications can be induced from IFML and ontologies. IFML allows representing abstractly the structure of user inter-faces, and dependencies between its elements in terms of interactions. However, ODM captures presentation features related to the UI and represent the basic concepts constituting a user interface. The major contribution in the proposed approach is the addition of the extension part to IFML and the definition of the GUI ontology that describe efficiently the graphical components of the application. Our challenge is to generate RIAs by using this Model Driven method without having to know all the technical specification of the execution platform.

Future works will cover the implementation of more refined code generator. Also, to obtain an enhanced result, this work can be extended to supplementary platforms like mobile Plateform starting from the same input models. Moreover, we can consider integrating other frameworks like JavaFX and GWT for Rich Internet Application.

References

1. Schmidt, D.C.: Guest editor's introduction: model-driven engineering. Computer **39**(2), 25–31 (2006)
2. OMG, MDA. Guide Version 1.0. 1. Object Management Group (2003)
3. Object Management Group: Ontology Definition Metamodel. Version 1.0, OMG (2009). http://www.omg.org/spec/ODM/1.0/
4. Paulheim, H., Probst, F.: A formal ontology on user interfaces yet another user interface description language. In: 2nd Workshop on Semantic Models for Adaptive Interactive Systems (SEMAIS) (2011)
5. Object Management Group: Interaction Flow Modeling Language. Version 1.0, IFML (2015). http://www.omg.org/spec/IFML/1.0/
6. OWL 2 Web Ontology Language Structural Specification and Functional-Style Syntax W3C Editor's Draft 14 September 2009 This version: http://www.w3.org/2007/OWL/draft/ED-owl2-syntax-20090914/
7. Brambilla, M., Fraternali, P.: Interaction Flow Modeling Language: Model-driven UI Engineering of Web and Mobile Apps with IFML. Morgan Kaufmann, San Francisco (2014)
8. Pan, J.Z., Staab, S., Aßmann, U., Ebert, J., Zhao, Y.: Ontology-Driven Software Development. Springer Science & Business Media, Heidelberg (2012)
9. Paulheim, H., Probst, F.: Ontology-enhanced user interfaces: a survey. Semantic-Enabled Advancements on the Web: Applications Across Industries: Applications Across Industries, p. 214 (2012)
10. Paulheim, H.: Ontology-Based Application Integration. Springer Science & Business Media, New York (2011)
11. Gašević, D., Djuric, D., Devedžic, V.: Model Driven Engineering and Ontology Development. Springer Science & Business Media, Heidelberg (2009)
12. Parreiras, F.S., Staab, S., Winter, A.: TwoUse: Integrating UML models and OWL ontologies. Inst. für Informatik (2007)
13. Brockmans, S., Haase, P., Hitzler, P., Studer, R.: A Metamodel and UML profile for rule-extended OWL DL ontologies. Springer, Heidelberg (2006)
14. Bumans, G.: Mapping between Relational Databases and OWL Ontologies: an example. Comput. Sci. Inf. Technol. **756**, 99–117 (2010). Scientific Papers, University of Latvia
15. Parreiras, F.S., Staab, S.: Using ontologies with UML class-based modeling: the TwoUse approach. Data Knowl. Eng. **69**(11), 1194–1207 (2010)
16. Canadas, J., Palma, J., Túnez, S.: Model-Driven Rich User Interface Generation from Ontologies for Data-Intensive Web Applications (2011)

17. Wysota, W.: Porting graphical user interfaces through ontology alignment. In: Ryżko, D., Rybiński, H., Gawrysiak, P., Kryszkiewicz, M. (eds.) Emerging Intelligent Technologies in Industry. SCI, vol. 369, pp. 91–104. Springer, Heidelberg (2011)
18. OMG, QVT. Meta Object Facility 2.0, Query/View/Transformation Specification, OMG (2008). http://www.omg.org/spec/QVT/1.0/PDF
19. Wagner, C.: Model-Driven Software Migration: A Methodology: Reengineering, Recovery and Modernization of Legacy Systems. Springer, Wiesbaden (2014)

Identity Judgments, Situations, and Semantic Web Representations

William Nick, Yenny Dominguez, and Albert Esterline[(⊠)]

North Carolina A&T State University,
1601 East Market Street, Greensboro, NC 27411, USA
{wmnick,ydomingu}@aggies.ncat.edu, esterlin@ncat.edu

Abstract. We present our framework for the identity of agents based on situation theory as developed by Barwise, Devlin, and others. Semantic Web standards are used to capture the information present in a constellation of situations ("id-case") that relate to identity attributions. We present examples of id-cases and discuss how they are encoded using RDF and other semantic-web standards. The examples include straightforward cases of identification using fingerprints and mugshots. They also include developing a profile from a set of documents. And they include a case with learning, specifically, learning a writing style so as to be able to identify the author. Using semantic-web standards, we can make SQL-like queries and execute rules to classify id-situations and entire id-cases. Our encodings also facilitate and account of how evidence accrues to identity judgments.

Keywords: Identity · Situation theory · Semantic web

1 Introduction

We are concerned with the identity of agents both in cyber and physical environments. We are particularly concerned with provenance of information and how a case fits together to support an identity judgment. While semantic-web standards are used for representation and inference, we are not concerned at this point with issues of how identity in general may be represented on the web (cf., e.g., owl:sameAs) as in, for example, [1]. We do, however, intend an equivalence relation, but referential opacity is not an issue since we are not concerned with modal contexts or propositional attitudes (where it is an issue).

The Superidentity Project [2], which represents the state of the art in identity, is developing a model of identity that connects elements from both the cyber and physical universes. An *element of identity* has a type, and a *characteristic*, is a multiset of elements of identity of the same type. A *superidentity* is a set of characteristics. An initial superidentity has at least one seed identity element and is enriched by deriving new elements of identity via functions that *transform* one or more elements of given types to an element of another type. For example, an email address may be transformed to a username.

When we attribute identity, we want something like a legal case. It became apparent that the elements of identity and transforms of the Superidentity project do not support

© Springer International Publishing Switzerland 2016
C. Dichev and G. Agre (Eds.): AIMSA 2016, LNAI 9883, pp. 237–246, 2016.
DOI: 10.1007/978-3-319-44748-3_23

the internal structure we require. We turned to situation theory developed by Barwise and his colleagues [3] and are developing a framework for identity based on situation theory. Barwise claimed that the "key insight" regarding situation theory in the form of situation semantics "is that speech, writing, thought, and inference are situated activities" [4]. Accordingly, "'[s]ituation' is our name for those portions of reality that agents find themselves in, and about which they exchange information" [4].

Situation theory has provided the foundation for several computational approaches; we mention only two. Several programming languages have been based on situation-theoretic notions [5], and several researchers have attempted to capture the notion of context in a computational setting in terms of situation [6].

Semantic Web standards including RDF and OWL are used here to capture the information present in a constellation of situations that relate to identity attributions. Kokar and Endsley applied their Situation Theory Ontology (STO) to situation awareness in [7]. See also Kokar et al. in [8] for using the STO in explaining human-computer collaboration. They produce an OWL ontology based on the same sources we use but have a somewhat different structure for the basic concepts.

Identity-related situations (*id-situations*) are those that include identity-relevant actions (*id-actions*), each asserting that two expressions denote the same agent. From our perspective, a superidentity is an equivalence class.

The remainder of this paper is organized as follows. The next section sketches situation theory. We represent situations using semantic web resources. Section 3 introduces these resources and explains why they are particularly appropriate for addressing identity here. To capture provenance of information and evidence, we consider id-situations within constellations of situations ("id-cases"). Section 4 presents an example with two straightforward cases: identifying one and the same person by fingerprint and by mugshot. Section 5 shows how we represent these cases with semantic-web resources (especially RDF). Section 6 sketches a more involved case, that of identifying someone from a set of documents, and Sect. 7 sketches another different kind of case, where we learn a writing style to identify the author; both sections discuss representations. Finally, Sect. 8 concludes. In other publications, we have extensively discussed how we can make SQL-like queries on our store of id-cases and how we can classify id-situations and id-cases [9]; we have also applied the Dempster-Shafer theory of evidence to explain how evidence accrues to an identity judgment [10].

2 Situation Theory

Situation theory was first articulated at length by Barwise and Perry in their 1983 work *Situations and Attitudes* [3] ([11] is a useful linguistically-oriented source). We consider Devlin's account of situation theory [12], which is faithful to the original.

An *infon* here is the basic item of information. It involves an n-place relation, R, and n objects appropriate for the corresponding argument places of R. For example, it might happen that a given book, denoted, say, by the identifier *book123*, is on the desk, denoted by the identifier *desk43*, in my office this evening. Here the two-place (binary) relation is *on*, and the two objects filling the argument positions are *book123* and *desk43*. We are talking about the relation *on* and the objects *book123* and *desk43*, not about the terms "on," "book123," and "desk43." An infon is not a linguistic entity, although the relation and objects involved are denoted by noun phrases and various other parts of speech.

A situation *s supports* an infon σ if the information available in *s* includes σ. For example, anyone in my office can see that said book is indeed on said desk. (What infons a situation supports depends in part on the sensory capacities of potential observers. In fact, what we conveniently take as a situation depends on how we address our world.)

An infon, however, encompasses more than a relation and the objects filling its argument positions since the relation is thus instantiated at a given time and place. In some situation *s*, it may be significant that certain objects a_1, a_2, ..., a_n are not related by R at a certain time and place. For example, in my office this evening, the mug (say, *mug73*) that is usually on my desk is not on it. We might express the infon in question as

$$<<on,\ mug73,\ desk43,\ l,\ t,\ 0>>,$$

where *l* is the location of my office and *t* is this evening. We include the '0' to indicate negative 'polarity'. In general, we handle such information by associating a polarity with an infon. Polarity 1 indicates that the objects in question are related as indicated (at the time and place in question) while 0 indicates otherwise.

Barwise and his colleagues emphasized that a situation is a *partial structure*: it does not support information, positive or negative, about many sets of objects and corresponding relations (at a given time and place), that is, there are many infons it does not support. A full expression of an infon, then, is as a structure $<<R, a_1, ..., a_n, l, t, i>>$, where R is an n-place relation, $a_1; ..., a_n$ are objects appropriate for the corresponding argument places of R, l is a location, t is a temporal location, and i is the polarity.

Situations and infons are real, but internal structure is imposed on them as agents get attuned to uniformities across situations. A *real situation* is a part of reality that supports indefinitely many infons while an *abstract* situation is a finite set of (possibly parameterized) infons. Think of abstract situations as types to classify real situations.

Situation theory arose as part of the situation semantics, where we identify an *utterance situation*, in which a speech act is performed, and a *described situation*, which the speech act is about. Besides supporting information, a situation may *carry information* about another situation. This is possible because of *constraints*. Some constraints are natural (as in smoke means fire), and some are conventional, such as those constraints by virtue of which a speech act carries information about a described situation.

3 Semantic Web Resources

The Semantic Web is built off W3C standards: the resource description framework (RDF) for representing simple facts, the RDF schema (RDFS) for extending the set of terms available, and the more expressive OWL (Web ontology language) for authoring ontologies, that is, conceptualizations of domains.

An RDF statement (triple) has the form *subject property object*, where *property* is a binary relation term. This asserts that *subject* has *object* as its value for *property* (e.g., *Fred hasFather Bill*). We use the N3 serialization of RDF, which expresses a triple as the three components separated by whitespace. When triples share a subject, as in *subject property$_1$ object$_1$* and *subject property$_2$ object$_2$*, we may abbreviate by listing the subject once and separating the property-object clauses by semi-colons: *subject property$_1$ object$_1$; property$_2$ object$_2$*.

To denote "resources" (things), RDF exploits the Web's uniform resource identifiers (URIs), which are unique across the Web. It in fact allows URIrefs: URIs with optional fragment identifiers. A URIref is written as a *qname*, **prefix:localPart**, where the namespace prefix stands for a URI. We have different prefixes for different vocabularies from different RDFS documents.

SPARQL is a query language for triple stores resembling SQL We use SPARQL provided by the Jena Semantic Web framework [13]. We issue SPARQL queries that navigate across situations connected by, say, shared individuals. In the semantic web rule language (SWRL), a rule is a conditional, with an antecedent ("body"), followed by an arrow ('->'), followed by a consequent ("head"). For classifying situations according to their types (abstract situations), we use SWRL rules. We also classify constellations of situations as instances of case types. We use the Pellet reasoner [14] within Jena and in the Protégé ontology editor [15]. For our use here of SPARQL and SWRL, see [9].

The semantic web is particularly apt for handling identity. Individuals are denoted by URIrefs, which are unique across the web, but there is no unique name assumption: two "names" (URIrefs) may refer to the same individual, and we must allow for this since it is impractical for all web authors to coordinate their naming efforts. This handling of names is one aspect of a basic tenet of the web in general, and of the semantic web in particular, the AAA slogan: Anyone can say Anything about Any topic. As a consequence of this, we have the open-world assumption: we can always include more facts, that is, from the absence of a statement, we cannot infer that it is false. This assumption fits well with situation theory as situations are partial information structures.

4   An Example: Identifying by Mugshot and Fingerprint

We present a constellation of situations involved in identifying an individual by mugshot and by fingerprint. Our running example is shown in Figs. 1 and 2 and involves six situations (within clouds), s_1-s_6. The situations on the right (s_1 and s_2) are id-situations coordinated in that they identify the same individual via their "name" (any identifier unique in the context). In a sense, we have one id-situation made of two coordinated id-situations. Id-situation s_1 matches fingerprints on file with those on a doorknob. The fingerprints on file were produced in s_3, and the fingerprints on the doorknob were produced in s_4. Situation s_4 is a (spatiotemporal) part of the situation portrayed in the ellipse in the representation of s_5 (call it s_{5a}), where people are socializing. s_{5a} is in turn part of s_5, where someone takes a picture of the group. In s_6, a mugshot is produced. It is used in s_2 to pick out the person in the group photo.

Combined id-situation s_1-s_2 is an utterance situation as the investigator in effect utters a judgment about the identity of the culprit. The described situation is s_{5a}, which has a structure imposed on it by the fact that the fingerprint-matching activity references s_4 as a part of s_{5a}. We assume that there were in fact situations s_{5a} and s_4 and that they were thus related. We also assume that the situation recorded in s_5 is indeed s_{5a}. These assumptions are embedded in a network of episodic and common-sense knowledge.

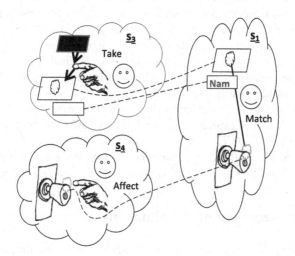

Fig. 1. The fingerprint case

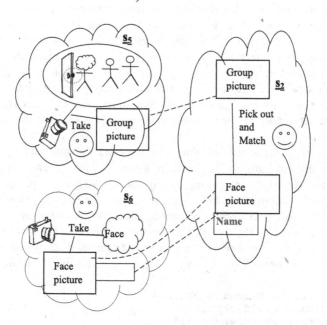

Fig. 2. The mugshot case

Situations s_3 and s_6 are neither utterance situations nor described situations, yet they play an essential role in the id action. We call them *support situations*. A support situation is is similar to what Devlin calls a *resource situation*, which helps us resolve references; for example, to identify, "the man", we might relate to another situation (a resource situation) we witnessed where there was a unique man. Our support situations, unlike resource situations (which relate to language use), tend to produce artifacts used in id-situations, but there remains a conventional reference to a previous situation. The dashed lines between situations on the left and id-situations connect things produced (left) and used (right). These dashed lines may relate to additional copying or rendering situations not shown in in the figures. For example, the mugshot matched against the group photo may be a copy of the original, and, in any case, what is produced in the situation where the mugshot is taken is in a medium that is different from that of the picture that is used.

To measure evidence of an identity from the supporting situations, we use the Dempster-Shafer theory of evidence enhanced in two ways to leverage the information available in a constellation of situations. One way exploits the structure within the situations [16], and the other way interprets the evidence-relationships in terms of argument schemes [17]. For our use of Dempster-Shafer theory, see [10].

5 RDF Representation of the Situations

We use the empty prefix, **:**, for the namespace in which we define the basic classes and properties. An instance *s* of class **:Situation** generally appears as subject in triples identifying the time and location of the situation, so time and spatial location need not be associated with infons. (This simplifies our representations.)

In RDF, our situations have local names **s1**, **s2**, and so on. We use namespace prefix **sit:**, but we usually use just local names. That a given situation *s* supports an infon *i* (an instance of class **:Infon**) is expressed as "*s* **:supports** *i***.**" Various relations are captured by subclasses of **:Infon**. If R is a relation with roles $r_1, r_2, \ldots, r_n$, we define subclass **:RInfon** of **:Infon** and properties **r1**, **r2**, ..., **rn** with domain **:RInfon**. This avoids RDF's restriction of relations to binary relations since an instance of **:RInfon** may be a subject of any number of triples with one of **r1**, **r2**, ..., **rn** as the property.

The fingerprint id-case involves three situations: **s1** (id-situation), **s3** (taking the polarity. We discuss only **i1** in detail. It is an instance of **:AnalystMatch-ingFpInfon**fingerprint), **s4** (leaving the fingerprint). We discuss only **s1** in detail. It has three important infons: **i1**, **i1a**, and **i14**. Assume (for simplicity) that infons have positive, that an analyst is matching a forensic fingerprint and a fingerprint on file (no suggestion of similarity). Three properties are recorded for it: **:fpObserved**, whose value is the URIref of the forensic fingerprint, **:fpRecorded**, whose value is the URIref of the fingerprint on file, and **:fpAnalyst** is for the officer making the match. In N3, this is

```
_:i1 a :AnalystMatchingFpInfon;
     :fpAnalyst officer:117;
     :fpObserved forensicfp:822;
     :fpRecorded fpfile:496 .
```

(**i1**, like all our infons, is represented by a bnode, indicated by an underscore prefix.) Infon **i1a** is an instance of **:SimilarFpInfon**, that the forensic fingerprint and the one on file have a similarity of 0.94 according to a similarity-measuring procedure. Infon **i14** is an instance of **:OnInfon**, that the forensic fingerprint is on the doorknob.

Situation **s3** has one infon that is important here, **i3**, an instance of **:TakeF-pInfon**, that a given officer takes the fingerprint of our suspect. This infon is the subject in triples identifying the officer, the fingerprint on file, and the person whose fingerprint is recorded. Situation **s4** has two important infons, one that the fingerprint is on the doorknob, which is also an infon in **s1**. The other is an instance of **:LeaveFpInfon**, that our suspect left the fingerprint on the doorknob. It is the subject in triples identifying the producer (our suspect), the forensic fingerprint, and the doorknob.

The photo id-case also involves three situations: **s2** (id-situation), **s5** (taking the forensic photo), and **s6** (taking the mug shot). **s2** is analogous to **s1** but lacks the analogue of the fingerprint on the doorknob. **s6** is analogous to **s3** (fingerprint on file). **s5** is only roughly analogous to **s4**. One of **s5**'s infons, **i5**, is an instance of **:ForensicPicInfon** and is the subject of triples identifying the photographer, the camera, the photo produced, and the described situation, **s5a**, caught on camera. **s5a** supports an infon that our suspect is touching the doorknob; it also has

```
sit:s5a :inSituation group:5342;
```

This says that this group is in the described situation but does not identify any information associated with the group. We also have the following, using the FOAF (friend of a friend) vocabulary for groups; **insys:201** is our suspect, and **insys:563** just happens to be in the group.

```
group:5342 a foaf:Group;
    foaf:member insys:201;
    foaf:member insys:563.
```

There is thereby in **i5** the information that **insys:201** is pictured in the photo; we do not necessarily have the information that **insys:201** is a member of **group:5342**. And we have the following, where **fshot:812** is the URIref of the group photo and **biom:GroupImage** is the class in the biometrics ontology for group photos.

```
fshot:812 a biom:GroupImage;
    foaf:depicts  sit:s5a .
```

foaf:depicts is an information relation. While the fingerprint and the group photo both ostensibly refer to our culprit, only the photo depicts him.

There is also a part-whole (mereological) relation between **s4** and **s5a**:

```
sit:s4 mereo:properPartOf sit:s5a.
```

where **mereo** is a prefix for a URI for mereological concepts. The spatial containment relations among these situations is clear, but the temporal relations are more involved: **s5** is roughly contemporaneous with **s4** and one brief action in the period of **s5a**.

6 Example: Identification from Documents

As a second example, suppose we are trying to identify a terrorist through multiple newspaper articles, email messages, social media postings, etc. A case that is structurally similar but more clearly illustrates some aspects is where an historian reads about a figure from the Dark Ages in multiple original documents. In both cases, the analyst reads several documents and builds a profile ("identity") of the person from references in the documents. The documents describe situations that are interrelated in time and space, drawing on resource situations when needed to resolve references.

There is a rich weaving of situations here. To get a handle on the structure, consider the simpler case where one inspects an email, for example, and notes that it is from someone. The id-action is the inspection, and the described situation is the sending of the email. On top of the described situation, we have a *referenced situation*, supporting real-world information that is the background content for the described situation, e.g., what the content of the email was about or what the characters in a chronicle discuss.

When an object is provided for reference, we take the referenced situation to be the situation producing the object. For example, when a mug shot is attached to an email, the referenced situation is the mugshot being taken. When an id-action identifies an individual, it generally (but not always) references a described situation. For establishing identity, we may focus on the id-situation itself, on a separate described situation, or on a referenced situation.

We have discussed the described and resource situations in the identification-by-documents case without mentioning utterance situations. In the straightforward case, the id-situation here itself is an utterance situation. Unlike in the mugshot and fingerprint cases, the described situation denoted by the id-situation in this case is simply an identity. The described situations that constitute the content of our subject's profile are each associated with an utterance situation consisting of the composition of a certain document. Each of these situations, like the mugshot and fingerprint-on-file situations, produces an enduring object, viz., the document. Unlike those situations, the objects here describe situations because a document is essentially an utterance frozen in time.

Regarding how these situations might be captured in RDF, and first the id-situation, this depends very much on how we represent the profile. In our prototype, we represent the id-situation as supporting infons involving the judgment, its support measure, and, for each document, a structure with identity-relevant information from it.

The infons that we consider being supported by a document-writing situation generally relate to metadata (perhaps better called meta-information), such as the date of composition, URIrefs for the author and location, and so on. For manuscripts from the Dark Ages, the URIrefs could dereference to dedicated articles. For emails or social media postings, much of this information would be electronically recorded and possibly accessible via URIs that dereference (i.e., URLs). The unique identifier for the document itself is a URI, which is the URL of the document if it is available online. How a situation supports meta-information requires a paper on its own, but such a case is quite natural when we consider intentions and subsequent uses of artifacts produced.

The profile referenced in the RDF for the id-situation has to pick out the identity-relevant passages and provide some machine-readable version of the information therein. One approach is to use RDF reification quads, but this assumes that we have a way (outside RDF) or associating URIs with statements.

7 Example: Identification by Writing Style

As a third example, we might learn to recognize a person's writing style from multiple examples of their writing. Then, given an example of their writing, we should be able to identify them as the author. The learning and matching could be done by man or machine. We suppose the latter.

The id-situation here is where one compares the learned characterization of the person's style against a writing sample and pronounces for or against the person's authorship. The described situation is where the person writes the sample being matched. The support situations include all the situations where the person produces writing samples used by the machine learning. The salient support situation is where these samples are provided to the machine-learning program. What is described in the writing sample is irrelevant, so there is no further described situation than that where the person writes the sample that is identified as belonging to them.

The RDF capturing these situations includes much metadata relating to provenance, but the id-situation is much like the id-situations in the mugshot or fingerprint cases. For the described situation, we have metadata for the sample, and the sample is stored in a text file. For each situation where a sample is produced, we have a unique URIref and metadata for the sample, which again is stored in a text file. For the situation with machine learning, we have metadata about the workstation used, where and when the learning took place, the person in charge, and the URIref of each sample used.

8 Conclusion

We have presented our framework for identity based on situation theory as developed by Barwise, Devlin, and others. Semantic Web standards are used to capture the information present in a constellation of situations ("id-case") that relate to identity attributions. We presented three examples of id-cases and discussed how they are represented using RDF and other semantic-web standards. Using semantic-web standards, we can make SQL-like queries on our store of id-cases and we can classify id-situations (in which identity judgments are made) and id-cases. And the Dempster-Shafer theory of evidence can be applied to explain how evidence accrues to identity judgments. These topics are discussed in papers that complement this one.

Future work will continue capturing in RDF and other semantic-web standards id-cases with as wide a variety of structure as possible. What we have found is that there is a great variety in the kinds of situations and the way they are related in id-cases. The forensic id-cases will be filled out to reflect protocols actually followed by law enforcement. We are consulting with the Criminal Justice Program at North Carolina A&T State University in this regard, and they will provide expert evaluation.

Acknowledgments. The authors acknowledge support from the NSF (grant no. 1460864) and the ARO (Contract No. W911NF-15-1-0524)

References

1. Halpin, H., Hayes, P.J., McCusker, J.P., McGuinness, D.L., Thompson, H.S.: When owl: sameAs isn't the same: an analysis of identity in linked data. In: Patel-Schneider, P.F., Pan, Y., Hitzler, P., Mika, P., Zhang, L., Pan, J.Z., Horrocks, I., Glimm, B. (eds.) ISWC 2010, Part I. LNCS, vol. 6496, pp. 305–320. Springer, Heidelberg (2010)
2. Creese, S., Gibson-Robinson, T., Goldsmith, M., Hodges, D., Kim, D., Love, O., Nurse, J.R., Pike, B., Scholtz, J.: Tools for understanding identity. In: 2013 IEEE International Conference on Technologies for Homeland Security (HST), pp 558–563. IEEE (2013)
3. Barwise, J., Perry, J.: Situations and Attitudes. The MIT Press, Cambridge (1981)
4. Barwise, J.: The Situation in Logic, vol. 17. Center for the Study of Language (CSLI) (1989)
5. Tin, E., Akman, V.: Computational situation theory. ACM SIGART Bull. 5(4), 4–17 (1994)
6. Akman, V., Surav, M.: The use of situation theory in context modeling. Comput. Intell. Int. J. 13(3), 427–438 (1997)
7. Kokar, M.M., Endsley, M.R.: Situation awareness and cognitive modeling. IEEE Intell. Syst. 3, 91–96 (2012)
8. Kokar, M.M., Matheus, C.J., Baclawski, K.: Ontology-based situation awareness. Inf. Fusion 10(1), 83–98 (2009)
9. Dominguez, Y., Nick, W., Esterline, A.: Situations, identity, and the semantic web. In: Proceedings of the International Multi-Disciplinary Conference on Cognitive Methods in Situation Awareness and Decision Support, San Diego, CA, March 2016
10. Nick, W., Dominguez, Y., Esterline, A.: Situations and evidence for identity using dempster-shafer theory. In: Proceedings of the Modern Artificial Intelligence and Cognitive Science Conference (MAICS 2016), Dayton, OH, April 2016
11. Kratzer, A.: Situations in Natural Language Semantics. The Stanford Encyclopedia of Philosophy, Fall 2011 edn. CSLI, Stanford University, Stanford, CA (2011)
12. Devlin, K.: Logic and Information. Cambridge University Press, Cambridge (1995)
13. The Apache Software Foundation (2013). Apache jena. http://jena.apache.org. Accessed 20 Mar 2014
14. Sirin, E., Parsia, B., Grau, B.C., Kalyanpur, A., Katz, Y.: Pellet: a practical OWL-DL reasoner. Web Semant. Sci. Serv. Agents WWW 5(2), 51–53 (2007)
15. Stanford Center for Biomedical Informatics Research (2015). Protégé ontology editor. http://protege.stanford.edu/. Accessed 6 Nov 2015
16. Lalmas, M., Van Rijsbergen, C.: Situation theory and Dempster-Shafer's theory of evidence for information retrieval. In: Alagar, V.S., Bergler, S., Dong, F.Q. (eds.) Incompleteness and Uncertainty in Information Systems, pp. 102–116. Springer, London (1994)
17. Tang, Y., Oren, N., Parsons, S., Sycara K.: Dempster-Shafer Argument Schemes. In: Proceedings of ArgMAS (2013)

Local Search for Maximizing Satisfiability in Qualitative Spatial and Temporal Constraint Networks

Jean-François Condotta[1]([✉]), Ali Mensi[1,3], Issam Nouaouri[2], Michael Sioutis[1], and Lamjed Ben Saïd[3]

[1] Université Lille-Nord de France - Université d'Artois - CRIL-CNRS UMR 8188,
Lens, France
{condotta,mensi,sioutis}@cril.fr
[2] Université Lille-Nord de France - Université d'Artois - LGI2A, Béthune, France
issam.nouaouri@univ-artois.fr
[3] Institut Suprieure de Gestion (ISG) - SOIE, Tunis, Tunisia
lamjed.bensaid@isg.rnu.tn

Abstract. We focus on the recently introduced problem of maximizing the number of satisfied constraints in a qualitative constraint network (QCN), called the MAX-QCN problem. We present a particular local search method for solving the MAX-QCN problem of a given QCN, which involves first obtaining a partial scenario S of that QCN and then exploring neighboring scenarios that are obtained by disconnecting a variable of S and repositioning it appropriately. The experimentation that we have conducted shows the interest of our approach for maximizing satisfiability in qualitative spatial and temporal constraint networks.

1 Introduction

Qualitative spatial and temporal reasoning (QSTR) is a major field of study in Knowledge Representation that abstracts from numerical quantities of space and time by using qualitative descriptions instead (e.g., precedes, contains, left of). The representational languages used in the qualitative approach have increasingly gained a lot of attention during the last decades, as they have the advantage of being conceptually concise and sufficiently expressive for a variety of applications in many areas, such as ambient intelligence, dynamic GIS, cognitive robotics, and spatiotemporal design [2,8,14].

The Interval Algebra (IA) [1] and a subset of the Region Connection Calculus (RCC) [13], namely RCC8, are the dominant calculi in QSTR for representing qualitative temporal and spatial information respectively. In particular, IA encodes knowledge about the temporal relations between intervals in the timeline (see Fig. 1a), and RCC8 encodes knowledge about the spatial relations between regions in some topological space. In addition to IA and RCC8, numerous other

This work is partially supported by PHC-UTIQUE program (project RESCUESYS-34942VF) managed by the CMCU.

C. Dichev and G. Agre (Eds.): AIMSA 2016, LNAI 9883, pp. 247–258, 2016.
DOI: 10.1007/978-3-319-44748-3_24

qualitative calculi have been proposed in the literature for representing spatial and temporal information [10].

The problem of reasoning about qualitative spatial or temporal information can be modelled as a qualitative constraint network (QCN), i.e., a network comprising constraints corresponding to qualitative spatial or temporal relations between spatial or temporal variables respectively. In this paper, we focus on a recently introduced problem in the context of QSTR, called the MAX-QCN problem [5]. Given a QCN $\mathcal{N}$, the MAX-QCN problem is the problem of obtaining a spatial or temporal configuration that maximizes the number of satisfied constraints in $\mathcal{N}$. Solving the MAX-QCN problem is clearly at least as difficult as solving the consistency problem, which is NP-complete in general. To solve this optimization problem, the authors in [5] propose a branch and bound algorithm based on state of the art techniques for checking the consistency of a QCN, viz., the use of a triangulation of the constraint graph of the considered QCN to reduce the number of constraints to be treated, the use of a tractable subclass of relations to reduce the width of the search tree, and the use of partial ◇-consistency to efficiently propagate constraints and consequently prune non-feasible base relations during search. In another approach, the authors in [6] view the MAX-QCN problem as partial maximum satisfiability problem (PMAX-SAT) and propose two related families of encodings. Each proposed PMAX-SAT encoding is based on, what is called, a forbidden covering with regard to the composition table of the considered qualitative calculus. Intuitively, a forbidden covering is a compact set of triples that express all the non-feasible configurations for three spatial or temporal entities.

We follow another approach for solving the MAX-QCN problem of a QCN and present a particular local search method [11] which involves first obtaining a partial atomic refinement S of that QCN and then exploring neighboring atomic refinements that are obtained by disconnecting a variable of S and repositioning it appropriately. The search for the best neighboring atomic refinement is guided by a combination of heuristics for minimizing the number of unsatisfied constraints in a given neighboring atomic refinement of a QCN, a tabu list for excluding certain already considered atomic refinements or atomic refinements that are known to not be candidates for best neighboring atomic refinement, and a particular restart policy to deal with local minima.

Some preliminaries on QSTR and the MAX-QCN problem are made in Sect. 2. Section 3 introduces the neighborhoods of scenarios which will be used in the proposed method and an algorithm to compute them. In Sect. 4, we define the local search based method proposed to solve the MAX-QCN problem, namely, the method QLS. In Sect. 5, we report some experimental results about QLS. Finally, we conclude.

2 Preliminaries

A (binary) spatial or temporal qualitative calculus [10] considers a domain D of spatial or temporal entities respectively and a finite set B of *jointly exhaustive and pairwise disjoint* (JEPD) relations defined on that domain called base relations. Each base relation of B represents a particular configuration between two spatial or temporal entities. The set B contains the identity relation Id, and is closed under the converse ($^{-1}$). A (complex) relation corresponds to a union of base relations and is represented by the set containing them. Hence, 2^B represents the set of relations. The set 2^B is equipped with the usual set operations (union and intersection), the converse operation, the complement operation and the weak composition operation. The converse of a relation is the union of the converses of its base relations. The complement of a relation r, denoted by $\bar{r}$, is the relation $\{b \in B : b \notin r\}$. The weak composition $\diamond$ of $b, b' \in B$ is the relation of 2^B defined by $b \diamond b' = \{b'' : \exists x, y, z \in D$ such that $x\ b\ y$, $y\ b'\ z$ and $x\ b''\ z\}$. For $r, r' \in 2^B$, $r \diamond r'$ is the relation of 2^B defined by $r \diamond r' = \bigcup_{b \in r, b' \in r'} b \diamond b'$. In the sequel, $\hat{B}$ will denote the smallest subset of 2^B containing the singleton relations of 2^B and the universal relation and, which is closed under the operations $^{-1}$, $\diamond$ and $\cap$. Consider the well known temporal qualitative calculus introduced by Allen [1] and called the Interval Algebra (IA). Allen represents the temporal entities by the intervals of the line and considers a set of 13 base relations $B_{IA} = \{eq, p, pi, m, mi, o, oi, s, si, d, di, f, fi\}$ (Fig. 1a).

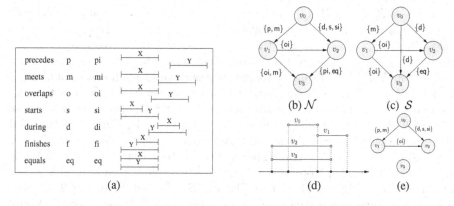

(a)   (b) $\mathcal{N}$ (c) $\mathcal{S}$ (d)   (e)

Fig. 1. The base relations of IA (a), a consistent QCN $\mathcal{N}$ of IA (b), a consistent scenario $\mathcal{S}$ of $\mathcal{N}$ (c), a solution of $\mathcal{N}$ and $\mathcal{S}$ (d), the QCN $\mathcal{N}^{\uparrow v_3}$, *i.e.* the relaxation of $\mathcal{N}$ w.r.t. v_3 (e). Notice that for the represented QCNs, the constraint between two variables v and v' is not represented when it is equal to B or $v = v'$ or when the constraint between v' and v is already represented.

Spatial or temporal information about a set of entities can be represented by a qualitative constraint network (QCN), which is a pair of a set of variables and a set of constraints. Each constraint is defined by a relation of 2^B and specifies the set of acceptable qualitative configurations between two spatial or temporal variables. Formally, a QCN is defined as follows: a QCN is a pair $\mathcal{N} = (V, C)$ where V is a non-empty finite set of variables, and C is a mapping that associates a relation $C(v, v') \in 2^B$ with each pair (v, v') of $V \times V$. Further, mapping C is such that $C(v,v) \subseteq \{\text{Id}\}$ and $C(v,v') = (C(v',v))^{-1}$. Given a QCN $\mathcal{N} = (V,C)$ and $v,v' \in V$, the relation $C(v,v')$ will also be denoted by $\mathcal{N}[v,v']$.

Concerning a QCN $\mathcal{N} = (V,C)$, we have the following definitions. A *solution* σ of $\mathcal{N}$ is a valuation σ of each variable V by an element of D such that for every pair (v,v') of variables in V, $(\sigma(v),\sigma(v'))$ satisfies a base relation belonging to the relation $C(v,v')$. $\mathcal{N}$ is *consistent* iff it admits a solution. $\mathcal{N}$ will be said trivially inconsistent iff one of its constraints is defined by the empty relation. A *sub*-QCN $\mathcal{N}'$ of $\mathcal{N}$, denoted by $\mathcal{N}' \subseteq \mathcal{N}$, is a QCN (V,C') such that $C'(v,v') \subseteq C(v,v')$ $\forall v,v' \in V$. A *scenario* $\mathcal{S}$ is a QCN that each contraint is defined by a singleton relation. A scenario $\mathcal{S}$ of $\mathcal{N}$ is a scenario which is a sub-QCN of $\mathcal{N}$. Given a variable $v \in V$, the relaxation of $\mathcal{N}$ w.r.t. v, denoted by $\mathcal{N}^{\uparrow v}$ is the QCN $\mathcal{N} = (V,C')$ defined by : for all $v',v'' \in V$, $C'(v',v'') = \mathsf{B}$ if $v' \neq v''$ and, $v' = v$ or $v'' = v$, $C'(v',v'') = C(v',v'')$ else. In the sequel, $\mathcal{N}_{[v,v']/r}$ with $r \in 2^{\mathsf{B}}$, denotes the QCN defined on V corresponding to $\mathcal{N}$ for which the relation defining the constraint between v and v' has been substitued by r.

Given two (undirected) graphs $G = (V,E)$ and $G' = (V',E')$, G is a subgraph of G', denoted by $G \subseteq G'$, iff $V \subseteq V'$ and $E \subseteq E'$. A graph $G = (V,E)$ is a *chordal (or triangulated) graph* iff each of its cycles of length > 3 has a chord [7]. The constraint graph of a QCN $\mathcal{N} = (V,C)$ is the graph (V,E), denoted by $\mathsf{G}(\mathcal{N})$, for which we have that $(v,v') \in E$ iff $C(v,v') \neq \mathsf{B}$.

Given a QCN $\mathcal{N} = (V,C)$ and a graph $G = (V,E)$, $\mathcal{N}$ is partially $\diamond$-consistent w.r.t. graph G or $\overset{\diamond}{G}$-consistent [4] iff for all $(v,v'),(v,v''),(v'',v') \in E$, we have that $C(v,v') \subseteq C(v,v'') \diamond C(v'',v')$. The closure under $\overset{\diamond}{G}$-consistency of $\mathcal{N}$, denoted by $\overset{\diamond}{G}(\mathcal{N})$, is the greatest $\overset{\diamond}{G}$-consistent sub-QCN of $\mathcal{N}$. This closure can be computed in $O(\delta|E|)$ time [4,15], where δ is the maximum degree of G. Note that if $\mathsf{G}(\mathcal{N}) \subseteq G$, $\overset{\diamond}{G}(\mathcal{N})$ is equivalent to $\mathcal{N}$, *i.e.* has the same solutions than $\mathcal{N}$.

Given a graph $G = (V,E)$, a *partial* scenario w.r.t. G, also called G-scenario, is a QCN (V,C) such that $C(v,v) = \{\text{Id}\}$ for all $v \in V$, $C(v,v') = \mathsf{B}$ for all $(v,v') \notin E$, and $|C(v,v')| = 1$ for all $(v,v') \in E$. As illustration, let us consider the inconsistent QCN $\mathcal{N}'$ represented in Fig. 2a. Its constraint graph $\mathsf{G}(\mathcal{N}')$ is represented in Fig. 2b. Moreover, a triangulated graph G such that $\mathsf{G}(\mathcal{N}') \subseteq G$ is illustrated in Fig. 2c. The four QCNs $\mathcal{S}_0$, $\mathcal{S}_1$, $\mathcal{S}_2$ and $\mathcal{S}_3$ in Fig. 2 are four $\overset{\diamond}{G}$-consistent (and consistent) G-scenarios. Now, we highlight a property useful in the sequel:

Definition 1. *The partial $\diamond$-consistency will be said complete for $\widehat{\mathsf{B}}$ iff for any triangulated graph $G = (V,E)$ and any $\overset{\diamond}{G}$-consistent $\mathcal{N}$ such that $\mathcal{N}[v,v'] = \mathsf{B}$ for any $(v,v') \in E$ and $\mathcal{N}[v,v'] \in \widehat{\mathsf{B}}$ for any $(v,v') \notin E$ we have $\mathcal{N}$ which is consistent.*

The partial $\diamond$-consistency is complete $\widehat{\mathsf{B}}$ for many qualitative calculi [9,15], in particular for IA and RCC8. Given a QCN $\mathcal{N} = (V,C)$, the MAX-QCN problem (MAX-QCN in short) is the problem of finding a consistent scenario over V that minimizes the number of unsatisfied constraints in $\mathcal{N}$ (or maximizes the number of satisfied constraints in $\mathcal{N}$). In order to more formally define MAX-QCN we introduce the operator α which takes as parameters two QCNs and returns the number of non overlapping constraints of these QCNs. Formally, given two $\mathcal{N} = (V, C)$ and $\mathcal{N}' = (V, C')$, $\alpha(\mathcal{N}, \mathcal{N}')$ is defined by $\alpha(\mathcal{N}, \mathcal{N}') = \frac{1}{2}.|\{(v, v') \in V \times V : v \neq v'$ and $C(v, v') \cap C'(v, v') = \emptyset\}|$. As illustration, for the QCNs in Fig. 2 we have $\alpha(\mathcal{S}_0, \mathcal{N}') = \alpha(\mathcal{S}_1, \mathcal{N}') = 3$, $\alpha(\mathcal{S}_2, \mathcal{N}') = 2$ and $\alpha(\mathcal{S}_3, \mathcal{N}') = 4$. Given $\mathcal{N} = (V, C)$, a solution of MAX-QCN for $\mathcal{N}$ is a consistent scenario $\mathcal{S}$ on V, said optimal scenario of $\mathcal{N}$, such that there is no consistent scenario $\mathcal{S}'$ on V with $\alpha(\mathcal{S}, \mathcal{N}) > \alpha(\mathcal{S}', \mathcal{N})$. Given a QCN $\mathcal{N} = (V, C)$, an optimal G-scenario of $\mathcal{N}$ is a consistent G-scenario $\mathcal{S}$ such that $G = (V, E)$ is a graph such that there is no consistent G-scenario $\mathcal{S}'$ with $\alpha(\mathcal{S}, \mathcal{N}) > \alpha(\mathcal{S}', \mathcal{N})$ and such that G is a triangulated graph (V, E) with $\mathsf{G}(\mathcal{N}) \subseteq G$. Note that all consistent scenario of an optimal G-scenario of a QCN $\mathcal{N}$ is an optimal scenario of $\mathcal{N}$. From [5,6], we have:

Property 1. *Let $\mathcal{Q}$ be a qualitative calculus for which the partial $\diamond$-consistency is complete for $\widehat{\mathsf{B}}$ and $\mathcal{N} = (V, C)$ a QCN in $\mathcal{Q}$. For any triangulated graph $G = (V, E)$ such that $\mathsf{G}(\mathcal{N}) \subseteq G$ and $\overset{\diamond}{_G}$-consistent G-scenario $\mathcal{S}$ we have : (1) $\mathcal{S}$ is an optimal G-scenario of $\mathcal{N}$, (2) any consistent scenario of $\mathcal{S}$ is an optimal scenario of $\mathcal{N}$ and (3) a consistent scenario of $\mathcal{S}$ can be computed in polynomial time.*

The method that we will define in the sequel is adapted for QCNs of a qualitative calculus $\mathcal{Q}$ that has the aforementioned property, *i.e.* a qualitative calculus $\mathcal{Q}$ for which the partial $\diamond$-consistency is complete for $\widehat{\mathsf{B}}$. From this, the method can consider partial scenarios rather than complete scenarios. The useful to consider partial scenarios is to render the treatment more fast by discarding some constraints of the considered QCNs.

3   Neighborhood of Partial Consistent Scenarios

The proposed search method that will be detailed in the next section, moves from partial consistent scenario to partial consistent scenario until a shut-off criterion is reached. Given a consistent partial scenario $\mathcal{S}$ *w.r.t.* a graph G, the candidate partial scenarios to be considered in the future step will be called neighbors of $\mathcal{S}$ and the set they comprise will be denoted by $\mathsf{Nb}(\mathcal{S}, G)$. A neighbor of the G-scenario $\mathcal{S}$ is a consistent G-scenario different from $\mathcal{S}$ that can be obtained by disconnecting a variable of $\mathcal{S}$ and repositioning it. A neighbor of $\mathcal{S}$ *w.r.t.* G is a consistent G-scenario of a relaxation of $\mathcal{S}$ with respect to one of its variables. The set of neighbors of a partial scenario is defined as:

Definition 2. *Let $\mathcal{S} = (V, C)$ be a consistent G-scenario, where $G = (V, E)$ is a graph. The set of neighbors of $\mathcal{S}$ w.r.t. G is the set:*

$$\mathsf{Nb}(\mathcal{S}, G) = \bigcup_{v \in V} \{\mathcal{S}' : \mathcal{S}' \neq \mathcal{S} \text{ and } \mathcal{S}' \text{ is a consistent } G\text{-scenario of } \mathcal{S}^{\uparrow v}\}.$$

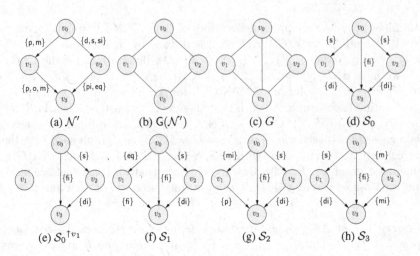

Fig. 2. An inconsistent QCN $\mathcal{N}'$ of IA (a), the constraint graph of $\mathcal{N}'$, *i.e.* $\mathsf{G}(\mathcal{N}')$ (b), a triangulation G of $\mathsf{G}(\mathcal{N}')$ (c), a consistent G-scenario $\mathcal{S}_0$ of $\mathcal{N}'$ (d), the relaxation of $\mathcal{S}$ *w.r.t.* variable v_1 (e). Three neighbors $\mathcal{S}_1$, $\mathcal{S}_2$, and $\mathcal{S}_3$ of $\mathcal{S}$ *w.r.t.* G (f, g, and h respectively).

As illustration, consider the QCN $\mathcal{N}' = (V, C)$ in Fig. 2a, the graph $G = (V, E)$ in Fig. 2c, and the consistent G-scenario $\mathcal{S}_0$ in Fig. 2d. Among the neighbors of $\mathcal{S}_0$ *w.r.t.* graph G, we have the three partial scenarios $\mathcal{S}_1$, $\mathcal{S}_2$, and $\mathcal{S}_3$ (depicted in Figs. 2f–h respectively). Note that $\mathcal{S}_1$ and $\mathcal{S}_2$ are consistent G-scenarios of the relaxation of $\mathcal{S}_0$ *w.r.t.* v_1 (Fig. 2e), whereas $\mathcal{S}_3$ results from the relaxation of $\mathcal{S}_0$ *w.r.t.* v_2.

At each step of the proposed method, among the neighbors of the current partial scenario, the future chosen partial scenario is selected among the best neighbors, *i.e.*, the neighbors satisfying a maximal number of constraints of the considered QCN and not belonging to a particular set of partial scenarios that can be seen as a tabu list. This set of best neighbors is formally defined by:

Definition 3. *Let* $\mathcal{N} = (V, C)$ *be a* QCN, $\mathcal{S} = (V, C')$ *a consistent* G-*scenario, where* $G = (V, E)$ *is a graph, and* $\mathcal{T}$ *a set of* G-*scenarios. The set of partial scenarios* $\mathsf{BestNb}(\mathcal{N}, \mathcal{S}, G, \mathcal{T})$ *is defined by:* $\mathsf{BestNb}(\mathcal{N}, \mathcal{S}, G, \mathcal{T}) = \{\mathcal{S}' \in \mathsf{Nb}(\mathcal{S}, G) \setminus \mathcal{T} : \text{there exists no } \mathcal{S}'' \in \mathsf{Nb}(\mathcal{S}, G) \setminus \mathcal{T} \text{ such that } \alpha(\mathcal{N}, \mathcal{S}'') < \alpha(\mathcal{N}, \mathcal{S}')\}$.

Now, we present function $\mathsf{bestNeighbors}$, which allows computing the set of best neighbors of a partial scenario. This function receives four parameters, namely, a QCN $\mathcal{N} = (V, C)$ for which we want to solve MAX-QCN, a triangulation $G = (V, E)$ of $\mathsf{G}(\mathcal{N})$, a consistent G-scenario $\mathcal{S} = (V, C')$ for which we want to compute the best neighbors, and a set $\mathcal{T}$ of G-scenarios that contains the partial scenarios to be excluded. The aim of $\mathsf{bestNeighbors}$ is to compute $\mathsf{BestNb}(\mathcal{N}, \mathcal{S}, G, \mathcal{T})$. In a first step, $\mathsf{bestNeighbors}$ initializes the global variables bestNb and $\mathsf{best}\alpha$. bestNb, an initially empty set, will contain the computed best neighbors. $\mathsf{best}\alpha$

Function bestNeighbors($\mathcal{N},\mathcal{S},G,\mathcal{T}$)

---

    in   : A QCN $\mathcal{N} = (V,C)$, a triangulation $G = (V,E)$ of $G(\mathcal{N})$, a consistent
              G-scenario $\mathcal{S} = (V,C')$, and a set of G-scenarios $\mathcal{T}$.
    output : The set of best neighbors of $\mathcal{S}$ w.r.t. $\mathcal{N}$, G and $\mathcal{T}$.

1 **begin**
2   bestNb $\leftarrow \emptyset$; best$\alpha \leftarrow +\infty$;
3   **foreach** $v \in V$ **do** bestNeighborsAux($\mathcal{N},\mathcal{S},G,\mathcal{T},\mathcal{S}^{\uparrow v}$) **return** bestNb;
4 **end**

Procedure bestNeighborsAux($\mathcal{N},\mathcal{S},G,\mathcal{T},\mathcal{N}'$)

---

    in   : Two QCNs $\mathcal{N} = (V,C)$ and $\mathcal{N}' = (V,C'')$, a triangulation $G = (V,E)$
              of $G(\mathcal{N})$, a consistent G-scenario $\mathcal{S} = (V,C')$, and a set $\mathcal{T}$ of
              G-scenarios of $\mathcal{N}$.

5 **begin**
6     $\mathcal{N}' \leftarrow \overset{\diamond}{G}(\mathcal{N}')$; $\alpha \leftarrow \alpha(\mathcal{N},\mathcal{N}')$;
7   **if** $\mathcal{N}'$ *is trivially inconsistant* **or** $\alpha >$ bestNb **then** **return**;
8   Select $(v,v') \in E$ such that $\mathcal{N}'[v,v']$ is not a singleton relation;
9   **if** *such a pair* (v,v') *exists* **then**
10   Select a base relation $b \in \mathcal{N}'[v,v']$;
11   bestNeighborsAux($\mathcal{N},\mathcal{S},G,\mathcal{T},\mathcal{N}'_{[v,v']/\{b\}}$);
12   bestNeighborsAux($\mathcal{N},\mathcal{S},G,\mathcal{T},\mathcal{N}'_{[v,v']/(\mathcal{N}'[v,v']\setminus\{b\})}$);
13   **return** ;
14   **if** $\mathcal{N}' \notin \mathcal{T}$ **then**
15   **if** $\alpha <$ bestNb **then**
16   bestNb $\leftarrow \{\mathcal{N}'\}$; best$\alpha \leftarrow \alpha$;
17   bestNb $\leftarrow$ bestNb $\cup \{\mathcal{N}'\}$;
18 **end**

---

corresponds to the number of constraints of $\mathcal{N}$ that are unsatisfied by a partial scenario of bestNb; this variable is initially assigned a value greater than the number of constraints of $\mathcal{N}$. In a second step (line 3), the relaxation of partial scenario $\mathcal{S}$ w.r.t. each variable $v \in V$ is addressed in order to characterize the consistent partial scenarios that are candidates to being best neighbors. This is done through a call to function bestNeighborsAux.

Function bestNeighborsAux receives five parameters. The first four parameters $\mathcal{N}$, $\mathcal{S}$, G, and $\mathcal{T}$ are similar to the parameters of bestNeighbors. The fifth parameter $\mathcal{N}'$ is a QCN defined on the set of variables of $\mathcal{N}$. bestNeighbors computes the consistent G-scenarios of $\mathcal{N}'$ that are candidates to being best neighbors of $\mathcal{S}$ by taking into account the candidate best neighbors of bestNb previously computed. In a first step, bestNeighborsAux prunes some non-feasible base relations of $\mathcal{N}'$ by enforcing $\overset{\diamond}{G}$-consistency on $\mathcal{N}'$. The integer α of non-overlapping constraints between $\mathcal{N}$ and $\mathcal{N}'$ is computed. When $\mathcal{N}'$ is detected as inconsistent or when the number of non-overlapping constraints between $\mathcal{N}$ and $\mathcal{N}'$ is greater than the number of unsatisfied constraints *w.r.t.* the neighbors of bestNb, we can assert that there exists no consistent G-scenario of $\mathcal{N}'$ that is better than those previously computed. Consequently, bestNeighborsAux terminates (line 3). In the contrary case, the treatment continues by selecting an edge (v, v') of G such that the corresponding constraint of $\mathcal{N}'$ is defined by a non-singleton relation. When such an edge exists, a base relation b is extracted from relation $\mathcal{N}'[v, v']$ and the treatment continues through two recursive calls to bestNeighborsAux. The first call (line 7) allows exploring the QCN resulting from $\mathcal{N}'$ by replacing the constraint between v and v' with relation $\{b\}$. The second call (line 8) concerns the QCN resulting from $\mathcal{N}'$ by removing b from the constraint between v and v'. When all the constraints of $\mathcal{N}'$ corresponding to edges of G are defined by a singleton relation, we can assert that $\mathcal{N}'$ is a G-scenario. As G is a triangulation of G($\mathcal{N}$) and $\mathcal{N}'$ is $\overset{\diamond}{G}$-consistent, we have that $\mathcal{N}'$ is a consistent G-scenario. Also, we know that the number of non-overlapping constraints between $\mathcal{N}$ and $\mathcal{N}'$ is smaller or equal to bestα. Consequently, if $\mathcal{N}'$ does not belong to the set of excluded partial scenarios $\mathcal{T}$, $\mathcal{N}'$ must be considered as a best neighbor of $\mathcal{S}$ and added to the set bestNb.

Theorem 1. *Let $\mathcal{N} = (V, C)$ and $\mathcal{S} = (V, C')$ be two QCNs of a qualitative calculus for which the partial $\diamond$-consistency is complete for $\widehat{\mathsf{B}}$, $G = (V, E)$ a triangulation of G($\mathcal{N}$), $\mathcal{T}$ a set of G-scenarios, and $\mathcal{S}$ a consistent G-scenario. Function bestNeighbors with parameters $\mathcal{N}$, $\mathcal{S}$, G, and $\mathcal{T}$ correctly computes the set BestNb($\mathcal{N}, \mathcal{S}, G, \mathcal{T}$).*

4   A Local Search Based Algorithm for the MAX-QCN Problem

In this section, we present a local search based algorithm, called QLS (Qualitative Local Search), for solving MAX-QCN. QLS takes as parameters a QCN $\mathcal{N} = (V, C)$ defined on a set of base relations for which partial $\diamond$-consistency is complete for partial scenarios and a triangulation $G = (V, E)$ of G($\mathcal{N}$). In short, to find a consistent G-scenario that minimizes the number of unsatisfied constraints of $\mathcal{N}$, QLS moves from consistent G-scenario $\mathcal{S}$ to consistent G-scenario $\mathcal{S}'$ according to the best neighbors presented in the previous section. QLS also includes a heuristic based on a weighting of the constraints to select at each iteration one of the best neighbors of the current G-scenario. A tabu list to exclude the G-scenarios already visited and a restart policy have been included. Now, we present the different steps realized in an iterative manner by QLS.

Initialization Step. An initial consistent G-scenario $\mathcal{S}$ on V to serve as the current partial scenario is randomly generated (see function randomScenario). The best partial scenario found, denoted by $\mathcal{S}^*$, is initialized with $\mathcal{S}$. Moreover, an integer weight denoted by $w(v, v')$ is associated with each edge $(v, v') \in E$. For each $(v, v') \in E$, $w(v, v')$ is initialized with 1 if $\mathcal{S}[v, v'] \subseteq \mathcal{N}[v, v']$, and

0 otherwise. In the case where the current partial G-scenario satisfies the constraint of $\mathcal{N}$ between v and v', the weight $w(v, v')$ will be incremented. Hence, $w(v, v')$ will represent the number of iterations for which the current partial G-scenario satisfies the constraint of $\mathcal{N}$ between v and v'. On the other hand, a tabu list $\mathcal{T}$ is introduced and initialized with $\{\mathcal{S}\}$. This tabu list will be used to avoid selecting new partial scenarios already selected.

Neighbor Generation and Selection Step. By using the function bestNeighbors presented earlier, the set of best neighbors $\mathsf{BestNb}(\mathcal{N}, \mathcal{S}, G, \mathcal{T})$ of $\mathcal{S}$ w.r.t. $\mathcal{N}$, G, and $\mathcal{T}$ is generated. Let us denote by BN this set. In the case where BN is an empty set, the procedure returns to the initialization step. In the contrary case, one partial scenario $\mathcal{S}'$ from the set BN is selected using a heuristic that uses assigned weights to each edge of E. More particularly, a weight $w(\mathcal{S}'')$ defined by $w(\mathcal{S}'') = 0 + \sum\{w(v, v') : (v, v') \in E \text{ and } \mathcal{S}''[v, v'] \subseteq \mathcal{N}[v, v']\}$ is associated with each partial scenario $\mathcal{S}''$ of BN. Intuitively, given a $\mathcal{S}'' \in$ BN, the greater the weight $w(\mathcal{S}'')$, the more the partial scenario $\mathcal{S}''$ corresponds to constraints used by the partial scenarios selected in the previous iterations. Consequently, to satisfy constraints of $\mathcal{N}$ which have been the least satisfied by the previous selected partial scenarios, the partial scenario $\mathcal{S}'$ is randomly selected among the elements of the set $\{\mathcal{S}'' \in \mathsf{BN} : w(\mathcal{S}'') = min\{w(\mathcal{S}''') : \mathcal{S}''' \in \mathsf{BN}\}\}$.

Acceptance and Restarts Step. The G-scenario $\mathcal{S}'$ replaces the current partial scenario $\mathcal{S}$ and the best partial scenario found $\mathcal{S}^*$ is possibly updated. Moreover, $\mathcal{S}'$ is added to the tabu list $\mathcal{T}$. For each $(v, v') \in E$ such that $\mathcal{S}^*[v, v'] \subseteq \mathcal{N}[v, v']$, $w(v, v')$ is incremented. When for the nbDivLoops last iterations (with nbDivLoops a parameter of QLS), the number of unsatisfied constraints of $\mathcal{N}$ by the selected partial scenario forms an increasing series, a diversification is realised by substituing the current partial scenario $\mathcal{S}$ by a partial scenario randomly selected in the tabu list $\mathcal{T}$. Concerning restarts, after a number of nbRestartDiv diversifications (with nbRestartDiv a parameter of QLS) a restart stage is realized by returning to the initialization step.

Termination Step. In this step, two additional integer parameters of QLS called maxLoops and expectedValue (with 0 as default value) are used. Whether the number of iterations is greater or equals to maxLoops or whether the $\alpha(\mathcal{N}, \mathcal{S}^*)$ is less or equals to expectedValue, the treatment terminates after returning the best partial scenario $\mathcal{S}^*$.

In the next section we will report some experimental results regarding the aforementioned method QLS and three variations of it denoted by $\mathsf{QLS}^{\mathfrak{r}}$, $\mathsf{QLS}^{\mathfrak{w}}$ and $\mathsf{QLS}^{\mathcal{T}}$. $\mathsf{QLS}^{\mathfrak{r}}$ is QLS without using restarts and $\mathsf{QLS}^{\mathfrak{w}}$ corresponds to QLS without using the weights. Hence, for $\mathsf{QLS}^{\mathfrak{w}}$, the selected neighbor is one of the elements of the whole set of best neighbors. $\mathsf{QLS}^{\mathcal{T}}$ corresponds to QLS without using a triangulation G of the considered QCN $\mathcal{N}$. In an equivalent manner, $\mathsf{QLS}^{\mathcal{T}}$ is QLS for which the parameter G is the complete graph over V, where V is the set of variable of the input QCN $\mathcal{N}$.

Function randomScenario(G)

in	: A graph $G = (V, E)$.	
output	: A consistent G-scenario on V randomly generated.	

19 **begin**
20     $\mathcal{N} \leftarrow B_V$; /**   */
21   [fB_V is the QCNon V whose each constraint is B
22   **foreach** $(v, v') \in E$ **do**
23   Select randomely a base relation $b \in \mathcal{N}[v, v']$ such that $\overset{\diamond}{G}(\mathcal{N}_{[v,v']/\{b\}})$ is not trivially inconsistent;
24        $\mathcal{N} \leftarrow \overset{\diamond}{G}(\mathcal{N}_{[v,v']/\{b\}})$
25   **return** $\mathcal{N}$;
26 **end**

---

5 Experiments

We considered the QCNs from IA used in the experiments reported in [6]. These 280 QCNs were randomly generated using the model $A(n, d, s)$ (proposed in [12]), with n being the number of variables of the generated QCNs, d the density of constraints defined by a relation other than the trivial relation (*i.e.* B), and s the average number of base relations in each constraint. The parameters used to generate the considered QCNs are $n = 20$, d varying from 8 to 14.5 with a step of 0.25, and $s = 6.5$. For each considered value of d, 10 instances were generated. Concerning triangulations of the constraint graphs of QCNs, we also use the same ones as those used in [6]. These triangulations were generated using a greedy triangulation algorithm (cf. the GreedyFillIn heuristic [3]). Moreover, QLS and its different variations have been implemented in Java. The main objective of our experiment is to validate the main different ingredients of the method QLS, *i.e.*, the use of (1) triangulations of the constraint graphs of the QCNs, (2) the proposed heuristic based on weights on the constraints, and (3) restarts. In order to do this, we compare QLS with its different variations. Note that for this comparison we used the value 5 for the parameter nbDivLoops and the same value for the parameter nbRestartDiv (the best values among the tested values). Moreover, we used the exact optimal number of unsatisfied constraints for the parameter expectedValue.

The first part of our analysis concerns the optimal number of unsatisfied constraints. Figure 3a presents the average of the optimal number of unsatisfied constraints found by the different methods for a maximal number of iterations equal to 4000 which corresponds approximatively to a 15-min timeout. The optimal values found are close enough to the exact optimal values. We note that QLS and QLS^T outperform QLS^R and QLS^W. Further, QLS is slightly better than QLS^T. This observation is reinforced when considering the number of exact optimal values found by each method. Indeed, for QLS, QLS^R, QLS^W, and QLS^T, this number is respectively 189, 58, 93, and 160. Moreover, note that by considering all the methods, 214 exact values are found.

The second part of our analysis concerns a more precise comparison between the two methods QLS and QLS^T. 280 QCNs for which QLS and QLS^T obtain the same optimal values. Clearly, in general QLS takes largely less time than

QLS^T to obtain the same optimal values. By examining the total number of scenarios computed during the neighbor generation step (not reported due to lack of space), we note that the respective numbers are close enough to one another regarding QLS and QLS^T. From this, we can explain the better performance of QLS considering time, due to the fact that the computation of the neighbors during the generation step is faster than that of QLS^T as the use of triangulations in the former case allows considering less constraints in general.

The last part of our analysis concerns a comparison between our approach and the one proposed in [6] which encodes MAX-QCN into the partial maximum satisfiability problem (PMAX-SAT) using a notion called *forbidden coverings*. We do not give details about this approach, nevertheless, we mention that we used the encoding referenced by $\mathcal{C}_{\mathsf{FCTEX}}^{5,8}$ in [6] for our experiments and that we will use the term of *complete method* to refer to the corresponding solving method. We ran the QLS method and the complete method with a 9000-second timeout. The number of QCNs for which we obtain the exact optimal value is 261 for QLS and 279 for the complete method. Regarding the solving time, in general the complete method outperforms the QLS method for the QCNs with a density of non trivial constraints less than 12.5. Focusing on the 80 instances with a density greater or equal to 13.0, for 27 instances the time needed by QLS to find the exact optimal value is less than that needed by the complete method.

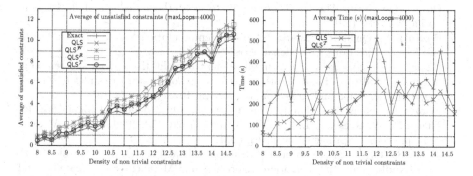

Fig. 3. (left) Optimal number of unsatisfied constraints; (right) average time for the instances of $A(20, d, 6.5)$ with $d \in \{8, \ldots, 14.75\}$.

6 Conclusions

To solve MAX-QCN, we propose a local search based method called QLS which is generic in the sense that it can be used in the context of numerous qualitative calculi. Preliminary experimentation shows the interest of our approach, in particular the useful of the different ingredients components of QLS: the using of traingulations of the constraint graphs, the using of restarts and the using of an heuristics based on weights on the constraints. Future work consists of conducting experiments with other calculi than IA and with large QCNs. Concerning

QLS, another perspective consists in the using of other neighbours that this one used currently. Particularly, we envisage to consider the neighbours computed from the moving of two or several variables.

References

1. Allen, J.F.: Maintaining knowledge about temporal intervals. Commun. ACM **26**(11), 832–843 (1983)
2. Bhatt, M., Guesgen, H., Wölfl, S., Hazarika, S.: Qualitative spatial and temporal reasoning: emerging applications, trends, and directions. Spat. Cogn. Comput. **11**, 1–14 (2011)
3. Bodlaender, H.L., Koster, A.M.C.A.: Treewidth computations I. Upper bounds. Inf. Comput. **208**, 259–275 (2010)
4. Chmeiss, A., Condotta, J.F.: Consistency of triangulated temporal qualitative constraint networks. In: Proceedings of 23rd International Conference on Tools with Artificial Intelligence (ICTAI 2011), Boca Raton, FL, USA, pp. 799–802 (2011)
5. Condotta, J.F., Mensi, A., Nouaouri, I., Sioutis, M., Saïd., L.B.: A practical approach for maximizing satisfiability in qualitative spatial and temporal constraint networks. In: Proceedings of 27th IEEE International Conference on Tools with Artificial Intelligence (ICTAI 2015). IEEE (2015)
6. Condotta, J., Nouaouri, I., Sioutis, M.: A SAT approach for maximizing satisfiability in qualitative spatial and temporal constraint networks. In: KR (2016)
7. Diestel, R.: Graph Theory. Graduate Texts in Mathematics, vol. 173, 4th edn. Springer, Berlin (2012)
8. Hazarika, S.: Qualitative Spatio-Temporal Representation and Reasoning: Trends and Future Directions. IGI Global, Hershey (2012)
9. Huang, J.: Compactness and its implications for qualitative spatial and temporal reasoning. In: Proceedings of 13th International Conference Principles of Knowledge Representation and Reasoning (KR 2012), Rome, Italy (2012)
10. Ligozat, G., Renz, J.: What is a qualitative calculus? A general framework. In: Zhang, C., W. Guesgen, H., Yeap, W.-K. (eds.) PRICAI 2004. LNCS (LNAI), vol. 3157, pp. 53–64. Springer, Heidelberg (2004)
11. Martí, R., Laguna, M., Glover, F.: Principles of tabu search. In: Gonzalez, T.F. (ed.) Handbook of Approximation Algorithms and Metaheuristics. Chapman and Hall, London (2007)
12. Nebel, B.: Solving hard qualitative temporal reasoning problems: evaluating the efficiency of using the ORD-horn class. Constraints **1**, 175–190 (1997)
13. Randell, D., Cui, Z., Cohn, A.: A spatial logic based on regions and connection. In: Proceedings of 3rd International Conference on Principles of Knowledge Representation and Reasoning (KR 1992), pp. 165–176 (1992)
14. Renz, J., Nebel, B.: Qualitative spatial reasoning using constraint calculi. In: Aiello, M., Pratt-Hartmann, I., Van Benthem, J. (eds.) Handbook of Spatial Logics, pp. 161–215. Springer, Berlin (2007)
15. Sioutis, M., Koubarakis, M.: Consistency of chordal RCC-8 networks. In: ICTAI (2012)

Forming Student Groups with Student Preferences Using Constraint Logic Programming

Grace Tacadao[✉] and Ramon Prudencio Toledo

Computer Science Department, Ateneo de Davao University, Roxas Avenue,
8016 Davao, Philippines
{gstacadao,rpstoledosj}@addu.edu.ph,
http://www.addu.edu.ph

Abstract. Forming student groups must be carefully planned for a successful collaborative work. Since there is no consensus in the literature and in practice as to the strategy and parameters to use, a strategy that takes the teachers' and the students' perspectives was developed in this preliminary study. Furthermore, a program based on this strategy was also written using Constraint Logic Programming (CLP). The parameters and conditions to use were obtained through a faculty survey and student interviews. Based on the results, the faculty does not regard teammate preferences as important while students prefer that these are given the utmost consideration. Thus, cohorts produced are not only evaluated based on satisfied constraints but also on satisfied teammate preferences. Hence, the study demonstrates not only that CLP can be applied in the field of computer-supported group formation but also that a grouping strategy can both include parameter constraints and teammate preferences.

1 Introduction

Working in groups for a class work or project is a critical part of a student's learning. The student should be comfortable enough with teammates so that team members are driven to contribute their ideas and eventually complete the required tasks.

Thus, the task of the teacher is to form groups whose members are well-matched with each other. Unfortunately, some teachers resort to form arbitrary groups by letting students choose their members or by selecting students at random to belong to the same group. But such strategies defeat the purpose since mismatched and sometimes opposing students end up on the same team. Those who consider the group formation really well, on the other hand, formulate a set of conditions inferred from student parameters, which can be their learning styles [1, 2], thinking styles [3], Myers-Briggs type indicators (MBTI) [4], programming styles [5], personality traits [6], Belbin team roles [3], grades in previous subjects [3, 6] and preferences to work with another [7]. Some teachers also include age, gender and nationality. Groups are then formed by imposing heterogeneity or homogeneity on each parameter used.

But the choices for the parameters to use also vary. This is intuitive since learning also varies relative to the culture and background of the students. Several researchers asserted that technical skills and grades in previous subjects are the most relevant factors [6, 8] while a few claimed that personal traits and relationships should have more

© Springer International Publishing Switzerland 2016
C. Dichev and G. Agre (Eds.): AIMSA 2016, LNAI 9883, pp. 259–268, 2016.
DOI: 10.1007/978-3-319-44748-3_25

bearing, particularly in the Eastern setting [9]. While there may be disparities in findings, there is however an agreement that any group formation strategy is better so long as it is not purely student-selected or based on random selection.

However, we further the above claim by asserting that a conceivably better strategy should not only consider the conditions imposed by the instructor but also the preferences of the students. Thus, in this paper, we present the generalized instructor-based model developed using Constraint Logic Programming (CLP) that is evaluated not only in terms of the constraints satisfied based on the parameters but also of the preferences of the students involved.

The rest of the paper is structured as follows. Section 2 describes the Group Formation Problem. Section 3 presents the methodology used in this research. Section 4 contains the survey and interview results, Sect. 5 describes the CLP model and Sect. 6 the experimental results. Lastly, the conclusions and future works are described in Sect. 7.

2   Group Formation Problem

The group formation problem generally involves a set of students $S = \{s_1, s_2, s_3, ..., s_{|s|}\}$ and a set of groups $G = \{g_1, g_2, g_3, ..., g_{|g|}\}$, where a predetermined number of students in S are allocated to a group in G [8] using predetermined parameters and are subject to constraints.

Group formation strategies depend on the initiator, the nature and duration of the activity and the group composition. Since most of the instances that appeared in the literature are those initiated by the instructor, strategies then vary according to the individual parameters that were factored in and the mixture method used for each parameter. These parameters may be categorized into the following:

- The *individual proficiency* may be obtained through the student's previous grades and test scores or through a pre-group work exam [10]. Technical skills, such as programming, can be similarly obtained for engineering projects [11].
- *Learning and thinking styles* are also natural choices to be included in the strategy. Mixing learning styles in a group may cause some tension [2] while groups with similar styles will likely produce the greatest individual progress [5].
- Groups who exhibited all the necessary skills might still fall short if the members do not take *team roles*. Falcão e Cunha et al. [12] classified the students based on their group role profiles while Ounnas et al. [3] used Belbin roles.
- Graf and Bekele [6] also considered *personality traits*, based on expert opinions and consultation with colleagues. Bekele [10] also showed that heterogeneous groups performed better than those formed randomly or using student selection.
- Thanh and Gillies [9] observed that *teammate preferences* is ideal in an Asian setting. 35 % of their students favored grouping based on pre-existing friendship since the students claimed they can work better in a positive group atmosphere.

The group formation problem is also two-fold. First, there is the issue on the appropriate grouping strategy to achieve the objectives of an activity. The second aspect deals with the computational modelling of the problem. Since most of the suggested strategies are too complex to do manually, most researchers include an automated solution to implement the strategy in their studies.

3   Methodology

The study proceeded in two phases where the group formation strategy was developed in the first phase and then modelled as a constraint logic program in the second phase.

The parameters and conditions that will be used in the strategy were initially obtained from a faculty survey, where 17 faculty members of the computer-related programs in the *Ateneo de Davao University* are involved. The questionnaire was divided into four parts: (1) the respondent profile, (2) the class profile, (3) the group formation practice of the faculty and (4) his/her perception on group formation.

The survey answers were then validated through student interviews with 15 students from classes that were taught by the faculty respondents. The main objective of the interview is to find out the parameters that students believe to have a positive influence to group formation for long-term student projects.

Faculty respondents were asked to rank the order in which of the five student parameters should be considered when forming groups for long-term collaborative activities. The said five parameters that figured most in the literature were (1) individual proficiency, (2) learning and thinking styles, (3) team roles, (4) personality traits and (5) student preferences. They were asked to choose between heterogeneous and homogenous mixture for each of the parameters and rank the resulting mixture methods. The results were then compared against those of the interviews.

To determine the ordinal ranking of both student parameters and mixture methods, Copeland's pairwise aggregation method was used. The agreement of the respondents was also measured using Kendall's Coefficient of Concordance W. To test the significance of W, Friedman's chi-square statistic and the F statistic were used.

A thematic approach was used in the analysis of the interview data. This will be sufficient for identifying and summarizing the recurring notions mentioned in the interviews.

In the next phase, the problem was modelled in ECLiPSe 6.1. Each cohort that the model produces is then evaluated using a metrics framework for group formation based on Ounnas et al. [14].

4   Group Formation Strategy

Based on the survey results, the respondents considered the individual proficiency as the most preferred characteristic that should be factored in when instructors form student groups. This is followed by the learning and thinking styles, team roles, personality traits and lastly, the student preferences.

Agreement between responses however is low at .34. The results also do not indicate a clear consensus yet for all parameters in general. However, for individual proficiency, data points tend to gather around ranks 1 and 2 only, which indicates that respondents most likely prefer individual proficiency over the three other parameters.

Respondents were also asked between heterogeneous and homogenous mixtures for each of the student parameters. Respondents prefer heterogeneous mixture for individual proficiency and team roles. It is also more preferred for learning and thinking styles but homogenous mixture is preferred for personality traits.

They were also asked to rank all the possible combinations of mixture methods and student parameters. The final ranking puts heterogeneous individual proficiency in the first place, heterogeneous learning and thinking styles in the second place, heterogeneous team roles in the third place and homogenous personality traits in the last place. The said results are consistent with the ranking of student parameters and the results of the mixture methods for each parameter. The degree of concordance between rankings is also very low at .18.

On the other hand, the interview results indicate that students still prefer student selection more than instructor selection. Although this is contrary to the thesis of the paper, this is in fact expected since students want to be in the same group as their peers who can potentially help them to either learn the course or get good grades.

However, if the instructors are to form the groups, a possible arrangement, as suggested by the frequency of the students' responses, may be as follows: (1) student preferences, (2) individual proficiency and (3) team roles. The students also prefer a heterogeneous grouping on individual proficiency and team roles. In fact, whatever parameters may be used, the students prefer that a heterogeneous mixture is imposed.

The lack of any decisive consensus indicates that the perceived differences between the parameters and mixture methods in terms of their effects on group dynamics are not yet generalized.

As such, the group formation strategy must be generic such that the instructor chooses the parameters and ranks them as he/she deems appropriate. But, as the interview results suggest, it must consider the teammate preferences of the students.

5   CLP Model

In the present study, the group formation problem was modeled as an assignment problem where each student is assigned a group.

The main predicate, named `form_groups`, accepts five arguments: the list of input files, the maximum group size, the new student list, group variable list and the cohort rating on each constraint.

5.1   Input Files

The input to the model includes the parameters file and the students file. Since the model is generic, the user must list the parameters as facts in the order that they should be considered in the strategy. Each parameter fact must contain the ID, name, output type (single or multiple) and data type (integer or symbolic). It must also contain the list of all possible values, the mixture method (heterogeneous or homogenous) that denotes its constraint, a yes/no to indicate whether it needs pre-processing and similarity index, which refers to the number of values that are considered similar. An example of the parameter fact is as follows:

```
parameter(1, "Individual proficiency", single, [below,
   average, above], heterogeneous, symbolic, yes, 0).
```

The students file, on the other hand, contains the three attributes: ID, the list of student's parameters and the student IDs of the student's preferred group mates. An example of the student fact is as follows:

```
student(0006, [84, [-9,1,3,3], shaper, [intuition,
   thinking, perceiving]], [34,18]).
```

Note that values in the parameters list do not correspond with those in the parameters file since the values will still undergo preprocessing.

5.2   Constraints

As mentioned, the first basic constraint restricts students from being assigned to multiple groups. This is achieved by a call to the predefined `alldifferent` constraint, which restrains the Team ID variables to be different for at most `GroupSize` times.

Parameters go through the optional preprocessing phase, as shown in the Fig. 1, where parameters that must be preprocessed are categorized and, in effect, will get new values based on the categorization. Some parameters have a lot of possible values that would make setting its constraint difficult since the solver will have to deal with all possible combinations. They are preprocessed using the corresponding facts in the category file.

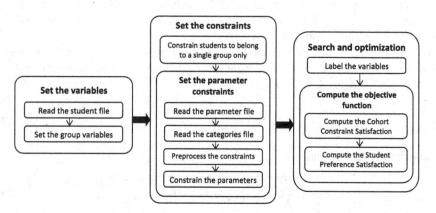

Fig. 1. The general flow of the `form_groups` predicate is shown in this diagram. It follows the "constrain and generate" methodology, where variables and their domains are first defined, then the constraints are modeled and lastly, a search and optimization strategy is used.

Each category fact contains the parameter ID, the category ID and category name. It also includes the inclusion condition, which indicates whether only *one* value, any value *between* a range or *all* values must be satisfied for the parameter to be categorized correctly. Lastly, it also contains the list of parameter values that belong to the category. It contains all possible values if the inclusion condition is *one*, the first and last values in a range if *between* and all values that must be included if *all*. The following are the categories for the individual proficiency parameter presented above.

```
category(1,1,below,between,[0,79]).
category(1,2,average,between,[80,89]).
```

Once parameter preprocessing is completed, the parameter is modified in the students list. The following is the categorized version of the sample student fact from above.

```
student(6, [average, [reflective, intuitive, verbal,
global], action, rational]),
```

Since the model is generic, users can define parameters using two possible types and two mixture methods. Thus, there are thus four constraint types in general: (1) symbolic homogeneity, (2) symbolic heterogeneity, (3) integer homogeneity and (4) integer heterogeneity. Either of the following two predefined constraints is then imposed: `all_eq` from the `gfd` module for homogenous mixture and `alldifferent` from the `ic_global` module for the heterogeneous mixture.

5.3 Objective Function

Cohort Constraint Satisfaction Quality. Defining an objective function appeared as an intuitive choice using the metric framework by Ounnas et al. [14]. The framework takes into account the number of constraints satisfied, the perceived satisfaction, group productivity in the task and the overall formation goal. Each group formed in the cohort as well as the entire cohort will be evaluated.

Since the formation strategy was not yet used in an actual class for a particular collaborative activity, only the cohort constraint satisfaction quality metric was measured for every constraint defined. It is computed as

$$f_{cG}(c_j) = \frac{1}{1+\sigma}. \tag{3}$$

where the standard deviation σ

$$\sigma = \sqrt{\frac{1}{N} \sum_{i=1}^{N} \left(f_{cg}(g_i, c_j) - \overline{f_{cg}} \right)^2}. \tag{4}$$

and the mean

$$\overline{f_{cg}} = \frac{1}{N} \sum_{i=1}^{N} f_{cg}(g_i, c_j). \tag{5}$$

Using this framework, the success of group formation is represented by the satisfaction of the goals, which were defined as constraints in the problem model.

To get the $f_{cG}(g_i, c_j)$ of every group g_i for constraint c_j, a point is added to the rating depending on the mixture method and whether the parameters matched. For integer parameters, the difference between two consecutive values is obtained and checked against the similarity index.

For parameters with multiple output, the sum of the individual ratings is obtained and is checked against the similarity index. If the sum is greater or equal to the similarity index, rating is set to one; otherwise, it is set to zero. But if the index was not indicated at all, the rating is only set to one if all the values are set to one.

The sum of all the $f_{cG}(c_j)$ values is then obtained and maximized for the optimal solution to be found.

Student Preferences Satisfaction. The constraint on student preferences was not included in the computation of the constraint satisfaction quality since the natures of both types of constraints are different. Student preferences are specified for each student while parameter constraints are applied on groups.

It was instead made an objective function that simply counts the average satisfied preferences for each student. It is thus defined as

$$f_{SP} = \sum_{i=1}^{N_s} \frac{1}{M} \sum_{j=1}^{M} f_{sp}(s_i, p_j). \tag{6}$$

where N_S is the number of students, M is the number of preferred teammates and $f_{sp}(s_i, p_j)$ is defined as

$$f_{sp}(s_i, p_j) = \begin{cases} 1 & \text{if } p_j \text{ is a teammate of } s_i \\ 0 & \text{if } p_j \text{ is not a teammate of } s_i \end{cases} \tag{7}$$

Regardless of the number of preferred group mates, the perfect score is always equal to the total number of students. The user-defined `rate_team` predicate counts the number of satisfied preferences. To do so, the team ID of the student is checked against that of the preferred team mates. If both match, a point is added.

6   Experimental Results

6.1   Cohort Satisfaction Quality

As indicated in Fig. 1, the final phase is the optimization and search. Each of the seven search methods provided by ECLiPSe was tested using 50 randomly generated student 50 records and 4 parameters for group sizes from 2 to 8. Each query is run ten times using each search method (Fig. 2).

One-Way Analysis of Variance (ANOVA) was then used to identify whether one of these search methods generated a solution faster than the rest of the methods. However, the p value is very close to 1.0 at .99, which connotes that the difference of the means

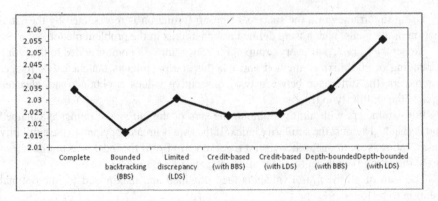

Fig. 2. Means plot of CPU time. This figure shows the average CPU time that each query takes after it was ran ten times.

of the populations is very close to 0. Thus, since there is no clear difference between the search methods, this implies that any of them can in fact be used for the problem context.

Since we are interested in the best grouping, the predefined bb_min predicate from the branch_and_bound library is called to query the solution for the constraint logic program. Given that there are no pronounced differences between the search methods, as concluded in the previous experiment, only the four search methods with the lowest mean CPU time were attempted. They were used to form groupings with sizes from 2 to 6 (Table 1).

Table 1. Experimental results of the cohort rating for group size = 2

Search method	CPU time	Cohort rating
LDS	11 041.77 s	3.54
CBS with LDS	–	3.67
CBS with BBS	–	3.59
BBS	–	3.48

There were queries however that did not return the most optimal solution even after running uninterrupted for a 24-hour period or longer. The CPU time for these queries was indicated with a dash (-) in the table above.

Moreover, it is worthy to note that the cohort ratings of the solutions shown are greater than 3.0, which implies that 3 of the 4 parameter constraints were satisfied.

6.2   Student Preferences Satisfaction

A similar experiment was also performed using the same group sizes and search methods but with the student preference metric minimized instead of the cohort satisfaction quality (Table 2).

Table 2. Experimental results of the student preferences for group size = 6

Search method	CPU time	Student preferences
LDS	172.72 s	8.00
CBS with LDS	–	10.50
CBS with BBS	–	11.50
BBS	–	8.50

The student preferences are quite low since they indicate the number of students whose preferences were satisfied. Based on these results, 11 student preferences were only satisfied out of 50.

7   Conclusions

In this paper, we presented the results of the survey conducted to the faculty of computer-related programs on the practice and perception regarding group formation. There was very little agreement among the faculty respondents in the ranking of student parameters and mixture methods.

Based on the interview results, students unsurprisingly preferred student selection more than instructor selection. However, if instructor selection will be used, students would prefer that their teammate preferences are given the highest consideration. The students also agree that heterogeneity is more aptly imposed on individual proficiency, should it be included as a factor in group formation.

Since both the survey and interview results do not agree, we thus suggest that instructors take the student preferences into consideration, no matter what parameters they may include in their strategy.

In this paper, we also presented the use of Constraint Logic Programming as an approach to computer-supported group formation using a number of student parameters as constraints. The experimental results using the search methods in ECLiPSe showed that there is no search method that can produce solutions faster than the others.

Since the two costs (i.e., constraint satisfaction quality metric and student preference metric) used in the model are disparate, one cost may be sacrificed for the other. An optimal solution in terms of the constraint satisfaction quality metric may not be optimal in terms of the student preference and vice versa. An objective function that includes both metrics would have been a better approach. Moreover, student preferences can also be modeled as soft constraints while maintaining the student parameters as hard constraints.

The other contribution of this study is the generic model for the instruction selection strategy that permits instructors to define the symbolic and integer parameter constraints. Initially, only parameters of type symbolic and integer can be defined using the model although other types, such as floating-point, may be added in the future.

Acknowledgments. This work is an extension of the paper originally reported in the *Proceedings of the 15th IEEE International Conference on Advanced Learning Technologies* [15].

This research has been funded by the University Research Council of the Ateneo de Davao University.

References

1. Paredes, P., Ortigosa, A., Rodriquez, P.: TOGETHER: an authoring tool for group formation based on learning styles. In: Proceedings of A3H 7th International Workshop on Authoring and Adaptable Hypermedia at EC-TEL 2009, Nice, France (2009)
2. Nielsen, T., Hvas, A., Kjaergaard, A.: Student team formation based on learning styles at university start: does it make a difference to the students? Reflecting Educ. 5(2), 85–103 (2009)
3. Ounnas, A., Millard, D., Davis, H.: A framework for semantic group formation in education. Educ. Technol. Soc. 12(4), 43–55 (2009)
4. Jensen, D., Feland, J., Bowe, M., Self, B.: A 6-Hats based team formation strategy: development and comparison with an MBTI based approach. In: Proceedings of the American Society for Engineering Education (ASEE) Annual Conference, St. Louis, Missouri (2000)
5. Adán-Coello, J., Tobar, C., de Faria, E., de Menezes, W., de Freitas, R.: Forming groups for collaborative learning of introductory computer programming based on students' programming skills and learning styles. Int. J. Inf. Commun. Tech. Educ. 7(4), 34–46 (2011)
6. Graf, S., Bekele, R.: Forming heterogeneous groups for intelligent collaborative learning systems with ant colony optimization. In: Ikeda, M., Ashley, K.D., Chan, T.-W. (eds.) ITS 2006. LNCS, vol. 4053, pp. 217–226. Springer, Heidelberg (2006)
7. Munson, S., Hutchins, G.: Maximizing satisfaction in group formation. In: Sean Munson's Website. http://www.smunson.com/portfolio/projects/groupform/paper.pdf (Accessed 2006)
8. Ani, Z., Yasin, A., Husin, M., Hamid, Z.: A method for group formation using genetic algorithm. Int. J. Comput. Sci. Eng. 2(9), 3060–3064 (2010)
9. Thanh, P., Gillies, R.: Group composition of cooperative learning: does heterogeneous grouping work in asian classrooms? Int. Educ. Stud. 3(3), 12–19 (2010)
10. Bekele, R.: Computer assisted learner group formation based on personality traits. In: Ph.D. Dissertations. University of Hamburg, Germany. http://www.sub.unihamburg.de/opus/volltexte/2006/2759 (Accessed 2005)
11. Strnad, D., Guid, N.: A fuzzy-genetic decision support system for project team formation. Appl. Soft Comput. J. (2008)
12. Falcão de Cunha, J., Borges, J., Dias, T.: Some results from managing the process of group formation and evaluation in student projects. In: Proceedings of the International Conference on Engineering Education (ICEE 2007), Coimbra, Portugal (2007)
13. Patton, M., Cochran, M.: A Guide to Using Qualitative Research Methodology (2002)
14. Ounnas, A., Millard, D., Davis, H.: A metrics framework for evaluating group formation. In: Proceedings of the 2007 International ACM Conference on Supporting Group Work, Sanibel Island, Florida, USA, pp. 221–224 (2007)
15. Tacadao, G., Toledo SJ, R.P.: A generic model for the group formation problem Using constraint logic programming. In: 2015 IEEE 15th International Conference on Advanced Learning Technologies (ICALT), Hualien, pp.307–308 (2015)

Intelligent Agents and Planning

InterCriteria Analysis of Ant Algorithm with Environment Change for GPS Surveying Problem

Stefka Fidanova[1]([✉]), Olympia Roeva[2], Antonio Mucherino[3],
and Kristina Kapanova[1]

[1] IICT, Bulgarian Academy of Sciences,
Acad. G. Bonchev Str., bl. 25A, 1113 Sofia, Bulgaria
`stefka@parallel.bas.bg, kkapanova@gmail.com`
[2] IBFBMI, Bulgarian Academy of Sciences,
Acad. G. Bonchev Str., bl. 105, 1113 Sofia, Bulgaria
`olympia@biomed.bas.bg`
[3] IRISA, University of Rennes 1, Rennes, France
`antonio.mucherino@irisa.fr`

Abstract. In this paper we apply InterCriteria Analysis (ICrA), which is based on the apparatus of Index Matrices and Intuitionistic Fuzzy Sets. We apply ICrA on the well-known Ant Colony Optimization (ACO) general framework including environment change. The environment is simulated by means of the Logistic map, that is used in ACO for perturbing the update of the pheromone trails. We compare different levels of perturbation of the one of the most important parameters in ACO – the pheromone. Based on ICrA we examine the obtained identification results and discuss the conclusions about existing relations and dependencies between defined criteria, defined, in terms of ICrA.

Keywords: InterCriteria Analysis · Ant Colony Optimization · GPS surveying

1 Introduction

InterCriteria Analysis (ICrA) is an approach [1] aiming to go beyond the nature of the criteria involved in a process of evaluation of multiple objects against multiple criteria, and, on this basis, to discover any existing correlations between the criteria themselves. Given in details in [1], ICrA has been further applied for the purposes of analyses of an economic case-study [5], bioprocess modelling [13], modular neural network preprocessing [16], etc.

In this paper ICrA has been applied for analysis of an Ant Colony Optimization (ACO) algorithm with environment change [12] to provide near-optimal solutions for Global Positioning System (GPS) surveying problem. Every metaheuristic is developed in order to find the best trade-off between two main concepts: *diversification* and *intensification* [17]. While the first one tries to widely

© Springer International Publishing Switzerland 2016
C. Dichev and G. Agre (Eds.): AIMSA 2016, LNAI 9883, pp. 271–278, 2016.
DOI: 10.1007/978-3-319-44748-3_26

extend the search in the domain of the optimization problem, the second one improves candidate solutions by focusing on local neighbors of the current best known solutions. ACO guarantes the diversification by the simulation of the typical ant behavior, while for intensification is applied a local search to a set of candidate solutions. In the considered ACO algorithm with environment changes [12] the pheromone updates are not supposed to be performed only on the basis of found solutions, but also on the basis of the current environment "surrounding" the ants. We will simulate environment changes by employing the Logistic map [19]. The Logistic map will be introduced with different parameters. The main purpose of this paper is ICrA of ACO with different perturbation parameters with the aim to learn the influence of perturbation parameters on ACO performance.

   This paper is organized as follows. In Sect. 2, we describe ACO approach for managing environment changes, which affects the fitness values used in the pheromone updating rule during the execution of ACO. We refer to the extended ACO as environmental ACO ($eACO$). In Sect. 3, we describe the problem we consider for testing our $eACO$, the GPS Surveying Problem (GSP). In Sect. 4 we apply ICrA and analyze the relations between perturbation parameters. Finally, in Sect. 5, we give some conclusions.

2   Including Environment Changes in ACO

Ants foraging for food deposit a substance named pheromone on the paths they follow. An isolated ant would just move randomly. The ants are stimulated to follow the paths with stronger pheromone concentration. During the time, the pheromone trails are reinforced around the optimal ones and the probability for the other ants to follow optimal paths increases. The repetition of this mechanism represents the auto-catalytic behavior of ant colonies in nature [6,8].

   ACO is inspired by this ant behavior. A colony of artificial ants working into a mathematical space is simulated. The ants search for candidate solutions of a given optimization problem, while possible paths are marked by artificial pheromone for guiding other ants in the regions of the search space where good-quality solutions were found. In ACO the ants generally create a sort of environment by themselves, by depositing the pheromone on marked paths. This environment is perturbed by means of the Logistic map [12,19]. The ants' search space is represented by graph, called graph of the problem, such that the solution to be represented by a path in the graph. Thus we look for a shorter path in the graph under some conditions, if we solve minimization problem. The ants decide on which edge to walk using probabilistic rule called transition probability. The transition probability p_{uv}, when the ants need to decide on which edge (u, v) to walk, is based on the heuristic information η_{uv} and on the current pheromone level τ_{uv}:

$$p_{uv} = \tau_{uv}^{\alpha} \eta_{uv}^{\beta} \left[\sum_{(u,w) \in E_S \, : \, w \not\subset X} \left(\tau_{uw}^{\alpha} \eta_{uw}^{\beta} \right)^{-1} \right], \tag{1}$$

where α and β are transition probability parameters. At the beginning the value of the pheromone on all elements is the same and is set to a small positive value between 0 and 1. After every iteration the pheromone is updated. First the value of the pheromone is decreased with a small constant called evaporation and after a new, proportional to the quality of the solution, pheromone is added.

The Logistic map is a quadratic dynamical equation proposed in 1938 as a demographic model [19].

$$x_{n+1} = r x_n (1 - x_n), \qquad n > 0, \tag{2}$$

where x_n represents the population size at time n and r is a constant, named growth coefficient. Given $x_0 \in [0,1]$ and a value for $r \in [0,4]$, this dynamical equation can either converge or be chaotic. In the first case, given any $x_0 \in [0,1]$, x_n tends to the so-called "attraction domain". In the second case, x_n never converges, but it can rather take, in an apparent random way, any possible value in the range $[0,1]$.

For small values of r, the Logistic map always converges to one single point, i.e. the attraction domain consists of one point only. The first bifurcation appears when $r = 3$, where the attraction domain consists of 2 points; then there is another bifurcation when $r = 1 + \sqrt{6}$, so that the attraction domain consists of 4 points. For larger values for r, the Logistic map experiences other bifurcations, and it can be chaotic for some subintervals of r. However, in these chaotic regions, it is still possible to identify regular attraction domains. Other regular attraction domains can be identified by looking at tighter subintervals of r, as well as other copies of the entire graphic.

We simulate regular and chaotic changes of environment in ACO by introducing the Logistic map in Eq. (2), which is used in ACO for updating the pheromone trails. In the hypothesis the objective function $F(Y)$, of the considered problem is positive and greater than 1, the term $1/f(Y)$ in Eq. (3) has always values ranging between 0 and 1, Y is the problem solution. It can therefore take the place of x_0 in the Logistic map, so that a perturbed value x_1 can be computed, for a given value of r in $[0,4]$. The equation for updating the pheromone therefore becomes:

$$\tau_{uv} = \tau_{uv} + r \cdot \frac{1}{f(Y)} \cdot \left(1 - \frac{1}{f(Y)} \right). \tag{3}$$

With this simple change in the rule for updating the pheromone, we artificially perturb the environment of the ants, which would otherwise only depend on the solution fitness values. Different values for r can produce different environment changes, depending on the behavior of the Logistic map. For values of r for which the Logistic map converges, the pheromone levels added to τ_{uv} tend to be constant, reducing in this way the effects of good-quality solutions, that might mislead the ants towards a local optimum. For values of r for which the Logistic map behaves instead chaotically, the environment is dominant on the choices of the ants, as the pheromone update mostly depend on the simulated environment, rather than on the actual fitness value. The idea is tested on GPS surveying problem.

3   GPS Surveying Problem

GPS consists of a certain number of satellites that constantly orbit around earth and that are provided with sensors able to communicate with machines located on earth [9]. Since such a machine lies over a sphere that does not contain the satellite, a very precise information about the distance between the earth surface and the satellite would allow for determining the precise location of the machine on earth [10,11].

GPS technology can in fact provide very accurate locations for all sensors forming a given sensor network. The related costs can be too high when the network is large. For this reason, researchers have been trying to design and install local ground networks having the task of recording satellite signals with the aim of decreasing the overall network functioning cost [7,15]. A network is composed by a certain number of *receivers* working in different *stations* at different times. Therefore, given a certain number of *sessions*, representing the temporarily assignment of a given number of receivers to a set of distinct stations, the problem is to find a suitable *order* for such sessions for reducing the overall cost. This cost is in fact strictly related to the order of the sessions, because receivers need to be moved from one station to another, when stepping from one session to another. Therefore, the distance between two involved stations is important for the computation of the costs. The session order is also named *session schedule* [18].

The GSP can be formalized as follows. Let $S = \{s_1, s_2, \ldots, s_n\}$ be a set of stations, and let $R = \{r_1, r_2, \ldots, r_m\}$ be a set of receivers, with $m < n$. Sessions can be defined by a function $\sigma : R \longrightarrow S$ that associates one receiver to one station. Considering that no more than one receiver should be assigned to the same station, σ can be represented by an m-vector $(\varsigma_1, \varsigma_2, \ldots, \varsigma_m)$ containing, for each of the m receivers, the labels of the chosen stations. Since $m < n$ (and generally fixed to 2 or 3 in the applications), the number of permutations of m objects from n distinguishable ones is $n!/(n-m)!$, which can be huge when the network is large.

Let C be an $n \times n$ matrix providing the costs $c(\varsigma_u, \varsigma_v)$ for moving one receiver from the station ς_u to the station ς_v. This matrix can be symmetric, however asymmetric case is more realistic.

An instance of the GSP can be represented by a weighted undirected multigraph $G = (V1, E, c)$ where vertices represent sessions σ_v and arcs (σ_u, σ_v) indicate the possibility to switch from session σ_u to session σ_v. The weight associated to the arcs provides the cost $c(\sigma_u, \sigma_v)$ for moving every receiver from the station $\varsigma_{u,i}$ to the station $\varsigma_{v,i}$, for each i:

$$c(\sigma_u, \sigma_v) = \sum_{i=1}^{m} c(\varsigma_{u,i}, \varsigma_{v,i}).$$

The graph G is not simple in general, because it might be feasible to switch from session σ_u to session σ_v, as well as from σ_v to σ_u, but with a different total cost. The problem consists in finding the path on G with minimal total cost, while covering the entire vertex set $V1$ [7].

4   InterCriteria Analysis

Following [1,4], an Intuitionistic Fuzzy Pair (IFP) [2] with the degrees of "agreement" and "disagreement" between two criteria applied on different objects is obtained. An IFP is an ordered pair of real non-negative numbers $\langle a, b \rangle$ such that: $a + b \leq 1$.

Let an Index Matrix (IM) (see [3]) whose index sets consist of the names of the criteria (for rows) and objects (for columns) be given. The elements of this IM are supposed to be real numbers. An IM with index sets consisting of the names of the criteria (for rows and for columns) with elements IFPs corresponding to the "agreement" and "disagreement" of the respective criteria will be obtained.

The set of all objects is denoted by O, and the set of values assigned by a given criteria C to the objects by $C(O)$, i.e.

$$O \stackrel{\text{def}}{=} \{O_1, O_2, \ldots, O_n\}, \ C(O) \stackrel{\text{def}}{=} \{C(O_1), C(O_2), \ldots, C(O_n)\}.$$

Let $x_i = C(O_i)$. Then the following set can be defined:

$$C^*(O) \stackrel{\text{def}}{=} \{\langle x_i, x_j \rangle | i \neq j \ \& \ \langle x_i, x_j \rangle \in C(O) \times C(O)\}.$$

In order to compare two criteria, the vector of all internal comparisons of each criteria which fulfill exactly one of three relations $R, \overline{R}$ and $\tilde{R}$ must be constructed. It is required that for a fixed criterion C and any ordered pair $\langle x, y \rangle \in C^*(O)$ it is true:

$$\langle x, y \rangle \in R \Leftrightarrow \langle y, x \rangle \in \overline{R}, \tag{4}$$
$$\langle x, y \rangle \in \tilde{R} \Leftrightarrow \langle x, y \rangle \notin (R \cup \overline{R}), \tag{5}$$
$$R \cup \overline{R} \cup \tilde{R} = C^*(O). \tag{6}$$

From the above it is seen that only a subset of $C(O) \times C(O)$ has to be considered for the effective calculation of the vector of internal comparisons $V(C)$, since from (4), (5) and (6) it follows that if the relation between x and y is known, the relation between y and x is known as well. Thus, only lexicographically ordered pairs $\langle x, y \rangle$ are considered. Let, for brevity, $C_{i,j} = \langle C(O_i), C(O_j) \rangle$. Then, for a fixed criterion C the following vector is constructed:

$$V(C) = \{C_{1,2}, C_{1,3}, \ldots, C_{1,n}, C_{2,3}, C_{2,4}, \ldots, C_{2,n}, C_{3,4}, \ldots, C_{3,n}, \ldots, C_{n-1,n}\}.$$

It can be easily seen that it has exactly $\frac{n(n-1)}{2}$ elements. Further, to simplify our considerations, the vector $V(C)$ is replaced with $\hat{V}(C)$, where for each $1 \leq k \leq \frac{n(n-1)}{2}$ for the k-th component it is true:

$$\hat{V}_k(C) = \begin{cases} 1 & \text{iff } V_k(C) \in R, \\ -1 & \text{iff } V_k(C) \in \overline{R}, \\ 0 & \text{otherwise} \end{cases}$$

Then, when comparing two criteria, the degree of "agreement" between the two is the number of matching components (divided by the length of the vector for normalization purposes). The degree of "disagreement" is the number of components of opposing signs in the two vectors (again normalized by the length).

The calculation of the degrees of "agreement" ($\mu_{C,C}$) and degrees of "disagreement" ($\nu_{C,C}$) between two criteria C and C' is implemented in Matlab environment according to the algorithm presented in [14].

We use 3 test problems from http://www.informatik.uni-heidelberg.de/groups/comopt/software/TSLIB95/ATSP.html: the test problems range 170 (instance ftv170), 403 (instance rgb403) and 443 (instance rgb443) sessions.

The average results (total cost), achieved over 30 runs of eACO, for every of the test problems and various values of the parameter r are shown in Table 1. The number of iterations is equal to the number of the sessions.

Table 1. GPS results over eACO 30 runs

r	ftv170	rgb403	rgb443
1	3314.20	3413.63	3749.93
2	3313.76	3392.23	3742.86
3	3319.83	3393.76	3742.43
3.25	3345.93	3392.90	3736.16
3.50	3336.16	3395.43	3747.90
3.75	3324.46	3392.16	3756.86
4	3338.53	3386.10	3754.50

We apply ICrA analysis on achieved results with different values of the perturbation parameter r in considered eACO. The objects are the values of the parameter r and the criteria are the values of the objective function. Results degree of "agreement" ($\mu_{C,C}$) are presented in Tables 2, 3 and 4.

Table 2. Degree of "agreement" ($\mu_{C,C}$) for ftv170

	GPS_1	GPS_2	GPS_3	$GPS_{3.25}$	$GPS_{3.50}$	$GPS_{3.75}$	GPS_4
GPS_1	1	0.77	0.775	0.602	0.513	0.508	0.497
GPS_2	0.77	1	0.699	0.517	0.437	0.451	0.600
GPS_3	0.775	0.699	1	0.722	0.462	0.494	0.600
$GPS_{3.25}$	0.602	0.517	0.722	1	0.602	0.639	0.515
$GPS_{3.50}$	0.513	0.437	0.462	0.602	1	0.823	0.453
$GPS_{3.75}$	0.508	0.451	0.494	0.639	0.823	1	0.490
GPS_4	0.497	0.600	0.600	0.515	0.453	0.490	1

Table 3. Degree of "agreement" ($\mu_{C,C}$) for rgb403

	GPS_1	GPS_2	GPS_3	$GPS_{3.25}$	$GPS_{3.50}$	$GPS_{3.75}$	GPS_4
GPS_1	1	0.520	0.579	0.544	0.554	0.577	0.598
GPS_2	0.520	1	0.607	0.497	0.561	0.510	0.506
GPS_3	0.579	0.607	1	0.584	0.490	0.457	0.655
$GPS_{3.25}$	0.544	0.497	0.584	1	0.524	0.526	0.547
$GPS_{3.50}$	0.554	0.561	0.490	0.524	1	0.791	0.499
$GPS_{3.75}$	0.577	0.510	0.457	0.526	0.791	1	0.533
GPS_4	0.598	0.506	0.655	0.547	0.499	0.533	1

Table 4. Degree of "agreement" ($\mu_{C,C}$) for ftv170

	GPS_1	GPS_2	GPS_3	$GPS_{3.25}$	$GPS_{3.50}$	$GPS_{3.75}$	GPS_4
GPS_1	1	0.579	0.549	0.455	0.517	0.561	0.526
GPS_2	0.579	1	0.701	0.529	0.494	0.508	0.529
GPS_3	0.549	0.701	1	0.559	0.531	0.510	0.584
$GPS_{3.25}$	0.455	0.529	0.559	1	0.684	0.575	0.607
$GPS_{3.50}$	0.517	0.494	0.531	0.684	1	0.671	0.582
$GPS_{3.75}$	0.561	0.508	0.510	0.575	0.671	1	0.483
GPS_4	0.526	0.529	0.584	0.607	0.582	0.483	1

Tables 2, 3 and 4 show level of agreement between the results achieved with different values of parameter r. We observe that the most of the parameter r value relations have strong dissonance. It means that the values of the parameter r are not related to each other. There is consonance only between the $r = 3.5$ and $r = 3.75$, because they are very close values and such similar performance is expected. The ICrA shows the correctness of the algorithm and that the perturbation is included in a correct way. There is meaningful to learn the different values of the perturbation, because they are independent.

5 Conclusions

In this paper we have applied ICrA on ACO algorithm with environmental changes – eACO. The test problem which we use is GPS surveying problem. Instances with 170, 403 and 443 sessions have been solved. The ICrA application shows the independence between the values of the parameter r and the correctness of the considered algorithm eACO and the way the environment changes are included.

Acknowledgments. This work was partially supported by two grants of the Bulgarian National Scientific Fund: DFNI-I02/5 "InterCriteria Analysis – A New Approach to Decision Making", and by the grant DFNP-176-A1.

References

1. Atanassov, K., Mavrov, D., Atanassova, V.: Intercriteria decision making: a new approach for multicriteria decision making, based on index matrices and intuitionistic fuzzy sets. Issues IFSs GNs **11**, 1–8 (2014)
2. Atanassov, K., Szmidt, E., Kacprzyk, J.: On intuitionistic fuzzy pairs. Notes IFS **19**(3), 1–13 (2013)
3. Atanassov, K.: On index matrices, part 1: standard cases. Adv. Stud. Contemp. Math. **20**(2), 291–302 (2010)
4. Atanassov, K.: On Intuitionistic Fuzzy Sets Theory. Springer, Berlin (2012)
5. Atanassova, V., Mavrov, D., Doukovska, L., Atanassov, K.: Discussion on the threshold values in the intercriteria decision making approach. Notes Intuitionistic Fuzzy Sets **20**(2), 94–99 (2014)
6. Atanassova, V., Fidanova, S., Popchev, I., Chountas, P.: Generalized nets, ACO-algorithms and genetic algorithm. In: Sabelfeld, K.K., Dimov, I. (eds.) Monte Carlo Methods and Applications, pp. 39–46. De Gruyter, Boston (2012)
7. Dare, P., Saleh, H.A.: GPS network design: logistics solution using optimal and near-optimal methods. J. Geodesy **74**, 467–478 (2000)
8. Dorigo, M., Birattari, M.: Ant colony optimization. In: Sammut, C., Webb, G.I. (eds.) Encyclopedia of Machine Learning, pp. 36–39. Springer, Heidelberg (2010)
9. Hofmann-Wellenhof, B., Lichtenegger, H., Collins, J.: Global Positioning System: Theory and Practice. Springer, Vienna (1993). 326 p
10. Leick, A.: GPS Satellite Surveying, 3rd edn. Wiley, Hoboken (2004). 464 p
11. Liberti, L., Lavor, C., Maculan, N., Mucherino, A.: Euclidean distance geometry and applications. SIAM Rev. **56**(1), 3–69 (2014)
12. Mucherino, A., Fidanova, S., Ganzha, M.: Ant colony optimization with environment changes: an application to GPS surveying. In: FedCSIS 2015, pp. 495–500 (2015). doi:10.15439/2015F33
13. Roeva, O., Vassilev, P., Angelova, M., Pencheva, T.: Intercriteria analysis of parameters relations in fermentation processes models. In: Núñez, M., Nguyen, N.T., Camacho, D., Trawiński, B. (eds.) ICCCI 2015. Lecture Notes in Computer Science, vol. 9330, pp. 171–181. Springer, Heidelberg (2015)
14. Roeva, O., Vassilev, P.: Intercriteria analysis of generation gap influence on genetic algorithms performance. Adv. Intell. Syst. Comput. **401**, 301–313 (2016)
15. Saleh, H.A., Dare, P.: Effective heuristics for the GPS survey network of Malta: simulated annealing and tabu search techniques. J. Heuristics **7**, 533–549 (2001)
16. Sotirov, S., Sotirova, E., Melin, P., Castilo, O., Atanassov, K.: Modular neural network preprocessing procedure with intuitionistic fuzzy intercriteria analysis method. Adv. Intell. Syst. Comput. **400**, 175–186 (2016)
17. Talbi, E.-G.: Metaheuristics: From Design to Implementation. Wiley, Hoboken (2009). 624 p
18. Teunissen, P., Kleusberg, A.: GPS for Geodesy, 2nd edn. Springer, Heidelberg (1998). 650 p
19. Verhulst, P.-F.: A note on the law of population growth. Correspondence Mathematiques et Physiques **10**, 113–121 (1938). (in French)

GPU-Accelerated Flight Route Planning for Multi-UAV Systems Using Simulated Annealing

Tolgahan Turker[✉], Guray Yilmaz, and Ozgur Koray Sahingoz

Computer Engineering Department, Turkish Air Force Academy, 34149 Istanbul, Turkey
{tturker,gyilmaz,sahingoz}@hho.edu.tr

Abstract. In recent years, Unmanned Aerial Vehicles (UAVs) have been preferred in different application domains such as border surveillance, firefighting, photography, etc. With the decreasing cost of UAVs, to accomplish the mission quickly, these applications facilitates the usage of multiple UAVs instead of using a single large UAV. This makes the trajectory planning problem of UAVs more complicated. Most of the users get help from the evolutionary algorithms. However, increased complexity of the problem necessitates additional mechanism, such as parallel programming, to speed up the calculation process. Therefore, in this paper, it is aimed to solve the path planning problem of multiple UAVs with parallel simulated annealing algorithms which is executed on parallel computing platform: CUDA. The efficiency and the effectiveness of the proposed parallel SA approach are demonstrated through simulations under different scenarios.

Keywords: Simulated Annealing · CUDA · Route planning

1 Introduction

Recent advances in aeronautic, electronic, and communication technologies have enabled the usage of Unmanned Aerial Vehicles (UAVs) in different application areas. Due to their low cost, usage of multi-UAV systems, which contain many numbers of UAVs in a coordinated form, is preferred in most of these application areas to decrease the mission completion time. Also, this approach increases the capability and efficiency of the system with the use of a large number of agents as autonomous UAVs.

As their pioneers, single-UAV systems, which have been used in practice for decades, are composed of large, expensive, and powerful UAV. On the other hand, multi-UAV systems consist of a group of small, low-cost, and mostly autonomous UAVs working together especially in collaboration with each other. Through their capability of simultaneous task execution and fault tolerant structure, multi-UAV systems promise more efficient ways to perform application-specific objectives. Due to these advantages, multi-UAV systems can be used

© Springer International Publishing Switzerland 2016
C. Dichev and G. Agre (Eds.): AIMSA 2016, LNAI 9883, pp. 279–288, 2016.
DOI: 10.1007/978-3-319-44748-3_27

in wide range of potential applications both in military and civil problem domains. However, putting these potential systems into practice, particularly for surveillance-based missions, requires computation of cost-effective and near-optimal flight routes in an acceptable period of time.

In a typical aerial surveillance mission, a number of predefined geographical locations, which is also called as *waypoints*, need to be traversed in a certain order by a UAV. The order of these waypoints that the UAV should follow is called as *flight route*. In order to minimize the energy consumption of the UAV, it is necessary to find the optimal flight route which provides the minimum total travel distance. However, for a large number of waypoints to be visited, searching for the optimal flight route *deterministically* requires an excessive amount of computational time because of combinatorial nature of the problem. Furthermore, in the case of planning routes for a multi-UAV system, there are additional parameters to be considered such as the number of UAVs in the system, take-off and landing points of each UAV, distribution of waypoints to each flight route in terms of different metrics, etc. Clearly, using multiple UAVs complicates the route planning problem further and necessitates a lot more computation time to obtain the optimal flight route. Therefore, computation of optimal route for surveillance missions is the main challenging issue in the flight route planning process and requires special optimization methods in practice.

To address this need, instead of brute-force search approaches, meta-heuristics such as *Simulated Annealing* (SA) [7] can be used to approximate the global optimum of a search space. Even these techniques do not assure to find the optimal flight route; they are able to provide acceptable local optimal flight routes in a reasonable amount of time. Additionally, it is quite possible to find flight routes with better quality in a shorter period of time by using high-performance parallel computing techniques. Today's many-threads processors, particularly graphics processing units (GPUs), allow significant increases in computing performance by executing data-parallel portions of a particular workload on GPU cores. In this context, SA algorithm can be modified and executed in parallel on a GPU in order to provide more efficient ways to obtain near-optimal flight routes.

Based on this background, the purpose of this research is to provide an efficient GPU-accelerated way to compute near-optimal flight routes for single depot multi-UAV systems. In practice, flight routes for each UAV are required to be as close as possible to each other in terms of travel distance, if the system is composed of homogeneous UAVs. This requirement is also considered as a constraint in this study. Because it is not feasible to use exact search algorithms due to the computational complexity of the route planning problem, SA algorithm, as a probabilistic local search technique, is preferred to use in this study. The algorithm is re-designed and implemented for parallel execution on NVIDIA's parallel computing platform, CUDA. Results obtained from the experiments are presented.

The rest of the paper is organized as four parts. In Sect. 2, some background information is given about SA algorithm and CUDA. Then, parallel implementation details are presented and techniques used are discussed in Sect. 3. Next, the findings are discussed and summarized. The paper concludes with a discussion of implications and directions for further research.

2   Background

2.1   Simulated Annealing

Simulated Annealing (SA) is a stochastic local search algorithm which is able to escape from local optima by allowing worsening moves probabilistically. It is originally proposed by Kirkpatrick et al. [7] and Cerny [3] to be used in combinatorial optimization problems. As it can be inferred from its name, SA is built on the basis of annealing in solids. In physical context, annealing is a process of increasing the temperature of the heat bath for a solid until it melts and then lowering the temperature sufficiently slow in order to obtain perfect structural integrity for the solid which is also called ground state. When the temperature is increased to the melting point of the solid, particles have higher energy, so the probability of re-arrangement for interior structure of the solid is high. During the cooling process, this probability starts to decrease because the particles start to have lower and lower energy with the intention of reaching the ground state. However, reaching the ground state is not a trivial task. It is required to increase the temperature sufficiently high and also decrease it sufficiently slow, otherwise, metastable states can be obtained, or a process called quenching may be observed.

SA provides a framework for optimization of complex systems through the analogy between this physical phenomenon and optimization problems. It is associated with a search process for the global optimum of an objective function. The terms mentioned in physical annealing; energy state, metastable state, ground state, quenching, and temperature are analogous to terms in SA as objective function, solution, local optimum, global optimal, local search, and control parameter respectively. SA algorithm is a series of Metropolis algorithm, which is proposed by Metropolis et al. [8], with gradually decreasing values of temperature. The most important algorithmic feature of SA algorithm is its capability of escaping from local optima by using Metropolis acceptance function. While better moves are accepted directly, it also provides acceptance of worsening moves with a probabilistic way. At high temperatures, the probability of accepting a worsening move is high, in contrast, at low temperatures, this probability is very low. It is, therefore, important to tune the control parameters of SA algorithm to approximate the global optimal.

2.2   CUDA: Compute Unified Device Architecture

CUDA is introduced by NVIDIA in 2006 as a general purpose parallel computing programming model and architecture [10]. It provides significant increases in computing performance owing to parallel compute mechanism of NVIDIA GPUs.

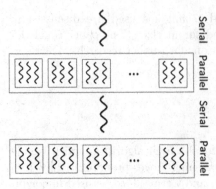
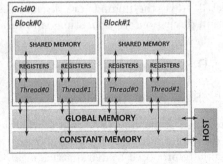

Fig. 1. CUDA execution model.   **Fig. 2.** CUDA device memory model.

Many-thread GPUs and general-purpose multicore CPUs have different design principles as pointed out by Kirk et al. in [6]. While CPUs are designed with the intent of providing minimum execution latency for a single thread, GPUs are designed as throughput oriented devices which maximize the total execution throughput of a large number of threads. So, GPUs cannot perform well on tasks demanding one or very few threads which are more sensible to perform on CPUs. It is, therefore, important for a program to be executed on both CPU and GPU for better resource utilization. To address this need, CUDA provides heterogeneous execution model, as presented in Fig. 1, which enables the execution of compute intensive parts on GPU and sequential parts on CPU.

A typical CUDA-capable GPU is formed as a set of highly threaded streaming multiprocessors (SMs) which perform the actual computations. The number of SMs in CUDA GPUs varies from generation to generation. There are a number of streaming processors (SPs) or in other words *CUDA cores* in each SM. Each SM has its own control units, execution pipelines, registers and caches which are shared by these CUDA cores. Another important element in a CUDA GPU is GDDR DRAM or global memory which is different from the system DRAMs and used as a high bandwidth off-chip memory for computation.

In CUDA programming model, threads are grouped into blocks, and blocks are grouped into a grid. In Fig. 1, the execution of two grids which are composed of a number of thread blocks is illustrated. As it is provided in Fig. 2, CUDA has additional techniques to access memory that can reduce the amount of data requests to the global memory in order to prevent the access traffic congestion. Global and constant memories can be accessed by the host using CUDA API functions. Also, these two memory types can be accessed by all threads in the system during the execution. Global memory, as mentioned before, is typically dynamic random access memory (DRAM) and tends to have long access latencies. On the other hand, the constant memory provides shorter latency read-only access by the device for the cases when all threads want to access to the same location. In addition to these two off-chip memories, registers and shared memory provides very high-speed access. Registers are used by individual threads for keeping frequently accessed variables and can be accessed by only associated thread. Lastly, shared memory is used by thread blocks for thread cooperation and sharing. In order to develop high-performance parallel applications, it is important to optimize the kernel's memory access model [9].

In the literature, there are a number of studies about UAV route planning problems using CUDA platform. Cekmez et al. provide efficient UAV path planning approaches on CUDA platform by using both Ant Colony Optimization [1] and Genetic Algorithms [2]. Another research on flight route planning using parallel genetic algorithm on NVIDIA GPUs is provided by [11]. Results obtained for all these studies show that parallel implementations of these algorithms using CUDA promise new efficient ways to solve route planning problem in the context of TSP. However, none of these researchers considers multi-UAV system within the context of mTSP using CUDA.

3   Implementation

3.1   Solution Representation

In this study, multi-chromosome genetic representation technique proposed by Király et al. [5] is used to represent solutions in the algorithm. Figure 3 illustrates this representation for 11 waypoints and 3 UAVs. Node H in the graph shows the *Home* which represents the single take-off and landing point for all UAVs in the system. It is, therefore, important to note that numbers 0, 1, and 2 used in the solution are reserved for 3 UAVs in the system, and each of them indicates the same take-off and landing point, *Home*. A solution is formed by individual routes for each UAV as it is illustrated by merging Route-0, Route-1, and Route-2 in Fig. 3, so each element in the solution denotes a geographical position to be visited.

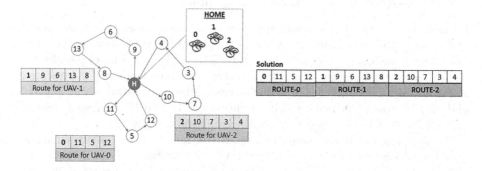

Fig. 3. Solution representation for a multi-UAV system with 3 UAV and 11 waypoints.

3.2 Initial Solution Construction

As it is mentioned in Sect. 1, in addition to try to find the route with minimum total travel cost, we want to make each individual route to have close cost values as much as possible. In this context, we set our initial solution using a simple heuristic. It is obvious that if the points representing the take-off and landing points of UAVs are located into the indices which are too close to each other in the solution vector, it is quite not possible to start with a solution with better cost value. However, we organize initial solution by locating each take-off point for UAVs equidistantly in the solution vector. With this way, we are hoping to have a solution with better quality.

3.3 Objective Function

Searching for the best route for a UAV typically means searching the shortest route. However, for a multi-UAV system, total travel cost is considered as to be minimum. Additionally, this study is looking for fair route distribution in terms of individual route costs of UAVs. That is, this research does not focus only to minimize the total travel cost for the multi-UAV system. Individual flight routes for UAVs in the system are also considered to be as close as possible to each other in terms of individual route costs. In order to ensure this additional requirement, the method provided by Hou et al. in [4] is used in this study. Given a solution with total travel cost C_T, the difference between the individual flight routes with minimum cost and maximum cost $C_{max} - C_{min}$ is added to C_T. With this way, a greater difference between the minimum and maximum individual routes cause a worsen solution with its greater penalty so these routes will be less preferable by the algorithm.

3.4 GPU-Accelerated Simulated Annealing

SA algorithm can be implemented using two different approaches, homogeneous and inhomogeneous. In this study, homogeneous version is used due to its capability over controlling the equilibrium state. The implementation of parallel SA algorithm for route planning consists of two major nested loops as it is presented in Fig. 4. These are *Cooling Loop* which controls the temperature of the system, and *Equilibrium State Loop* which all threads in the GPU are responsible to execute it as a CUDA kernel to reach the equilibrium state.

Cooling loop, as being the outer loop, is controlled by the host using static geometric cooling schedule. In each iteration, the temperature, as the main control parameter of the algorithm, is decreased by multiplying a predefined cooling factor. In addition to cooling scheduling, this loop is also responsible for selection of the best solution within the solutions found and returned by all CUDA threads. The best solution is cloned into an array with size of number of threads in order to comply with data-parallel computation mechanism of CUDA. This is because all CUDA threads should perform their computation on their own data

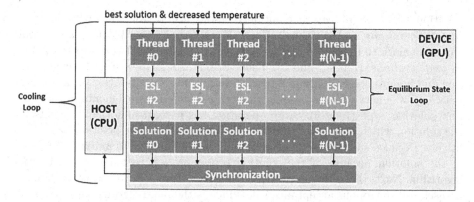

Fig. 4. Overview of parallel simulated annealing.

portion to exploit data parallelism. Then, this *solution array*, which is composed of the same best solutions, and decreased temperature are sent to device global memory (GPU DRAM) by passing them as CUDA kernel parameters, as it is illustrated in Fig. 4. For each iteration of cooling loop these operations are repeated until reaching the defined stopping criteria.

Algorithm 1. Cooling Iteration

1: **procedure** ANNEAL(*initialTemperature*, *initialConfiguration*) ▷ Runs on CPU
2:     *currentTemperature* ← *initialTemperature*
3:     *currentConfig* ← *initialConfiguration*
4:     *minConfig* ← *currentConfig*; k ← 0;
5:     **while** k < *noOfOuterIteration* **do** i ← 0;
6:         **while** i < *noOfThreads* **do**
7:             *configs*[i] ← *currentConfig*; i ← i + 1;
8:         **end while**
9:         *Kernel* <<< *DimGrid*, *DimBlock* >>> (*configs*, *currentTemperature*)
10:        *currentConfig* ← *GetConfigWithMinCost*(*configs*)
11:        **if** *currentConfig.Cost* < *minConfig.Cost* **then**
12:            *minConfig* ← *currentConfig*
13:        **end if**
14:        *currentTemperature* ← *currentTemperature* * *coolingFactor*; k ← k + 1;
15:     **end while**
16: **end procedure**

Equilibrium State Loop (ESL), as a CUDA kernel, is executed by an array of threads in parallel on the GPU. This kernel is invoked at each cooling iteration as it is provided in Algorithm 1, (Line 9). The ESL has two main responsibilities: neighbour generation and acceptance evaluation. As it is presented in Algorithm 2, *configurations* and *temperature* parameters are provided by cooling iteration. *The solution array* mentioned above is represented as the parameter *configurations* in Algorithm 2. Each thread gets its own solution *conf* from global memory to its local memory by using its unique id *threadId* as the index of *configurations*, (Lines 2 and 3). Then, ESL starts for new neighbour generation and acceptance evaluation. At each equilibrium state iteration, a neighbour *newConf* is generated by swapping randomly selected two elements in the current solution *conf*, (Line 6). Swap operation has a constraint that elements in the solution which represents the *Home* point should not be positioned consecutively. Because such a case leads a missing in the number of individual routes in the solution because of the solution representation technique used in the algorithm. Next, the difference between the cost of two solutions is calculated to be used for Metropolis acceptance rule which is the main algorithmic feature of SA to escape from local optima. If the cost of the new solution is lower than the current solution, it is directly accepted as the current solution *conf*. However, if the newly generated neighbour is worsen than the current solution *conf*, then, according to cost difference between two solutions and the temperature value, it can still be accepted as the current configuration *conf* with a gradually decreasing probability through the Metropolis acceptance function. These operations are repeatedly performed in the ESL by all threads until a predefined number of iteration, as referred to *noOfInnerIteration*, is reached. At the end of the ESL, each thread has its own candidate solution (*conf*) on the device global memory as provided in Fig. 4 and should wait for each other to synchronize for *deviceTo-Host* data transfer. After these candidate solutions generated at the end of the ESL by all threads transferred from GPU tp CPU, the host continues its own operations in the cooling iteration by decreasing the temperature and selecting the best solution, and so on.

Algorithm 2. Equilibrium State Loop as a CUDA Kernel

```
 1: procedure KERNEL(configurations, temperature)          ▷ Runs on GPU
 2:     threadId ← blockDim.x * blockIdx.x + threadIdx.x
 3:     conf ← &(configurations[threadId])
 4:     i ← 0
 5:     while i < noOfInnerIteration do
 6:         newConf ← Swap(conf);
 7:         ΔE ← newConf.Cost − conf.Cost
 8:         if IsAccepted(ΔE, temperature) then
 9:             conf ← newConf
10:         end if
11:         i ← i + 1
12:     end while
13: end procedure
```

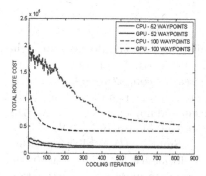

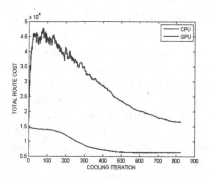

Fig. 5. Comparison of CPU and GPU performance for 52 and 100 waypoints using 3 UAVs.

Fig. 6. Comparison of CPU and GPU performance for 225 waypoints using 3 UAVs.

4 Experimental Results

SA algorithm is implemented both on CPU and GPU for performance comparisons. The hardware specifications are presented in Table 1. The initial temperature of SA algorithm is selected as 40 because of its average initial acceptance percentage (60 %) and cooling iteration is set to 1000. All GPU tests are executed with 128 threads. Additionally, the equilibrium state iteration which each thread execute is set to 10 for all tests.

Table 1. Hardware specifications

	CPU	GPU
Manufacturer	Intel	Nvidia
Model	i5	GeForce 840 M
Architecture	Broadwell-U	Maxwell
Clock frequency	2.2 GHz	1029 MHz
Cores	2	384
DRAM	8 GB DDR3	4 GB DDR3

Multi-UAV system is considered as consisting 3 UAVs. Data sets with 52, 100, and 225 waypoints are used in the experiments. As it is presented in Figs. 5 and 6, GPU implementation provides better total travel cost values for the similar number of iterations. Moreover, it can be inferred form the figures above, as the number of waypoints increases, the difference between the total travel costs provided by CPU and GPU at a certain iteration also increases at the begining of the iterations. It is therefore important to note that GPU implementation of SA algorithm provides better solutions as compared to serial CPU implementation. So, experiments show the efficiency of GPU implementation of SA algorithm.

5 Conclusion

In conclusion, this paper presents an alternative way to compute cost-fair flight routes for single station multi-UAV systems efficiently. Simulated Annealing algorithm is used to deal with exponentially increasing computation time of route planning problem due to its combinatorial nature. The algorithm is redesigned by making small modifications for GPU acceleration and implemented using CUDA platform in order to exploit data-parallel compute mechanism of NVIDIA GPUs. Experimental results show that GPU-accelerated Simulated Annealing algorithm provides significant increases in computing performance for flight route planning problem. As a future work, in order to simulate more realistic scenarios, additional environmental constraints which should be avoided by UAVs such as radars or missiles in a pre-known flight region are planned to be involved in this study.

References

1. Cekmez, U., Ozsiginan, M., Sahingoz, O.K.: A UAV path planning with parallel ACO algorithm on CUDA platform. In: 2014 International Conference on Unmanned Aircraft Systems (ICUAS), pp. 347–354, May 2014
2. Cekmez, U., Ozsiginan, M., Aydin, M., Sahingoz, O.K.: UAV path planning with parallel genetic algorithms on CUDA architecture. In: Proceedings of the World Congress on Engineering, pp. 347–354. IAENG (2014)
3. Černý, V.: Thermodynamical approach to the traveling salesman problem: an efficient simulation algorithm. J. Optim. Theor. Appl. **45**(1), 41–51 (1985). http://dx.doi.org/10.1007/BF00940812
4. Hou, M., Liu, D.: A novel method for solving the multiple traveling salesmen problem with multiple depots. Chin. Sci. Bull. **57**(15), 1886–1892 (2012)
5. Király, A., Abonyi, J.: A novel approach to solve multiple traveling salesmen problem by Genetic algorithm. In: Rudas, I.J., Fodor, J., Kacprzyk, J. (eds.) Computational Intelligence in Engineering. SCI, vol. 313, pp. 141–151. Springer, Heidelberg (2010). http://dx.doi.org/10.1007/978-3-642-15220-7_12
6. Kirk, D.B., Wen-mei, W.H.: Programming Massively Parallel Processors: A Hands-on Approach. Newnes, Oxford (2012)
7. Kirkpatrick, S., Gelatt, C.D., Vecchi, M.P.: Optimization by simulated annealing. Science **220**(4598), 671–680 (1983). http://science.sciencemag.org/content/220/4598/671
8. Metropolis, N., Rosenbluth, A.W., Rosenbluth, M.N., Teller, A.H., Teller, E.: Equation of state calculations by fast computing machines. J. Chem. Phys. **21**(6), 1087–1092 (1953). http://scitation.aip.org/content/aip/journal/jcp/21/6/10.1063/1.1699114
9. NVIDIA Corporation: CUDA C best practices guide, version 7.5, September 2015
10. NVIDIA Corporation: CUDA C programming guide, version 7.5, September 2015
11. Sancı, S., İşler, V.: A parallel algorithm for UAV flight route planning on GPU. Int. J. Parallel Program. **39**(6), 809–837 (2011). http://dx.doi.org/10.1007/s10766-011-0171-8

Reconstruction of Battery Level Curves Based on User Data Collected from a Smartphone

Franck Gechter[1]([✉]), Alastair R. Beresford[2], and Andrew Rice[2]

[1] University of Technologie of Belfort Montbliard, UBFC,
F-90010 Belfort Cedex, France
franck.gechter@utbm.fr
[2] Computer Laboratory, University of Cambridge, Cambridge, UK
{arb33,acr31}@cam.ac.uk

Abstract. We demonstrate how a multi-agent top-down approach can be used to interpolate between battery level measurements on a phone handset. This allows us to obtain a high fidelity trace whilst minimising the data collection overhead. We evaluate our approach using data collected by the Device Analyzer project which collects handset events and polled measurements from Android devices. The value of the multi-agent approach lies in the fact that it is able to incorporate implicit information about battery level from operating system events such as network usage. We compare our approach to interpolation using Bezier curves and show a 50 % improvement in mean error and variance.

Keywords: Power modelling · Smartphones · Device Analyzer · Physics inspired artificial intelligence

1 Introduction

Researchers are increasingly interested in how consumer devices are used in the real world. Projects such as Device Analyzer attempt to help with this by collecting usage data from Android devices. Device Analyzer has collected data from more than 30,000 users worldwide with some participants providing more than two years of continuous usage data [1]. It contains operating system events, such as screen-on and incoming-call, and also periodically polls for information, such as battery level and network byte counters.

Polled data in Device Analyzer is only collected every 5 min. This is necessary in order to minimise the energy footprint of the data collection itself on the participant's phone. However, this means that the trace lacks fidelity and so being able to accurately interpolate between readings would be a valuable benefit.

In this paper we focus on estimating the power consumption of a handset. These estimates are the basis of research into understanding (and subsequently reducing) the energy consumption of smartphones. The most accurate way to determine a device's power consumption is to measure it directly. This can be

© Springer International Publishing Switzerland 2016
C. Dichev and G. Agre (Eds.): AIMSA 2016, LNAI 9883, pp. 289–298, 2016.
DOI: 10.1007/978-3-319-44748-3_28

done by intercepting the current flow between the phone and its battery [2]. This overall consumption can then be used in a power model to estimate the consumption of different components on the phone [3]. Researchers have also been successful in estimating energy consumed by monitoring the usage of hardware at the operating system level [4] or instrumenting application binaries and logging system calls [5]. These techniques all suffer from the limitation that they require some sort of specialist modification to the device and so are only suitable for small-scale studies. These studies are unlikely to capture the full range of conditions and use-cases experienced by real users.

Wide-scale deployments of power estimation have been forced to rely on coarse-grained estimates of energy use collected by polling the state-of-charge of the battery directly. Drawing useful inferences from these measurements is difficult. In the Carat project for example the authors were able to identify energy-wasting applications ('energy hogs') by aggregating measurements from many thousands of devices [6]. Our intention is similar in that we seek to make use of additional information to improve the quality of data from coarse-grained measurements.

The energy consumption of one smartphone component is not independent of other components [3] and so we take a holistic, top-down approach. Our approach is similar to the top-down operating system approaches described above, however we apply a multi-agent system method and use coarse-grained events which relate to user actions rather than system calls or hardware usage.

A multi-agent system is composed of a set of *agents* in *interaction* with each other within an *environment*. The result of the three main components (agents, interaction, environment) leads to a *collective organisation*.

There are two main approaches in the multi-agent community. The first, classical, approach uses agents with a high-level decision process. These are called *cognitive agents* and are commonly found in the literature [9]. The second approach is to use agents with small (or no) cognitive ability and whose behaviours stem from a stimulus-response or influence-reaction scheme. These are called *reactive agents*. In this approach intelligence is not in the agents themselves but emerges from their interaction with each other and with their environment. These methods are now also widely used in some domains, such as cyber-physical systems control [10,14] and (distributed) problem solving [11]. The main difference between cognitive and reactive agents is in the role of the interaction processes that act upon the agents. For reactive agents, the environment plays an essential role since it formalises the constraints on the system's evolution [13]. We take a reactive agent paradigm using forces inspired by Coulomb's law as an interaction model.

The rest of the paper is structured as follows: we first characterise the data available from Device Analyzer (Sect. 2) and then explain the physics-inspired multi-agent model we developed (Sect. 3). In Sect. 4 we show the accuracy of our approach and discuss their validity range.

2 Device Analyzer

We use data collected by the Device Analyzer [1] project. This is coarse-grained information but with the benefit that is has been collected from a large number of different users over an extended time period. The device data consists of events of three broad types:

Immediate events are directly stored with their occurrence time. These correspond to events delivered by the Android operating system. Examples include *screen on* and *screen off* events.

Polled events are collected periodically (normally) every 5 min. These are split into two categories depending on whether they are continuous data (such as number of bytes transmitted over a network interface) or discrete data (such as phone notification settings).

Static events are collected at the first connection and which are not supposed to change during the period of analysis. These generally correspond to device information such as the OS version or the Hashed ID of the SIM card.

Table 1. Classification of the data collected by Device Analyzer

Immediate event	Polled event	
	Continuous	Discrete
Phone on/off	Local system time	Phone alert status
Charging time	Amount of free storage	Device Analyzer version
Take picture	Audio volume settings	Total number of photos
Screen on/off	Battery level and voltage	Roaming status
Airplane mode on/off	Screen brightness	
Network connectivity	Cellular signal strength	
Incoming/outgoing call	Amount of 3G data received	
Bluetooth/Wifi on/off	Amount of 3G data transmitted	
Tethering	Amount of Wifi data received	
Bluetooth scan	Amount of Wifi data transmitted	

We give some examples in Table 1. Note that one specific piece of data, such as number of bytes received on the 3G network, belongs to one category (*3G data received*) and to one type (*polled continuous event*).

3 The Particle Model

3.1 Principles

The general idea of our model is to use the event information collected by Device Analyzer to estimate the battery level of a device over a time period.

We consider collected events as disturbances which deform the ideal battery curve. The challenge is thus to determine how each event influences the battery curve. In the rest of this paper, the battery curve is made of virtual charged particles that will be influenced by the presence of events through attraction forces based on Coulomb's law. This can be considered as a beam. We make use of an existing linear power model [3] expressed as follows:

$$P = \sum_i (\beta_i \cdot x_i) + P_{base} + P_\epsilon \tag{1}$$

where P is the power used by the device, $\{x_i\}_i$ is the set of state variables of the system (e.g. when dealing with hardware, each element of this set corresponds to a hardware component and the associated value to its utilisation), P_{base} is the base power consumption, P_ϵ is a noise factor and β_i are the linear coefficients that must be determined in order to estimate the influence of the x_i component on the overall power model.

To estimate the energy consumption we must include the whole time period during which the components are used. This leads to the following equation:

$$E = \sum_i \sum_j (\beta_i \cdot x_i^j) \cdot d_i^j + (P_{base} + P_\epsilon) \cdot D \tag{2}$$

where E is the global energy consumption, x_i^j is the utilisation of the component i during the window d_i^j, and D is the duration of the experiment.

Since we are focusing on user events collected by Device Analyzer, we have to determinate the β_i and the d_i^j values for each type of event x_i^j.

Let S_E be a finite set of events collected from the device defined by $S_E = \{Ev_{i,t_i}\}_{i \in 1..N}$ and ordered relatively to a time line (if $i < j$, $t_i < t_j$). Each event Ev_{i,t_i} belongs to a category C_k of events which corresponds to those presented in Sect. 2.

Considering an ideal curve $\mathscr{C}$ of the battery and assuming that we can find out, for each category C_k, a representative Event Agent, EA_{C_k}, the battery consumption for the device, over a time period $\Delta t = [T_a, T_b]$, is the result of the influence of the set of event agents, occurring during this time period $S_{EA} = \{EA_{C_k,t_p}\}_{t_p \in [T_a, T_b]}$ on $\mathscr{C}$. The next section will detail the agent model used, starting with the design of the particle beam (the battery curve) and continuing with the event agents.

3.2   Agent Models

The Particle Beam. The particle beam is made of small particles that can be considered as either electrons or positrons depending on their charge. The system environment contains an electrical field in order to ensure that the beam moves from left to right. If one assumes that the charge of the particles is positive, the electrical field is oriented from left to right. Thus, the forces applied to each particle can be defined as follows:

$$\boldsymbol{F}_{\text{Field}} = q \cdot \boldsymbol{E} \tag{3}$$

where q is the charge of the particle of the beam (we use a charge of 1) and E is the electrical field. We avoid superposition of particles by introducing a repulsion force between beam particles. This force is a classical force, following Coulomb's law, of $1/d^2$ where d is the distance between two nearby particles. This force, applied to the particle P_1 by the particle P_2 is then defined by the following equation:

$$F_{P_1 P_2} = -\kappa \cdot \frac{q_1 \cdot q_2}{\|P_1 P_2\|^3} \cdot P_1 P_2 \tag{4}$$

where q_i is the charge of the particle P_i, and κ is a scalar, which corresponds to Coulomb's coefficient for real particles, that can be used in this virtual particle world as a tuning parameter. Combining these two types of forces using Newton's second law of motion allows us to obtain a virtual particle beam.

As an alternative to a particle beam, we might have used other models such as a single particle with a trace corresponding to its trajectory or a rope with nodes influenced by the event. The benefit of using a beam is that it provides better interaction when tuning parameters. Instead of waiting for the particle to travel through the whole time period to check the quality of the result, one can observe in real time the influence on the result of any modification on the parameters set.

The Event Agents. An Event Agent (EA) is a representation of a real event collected by Device Analyzer. In the curve reconstruction model they influence the trajectory of the particle beam depending on the values they are associated with. We define one EA for each event. An Event Agent for one category, denoted EA_{C_k} (where C_k is the category), is defined by the following two parameters:

→ A range R_{C_k} which corresponds to the influence range of the Event Agent in time. This range is associated with the coefficients d_i^j in Eq. 2.
→ A strength S_{C_k} which corresponds to the influence of the category of events. In this paper, the strength is a scalar value. It is associated with the coefficients β_i in Eq. 2.

The values of these parameters define the influence of each event as a local electrical field which deforms the particle beam. Thus, for each particle of the beam, if its position along the time axis is in the following interval $\left[t_{EA_{C_k}} - \frac{R_{C_k}}{2}, t_{EA_{C_k}} + \frac{R_{C_k}}{2} \right]$ its trajectory will be influenced by the following force:

$$F_{EA_{C_k}} = -q \cdot S_{C_k} \cdot u_y \tag{5}$$

where q is the particle charge, S_{C_k} the strength of corresponding category of events and u_y is the unity vector on the vertical axis which represents the battery level. The values of these parameters are the tuning parameters of the multi-agent model which have to be chosen so as to make the curve fit the control points. For most categories of event both parameters have to be chosen. However, there are some exceptions:

Screen on and screen off events These events are twinned events which correspond to a 0 to 1 state transition (screen on) and to a 1 to 0 state transition (screen off). Their influence on the curve appears only when the screen is at the upper level (equal to 1). The range of this twinned event is the time distance between screen on and screen off event.

General usage event This is a synthetic event which models the general trend of the battery to leak energy over time due its internal construction. This event has a range which corresponds to the size of the time window of the analysis and a constant strength.

For the remaining events, the range is computed using this equation:

$$R_{E_i} = \alpha_i \cdot R_{C_k} \tag{6}$$

where R_{E_i} is the range of the event i, R_{C_k} is the range of the category of the event and α_i is a coefficient which includes the values associated to the event collected by Device Analyzer. For instance, if the event corresponds to data transmitted over Wifi, α_i incorporates the number of bytes that are transmitted during this event. Consequently, the number of transmitted bytes has an influence over the range of the Event Agent and not over its strength which is the same for all events of the category. This is correlated to the way the Wifi hardware chip behaves (we are assuming it consumes constant power whilst transmitting).

3.3   Simulation Model

We prototyped the simulation model using the Tiny MultiAgent Java Platform (tinyMAS).[1] Once the events are filtered from the data file provided by Device Analyzer, each event is transformed into the relevant *EventAgent* depending on the category that it belongs to. The *BatteryEvent*s, which are used in the curve reconstruction process only as checkpoints, are placed first. The measured level gives their position on the y-axis and the recorded time gives the position on the x-axis. Other EventAgents are then placed into the *WorldModel* according to their timestamp. They are placed along the y-axis at the minimum battery level over the studied time window. The particle beam made of *ParticleAgent*s is initiated with a starting position which corresponds (in y) to the level of the first (in time) *BatteryEvent*. We then start the simulation. At each time step, each *ParticleAgent* finds out the subset S_L of *EventAgent*s that have influence according to its position along the time axis. For each element of the set, the associated force is computed (Eq. 5). Finally, using the second law of motion the acceleration, speed and future position of the particle is computed.

4   Results and Discussions

We focused first on a small set of possible categories which includes: screen on, screen off, network received, network sent and application usage (i.e. the foreground or active application). We ran the simulation over several test sets which

---

[1] http://www.arakhne.org/tinymas/.

are made of subsets of the selected event categories. For each of them we found out, using a training set, the corresponding sets of parameters which minimised the error made on the battery measurement points which have been taken as checkpoints with no influence on the particle beam. After this tuning step, we ran the simulation on a non-overlapping test set using only the first battery measured value to initiate the y-axis particle beam position. The other battery measurement points are then used so to compute the error of the estimation. For reference we also compare this error to one obtained using a Bezier-spline as an estimator applied to a subset of battery measurement points.

4.1   Estimation Results

Figure 1 shows a comparison between the agent estimation and a standard Bezier-spline curve using a test set comprising 8 635 collected events. The Bezier-spline used is a cubic Bezier curve with four control points taken regularly over the time period: the first and the last battery measurements together with battery points nearest to one-third and two-thirds of the way through the time period. The coarse-grained events from Device Analyzer are only collected once every 5 min and so one cannot know where in the 5 min period the recorded activity actually occurred. For this reason, we tried several strategies to place the event agents: (1) placing polled events *directly* at the time they were collected; (2) placing *network* events half-way through the poll window; (3) placing *all polled* events half-way through the poll window.

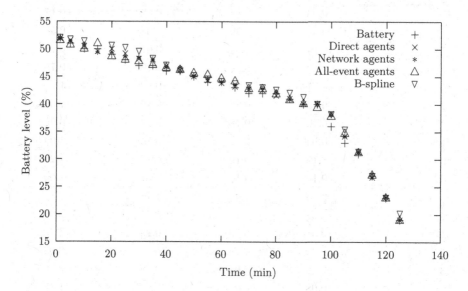

Fig. 1. Estimation results showing measured battery levels (+), agents with the direct placement strategy (×), with the network events strategy (∗), with the all-polled events strategy (△) and the cubic Bezier-spline with 4 control points (∇).

We can see that the agent approaches (see symbols ×, *, △ in Fig. 1) are similar to each other. Strategy 3 (×) shows the least error compared to the measured battery levels (+). Recall that the agent approach is estimating this curve without using any battery measurements except for the first one to position the beam. The Bezier-spline (▽) shows more error—it converges on the measured trace at the 4 control points and then diverges between them.

Table 2. Agent and Bezier estimation for specific battery measurement points.

Battery measurements	Estimation error			
	Direct agents (Strategy 1)	Network agents (Strategy 2)	All-events agents (Strategy 3)	B-spline
52	0.12	0.19	0.44	0.00
51	0.53	0.27	0.29	0.72
50	0.51	0.78	0.00	1.36
49	1.01	0.56	0.28	1.98
49	1.01	0.39	0.37	1.60
48	1.16	0.51	0.01	2.23
47	1.16	1.39	0.73	2.52
47	1.06	1.11	0.07	1.24
46	0.95	0.59	0.59	1.06
46	0.15	0.59	0.2	0.00
45	0.03	0.14	0.62	0.17
44	0.34	0.59	1.33	0.58
44	0.14	0.04	0.67	0.13
43	0.17	0.47	1.19	0.75
42	0.71	0.88	0.82	1.41
42	0.60	0.85	0.80	1.04
42	0.42	0.18	0.14	0.60
41	0.03	0.15	0.16	1.04
40	0.19	0.16	0.10	1.26
40	0.15	0.15	0.61	0.00
36	2.35	1.95	1.78	2.15
33	1.22	1.22	1.71	2.37
31	0.39	0.39	0.23	0.36
27	0.22	0.26	0.13	0.03
23	0.19	0.08	0.03	0.00
19	0.08	0.08	0.03	1.15
Mean error	0.57	0.53	0.51	0.99
Variance	0.29	0.22	0.26	0.64

Table 2 shows the errors in estimation for the different approaches. The results obtained show that the agent methods have both a smaller error and a smaller variance than the Bezier solution. Placing polled events in the centre of the polling window (Strategy 3) gives the best error performance but moving just network events (Strategy 2) gives better variance. The mean error and variance from the agent approaches is less than 50 % of the Bezier-spline approach. This is in spite of the fact that the Bezier approach relies on four times as many samples as the agent method.

5   Conclusion

This paper described a physics-inspired agent-based algorithm for battery curve estimation on smartphones. The algorithm relies on a top-down approach based on an analysis of event collected by an application installed on the target handset. The main interest of this integrative point of the view is to consider the smartphone as a complex system in which one does not neglect interactions between hardware components, software, infrastructure and user. The results obtained show that we can make good predictions of battery level as compared to a standard Bezier estimation. The next step will be to compare our approach to other alternative methods such as particle filters, Kalman filters and Gaussian process regression.

Our original goal was to improve the fidelity of coarse-grained measurements and so we now need to apply our approach to interpolating between battery measurements rather than predicting them. This is essentially the same model that we have described but on a smaller scale. Occasional periods of high-rate sampling could be used to provide a ground truth to measure the accuracy of the approach. One drawback of our approach is the need to tune the model parameters for each event. Occasional periods high-rate sampling would assist with this problem too by providing base-line data for an automated parameter search.

Acknowledgements. The Device Analyzer project was supported by a Google Focussed Research Award.

References

1. Wagner, D.T., Rice, A., Beresford, A.R.: Device analyzer: understanding smartphone usage. In: Stojmenovic, I., Cheng, Z., Guo, S. (eds.) Mobile and Ubiquitous Systems: Computing, Networking, and Services. Lecture Notes of the Institute for Computer Sciences, Social Informatics and Telecommunications Engineering, vol. 131, pp. 195–208. Springer, Heidelberg (2014)
2. Rice, A., Hay, S.: Measuring mobile phone energy consumption for 802.11 wireless networking. Pervasive Mob. Comput. **6**, 593–606 (2010). Elsevier
3. Zhang, L., Tiwana, B., Qian, Z., Wang, Z., Dick, R.P., Mao, Z.M., Yang, L.: Accurate online power estimation and automatic battery behavior based power model generation for smartphones. In: The Proceedings of the 8th IEEE/ACM/IFIP International Conference on Hardware/Software Co-design and System Synthesis (CODES/ISSS), pp. 105–114. ACM Press (2010)

4. Yoon, C., Kim, D., Jung, W., Kang, C., Cha, H.: AppScope: application energy metering framework for Android smartphones using kernel activity monitoring. In: The Proceedings of the USENIX Annual Technical Conference (ATC), pp. 387–400. USENIX Association (2012)
5. Pathak, A., Hu, Y.C., Zhang, M.: Where is the energy spent inside my app? Fine grained energy accounting on smartphones with Eprof. In: The Proceedings of the 7th European Conference on Computer Systems (EuroSys), pp. 29–42. ACM Press (2012)
6. Oliner, A.J., Iyer, A.P., Stoica, I., Lagerspetz, E., Tarkoma, S.: Carat: collaborative energy diagnosis for mobile devices. In: The Proceedings of the 11th Conference on Embedded Networked Sensor Systems (SenSys), pp. 1–14. ACM Press (2013)
7. Ferreira, D., Dey, A.K., Kostakos, V.: Understanding human-smartphone concerns: a study of battery life. In: Lyons, K., Hightower, J., Huang, E.M. (eds.) Pervasive 2011. LNCS, vol. 6696, pp. 19–33. Springer, Heidelberg (2011)
8. Rahmati, A., Qian, A., Zhong, L.: Understanding human-battery interaction on mobile phones. In: Proceedings of the 9th International Conference on Human-Computer Interaction with Mobile Devices and Services (MobileHCI), pp. 265–272. ACM Press (2007)
9. Wooldridge, M.: Reasoning About Rational Agents. MIT Press, Cambridge (2000)
10. Dafflon, B., Gechter, F.: Making decision with reactive multi-agent systems: a possible alternative to regular decision processes for platoon control issue. Res. Comput. Sci. **86**, 101–112 (2014)
11. Cox, J.S., Durfee, E.H., Bartold, T.: A distributed framework for solving the multiagent plan coordination problem. In: The Proceedings of the 4th International Joint Conference on Autonomous Agents and Multiagent Systems (AAMAS), pp. 821–827. ACM Press (2005)
12. Dunin-Kęplicz, B., Verbrugge, R.: A tuning machine for cooperative problem solving. Fundamenta Informaticae **63**, 283–307 (2004). IOS Press
13. Simonin, O., Gechter, F.: An environment-based methodology to design reactive multi-agent systems for problem solving. In: Weyns, D., Van Dyke Parunak, H., Michel, F. (eds.) E4MAS 2005. LNCS (LNAI), vol. 3830, pp. 32–49. Springer, Heidelberg (2006)
14. Gechter, F., Chevrier, V., Charpillet, F.: A reactive agent-based problem-solving model: application to localization and tracking. Trans. Autonom. Adapt. Syst. (TAAS) **1**(2), 189–222 (2006). ACM Press

Possible Bribery in k-Approval and k-Veto Under Partial Information

Gábor Erdélyi and Christian Reger[(⊠)]

University of Siegen, Unteres Schloß 3, 57072 Siegen, Germany
{erdelyi,reger}@wiwi.uni-siegen.de

Abstract. We study the complexity of possible bribery under nine different notions of partial information for k-Approval and k-Veto. In bribery an external agent tries to change the outcome of an election by changing some voters' votes. Usually in voting theory, full information is assumed, i.e., the manipulative agent knows the set of candidates, the complete ranking of each voter about the candidates and the voting rule used. In this paper, we assume that the briber only has partial information about the voters' votes and ask whether the briber can change some voters' votes such that there is a completion of the partial profile to a full profile such that the briber's preferred candidate (or most despised candidate in the destructive case) is a winner (not a winner) of the resulting election.

Keywords: Computational social choice · Voting · Algorithms and complexity

1 Introduction

Voting provides a natural framework for preference aggregation and collective decision making. Voting rules have been applied in many settings ranging from politics to computer science. Take automated, large-scale computer settings for example, in which voting rules have been used in the design of ranking algorithms [3,8] to lessen the spam in meta-search web-page rankings. Another application of voting is the movie recommender system designed by Ghosh et al. [17], which recommends movies to users based on their and other users' previous preferences. Beyond that, voting rules have been used in planning [9–11], similarity search [12], machine learning [25], and the list of examples could be continued. Generally speaking, voting is applied each time certain options are to be selected based on certain preferences, so the range of applications is virtually limitless.

In applications like recommender systems, machine learning, and meta-search engines, we are usually dealing with huge data volumes, thus it is worth studying the computational aspects of problems related to voting and elections. In the nearly three decades there has been much research about the complexity of

This work is supported in part by the DFG under grant ER 738/2-1.

C. Dichev and G. Agre (Eds.): AIMSA 2016, LNAI 9883, pp. 299–309, 2016.
DOI: 10.1007/978-3-319-44748-3_29

voting problems in various settings, especially on the complexity of manipulative actions. One of such manipulative actions is *bribery*—introduced by Faliszewski et al. [13]—where an external agent changes some voters' votes in order to change the outcome of the election.

Traditionally, the complexity of voting problems is studied under the full information assumption, i.e., the manipulative agent knows the candidate and voter sets, each voter's complete preference over all candidates, and the voting rule used. However, in many real-world settings we are dealing with incomplete information. Recently, various papers analyzed the complexity of voting problems under some kind of uncertainty (see the Related Work Section for a detailed overview). In this paper, we systematically study the complexity of an optimistic bribery variant, called possible bribery, under nine different models of partial information. Informally, the possible bribery problem is the following: Given an election (in our case in form of a partial profile according to one of the partial information models), a number ℓ, and a distinguished candidate c, can we make c a winner in at least one extension of the partial profile to a complete profile by changing at most ℓ votes? We provide a complexity study of the possible bribery problem both in the constructive and destructive variants under nine notions of partial information of the k-Approval and k-Veto families of voting rules. We chose these two classes of voting rules since bribery under full information is easy for k-Approval, with $k \leq 2$, and k-Veto, with $k \leq 3$.

Related Work. First of all, bribery under full information was introduced by Faliszewski et al. [13,14], where the authors studied the complexity of bribery for Plurality, Veto, and Approval, amongst others. The bribery results for k-Approval and k-Veto ($k \geq 2$) under full information were published by Lin [20]. Destructive bribery has been investigated for several voting rules, e.g., for Copeland [15], Schulze [21], cup elections [23], and for Bucklin and Fallback voting [16]. A closely related problem to destructive bribery is the *margin of victory*, a critical measure of the robustness of voting rules in terms of changing the outcome of elections by errors or fraud. The complexity theoretic analysis of the margin of victory was initiated by Xia [24] (amongst others, for all scoring rules).

Regarding the combination of partial information and winner determination, Konczak and Lang have introduced the possible and necessary winner problems [19]. The possible winner problem was further studied in [2,4,26]. Baumeister et al. introduced *top-truncated*, *bottom-truncated*, and *doubly-truncated orders*, where each voter only specifies some top-ranked, bottom-ranked, or both top- and bottom-ranked candidates, respectively [1]. Chevaleyre et al. used a partial information model, where each voter completely ranks a subset of candidates and after that some new candidates join the election [6]. The final partial information model we have to mention was introduced by Conitzer et al. [7], where they do not define a special model of partial information, but in their setting, an information set contains all complete profiles that can be obtained by completing the partial profile.

The complexity of bribery under partial information has been studied by Briskorn et al. [5] for k-Approval and k-Veto, however, their model differs from ours in a sense that while we are only looking for at least *one* successful completion, they wanted the briber to be successful under *each* completion of the partial profile to complete profiles. Furthermore, they have pointed out the interrelations of nine different partial information models. For the voting rule families k-Approval and k-Veto, the complexity of control has been studied for the same partial information models we use by Reger [22].

2   Preliminaries

Formally, an *election* is defined by a pair $E = (C, V)$ with the finite set of *candidates* C and a finite collection of *voters* V. Each voter v_i has a strict (i.e., total, asymmetric and transitive) *order* $\succ_i$ over the set of candidates. We will simultaneously use the terms *ranking* and *preference order*. Sometimes we write $c \succ d$ instead of $c \succ_i d$ if the voter is clear from the context. An *n-voter profile* P on C consists of n strict linear orders $P = (v_1, \ldots, v_n)$. Analogously, one can define a *partial profile* where all votes v_i are partial according to the same partial model X. A *completion* or *extension* of a partial vote (profile) is a complete vote (profile) which does not contradict the partial vote (profile). We use the game-theoretic notion *information set* $I(P)$ for a partial profile P containing all complete profiles P' not contradicting P.

A *voting rule* $\mathscr{E}$ maps an election E to a subset $W \subseteq C$, the set of *winners* of the election. Throughout this paper, we use the *nonunique-winner model* (i.e., every candidate in W is a winner). An important class of voting rules are *scoring rules* specified by their *scoring vectors* $\alpha = (\alpha_1, \ldots, \alpha_m)$ (with $m = |C|$) with $\alpha_1 \geq \ldots \geq \alpha_m$ such that each voter gives α_1 points to his favorite candidate, α_2 points to his second most preferred candidate and so on. The *score* of a candidate c, denoted by $score(c)$, is the sum of points he receives from each voter. The winners are the candidates with the highest overall score. We will consider the following scoring rules in this paper.

- k-Approval is the scoring rule with the scoring vector $\alpha = (\underbrace{1, \ldots, 1}_{k}, 0, \ldots, 0)$.

  1-Approval is also known as Plurality.
- k-Veto is the scoring rule with the scoring vector $\alpha = (1, \ldots, 1, \underbrace{0, \ldots, 0}_{k})$. We

  will write Veto instead of 1-Veto. Sometimes, in our proofs we will rather count the vetoes a candidate receives than the approvals. $vscore(c)$ denotes the number of *vetoes* for c, i.e., the number of voters assigning zero points to c.

Note that for a fixed number m of candidates, k-Approval equals $(m-k)$-Veto. In our setting, however, k is always fixed, but m is not fixed.

In our proofs we will use the following notations: $pscore(d)$ sums up all potential (definite and possible) approvals for candidate d. Sometimes we write $score_{(C,V_1)}(d)$ which counts the number of points for d in subelection (C, V_1). In a similar manner, we may write $pvscore(c)$ or $vscore_{(C',V)}(d)$.

In the remainder of this section, we give a short survey about all nine models of partial information for which we will study the complexity of the possible bribery problem. For each model of partial information, we will specify the structure of data given. Note that this is exactly the information given in our problem instances. The "real" vote, i.e., the actual complete ranking of the regarded voter, is among all potential completions of the given partial vote. Moreover, one can easily verify that full information can be displayed by each model. We let $m = |C|$.

- **Gaps (GAPS)** [5] **and One Gap (1GAP)** [1]. Our first model GAPS handles the case where there are *gaps* in a vote, i.e., each vote has some fully ranked blocks and some blocks in between where it is only known which candidates belong to these blocks, but not how they are ordered. Formally, for each vote v, we have a partition $C_1^v, \ldots, C_{2m+1}^v$ of the set of candidates and a total order for each C_k^v with k even. Note that possibly $C_k^v = \emptyset$ for some k. If $C_k^v = C_{k+1}^v = \emptyset$, we can drop both partite sets without changing the information set. Therefore, we can restrict ourselves to at most $2m+1$ partite sets. A special case is 1GAP, where in each vote some candidates are ranked at the top and at the bottom of the votes, and there is at most one hole in between. Formally, 1GAP refers to the special case of GAPS with $C_k^v = \emptyset$, for each $k \in \{1, 5, 6, \ldots, 2m + 1\}$ and each voter v.
- **Top- and Bottom-Truncated Orders (TTO,BTO)** [1]. TTO equals GAPS with $C_1^v = C_4^v = \ldots = C_{2m+1}^v = \emptyset$ for each voter v. BTO refers to the special case of GAPS with $C_3^v = \ldots = C_{2m+1}^v = \emptyset$ for each voter v.
- **Complete or empty votes (CEV)** [19]. CEV handles the case where each vote is either complete or empty. Formally, CEV is a special case of TTO with either $C_2^v = \emptyset$ or $C_3^v = \emptyset$ for each voter v.
- **Fixed Positions (FP)** [5]. For each vote v we have a subset of candidates C^v with distinct positions in range between 1 and m assigned.
- **Pairwise Comparisons (PC)** [19]. PC (aka *partial orders*) is probably the most natural way to display partial preferences. Formally, for each vote v we have a subset Π^v of $C \times C$ which we restrict to be asymmetric and transitive for matters of convenience.
- **Totally Ordered Subset of Candidates (TOS)** [5]. In TOS, for each vote v, there is a complete ranking about a subset $C^v \subseteq C$.
- **Unique Totally Ordered Subset of Candidates (1TOS)** [6,19]. An important special case of TOS, 1TOS, requires that $C^v = C'$ for each voter $v \in V$.

Briskorn et al. gave a complete picture of the interrelations of the above described partial information models [5]. The interrelations can be visualized with a Hasse diagram, see Fig. 1. For a detailed overview, motivation, and examples for these partial information models, we refer the reader to the work of Briskorn et al. [5].

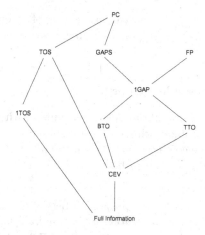

Fig. 1. Hasse diagram of the nine partial information models and full information.

2.1   Problem Settings

In this section, we define the two possible bribery problems we will investigate in this paper. Let PIM := {GAPS, FP, TOS, PC, CEV, 1TOS, TTO, BTO, 1GAP} and $\overline{\mathrm{PIM}}$:= PIM $\cup$ {FI} where FI is the model of full information. If $X = FI$, we obtain the standard definition of bribery under full information known from [14]. In the following, we let $X \in \overline{\mathrm{PIM}}$.

CONSTRUCTIVE $\mathscr{E}$-X-POSSIBLE BRIBERY

Given: An election (C, V), a designated candidate $c \in C$, a non-negative integer $\ell \leq |V|$, and a partial profile P according to X.

Question: Is it possible to make c a winner of the election under voting rule $\mathscr{E}$ for at least one complete profile in $I(P)$ by changing up to ℓ votes?

In the destructive variant we ask whether it is possible to prevent c from being a winner of the election. It shall be pointed out that the bribed votes are complete after the bribery, although they may have been partial before the bribery.

Finally, we mention that the $\mathscr{E}$-X-POSSIBLE WINNER problem asks whether for a given election (C, V), a designated candidate $c \in C$, and a partial profile P according to X, c is a winner under voting rule $\mathscr{E}$ for at least one complete profile in $I(P)$ [19].

3   Results for Possible Bribery

In this section, we will perform a complexity analysis both for the constructive and destructive case. If not mentioned other, we are given an election (C, V) with m candidates $C = \{c, c_1, \ldots, c_{m-1}\}$, n voters $V = \{v_1, \ldots, v_n\}$, a designated candidate c, and an $\ell \in \mathbb{N}_0$ as the maximum number of voters that can be bribed. Sometimes we also use d for a general, not further specified (non-distinguished) candidate.

3.1   Constructive Possible Bribery

In this subsection, we will provide complexity results for the constructive variant of the possible bribery problem which are summarized in Table 1. We start with a lemma, pinpointing the connections between the possible bribery problem and the possible winner and standard bribery problems.

Table 1. Summary of results for the possible bribery problem. Column FI displays the results for the case with full information. Results in italic are hardness results that follow from already existing hardness results for bribery under full information or the possible winner problem. Results in boldface are new.

Voting rule	FI	Gaps	FP	TOS	PC	CEV	1TOS	1Gap	TTO	BTO
Plurality	P	**P**	**P**	**P**	**P**	**P**	**P**	**P**	**P**	**P**
2-Approval	P	**P**	**P**		*NPC*	**P**	**P**	**P**	**P**	**P**
(≥ 3)-Approval	NPC	*NPC*	*NPC*	*NPC*	*NPC*	*NPC*	*NPC*	*NPC*	*NPC*	*NPC*
Veto	P	**P**	**P**	**P**	**P**	**P**	**P**	**P**	**P**	**P**
2-Veto	P	**P**	**P**		*NPC*	**P**	**P**	**P**	**P**	**P**
3-Veto	P	**P**	**P**		*NPC*	**P**	**P**	**P**	**P**	**P**
(≥ 4)-Veto	NPC	*NPC*	*NPC*	*NPC*	*NPC*	*NPC*	*NPC*	*NPC*	*NPC*	*NPC*

Lemma 1. *1. $\mathscr{E}$-X-Possible Bribery is at least as hard as $\mathscr{E}$-X-Possible-Winner.*
2. $\mathscr{E}$-X-Possible Bribery is at least as hard as $\mathscr{E}$-FI-Bribery.

Proof.

1. The possible winner problem is a special case of the possible bribery problem when $\ell = 0$.
2. For $X = $ FI (each vote is complete), we obtain the definition of standard bribery.   ❑

We obtain the following results as an immediate consequence of Lemma 1.

Corollary 1. *(a) For every model $X \in PIM$ and $k \geq 3$, k-Approval-X-Possible Bribery is NP-hard.*
(b) For each model $X \in PIM$ and $k \geq 4$, k-Veto-X-Possible Bribery is NP-hard.
(c) k-Approval-PC-Possible Bribery and k-Veto-PC-Possible Bribery are NP-hard for $k \geq 2$.

The first two parts of Lemma 1 follow from the bribery results under full information from [20], the last part follows from the possible winner results from [4,26].

Both the necessary winner problem and bribery under full information in Plurality are in P [14,19]. Briskorn et al. showed that combining these two problems increases the complexity to NP-completeness [5]. The following theorem shows that combining bribery and the possible winner problem does not increase the complexity for Plurality.

Theorem 1. PLURALITY-X-POSSIBLE BRIBERY *is in* P *for every model* $X \in$ *PIM.*

For 2-Approval, we still obtain P results for seven models. Due to space restrictions we have to omit the proofs, which work by a reduction to the generalized b-edge matching problem (with upper and lower capacity restrictions both for edges and vertices). The generalized weighted b-edge matching problem was introduced and proved to be in P by Grötschel et al. [18].

Theorem 2. 2-APPROVAL-X-POSSIBLE BRIBERY *is in* P *for each model* $X \in$ $\{GAPS, FP, CEV, 1GAP, TTO, BTO, 1TOS\}$

For Veto, possible bribery is easy under each partial information model. We omit the proof due to space restrictions.

Theorem 3. VETO-X-POSSIBLE BRIBERY *is in* P *for each* $X \in PIM$.

For 2-Veto, we obtain a P result for seven models.

Theorem 4. 2-VETO-X-POSSIBLE BRIBERY *is in* P *for every model* $X \in$ $\{GAPS, 1GAP, FP, BTO, TTO, CEV\}$.

Proof. It suffices to regard $X \in \{GAPS, FP\}$. For both models, it is easy to determine all definite, potential and excluded veto candidates for each voter. We rewrite each vote by fixing c on the highest possible position unless the position is already fixed.

The following types of votes may occur: Either two, one or no veto for non-distinguished candidates is definite and the other zero, one or two vetoes for non-distinguished candidates are in jeopardy. Or c is vetoed together with candidate $d \neq c$ or the other veto aside from c is in jeopardy. We let $vscore(c)$ be the number of definite vetoes that c receives. First we assume that $|C| \geq 3$ and $vscore(c) > \ell$. Otherwise c is or can be trivially made a possible winner. Clearly the briber only bribes voters initially vetoing c. Thus c has $vscore(c) - \ell$ vetoes after the bribery. To determine if a successful bribery exists, we will solve a flow maximization problem defined as follows (for our purposes, we assume that $v_1, \ldots, v_{n-vscore(c)}$ do not definitely veto c and $w_1, \ldots, w_{vscore(c)}$ definitely veto c):

The vertices include a source s, a sink t, nodes c_i ($i = 1, \ldots, m-1$), $v_1, \ldots, v_{n-vscore(c)}$, $w_1, \ldots, w_{vscore(c)}$, K_i ($i = 1, \ldots, n - vscore(c)$), and two additional nodes N and B. The edges are defined as follows:

- There is an edge from s to c_i $(i = 1, \ldots, m - 1)$ with capacity $vscore(c) - \ell$. These edges later will assure that every candidate receives at least as many vetoes as c, namely $vscore(c) - \ell$.
- From c_i to v_j, there is an arc with capacity one if and only if c_i is definitely vetoed by v_j $(i = 1, \ldots, m - 1, j = 1, \ldots, n - vscore(c))$. This and the next three groups of edges ensure that each candidate can only be vetoed by voters giving them at least possibly a veto. Moreover, these edges fix definite vetoes and admissibly assign vetoes in jeopardy to non-distinguished candidates.
- There is an arc from c_i to K_j $(i = 1, \ldots, m - 1, j = 1, \ldots, n - vscore(c))$ with capacity one if and only if c_i is only potentially but not definitely vetoed by v_j.
- There is an edge from K_j to v_j $(j = 1, \ldots, n - vscore(c))$ with capacity equal to the number of indefinite veto positions in this vote v_j.
- There is an edge from c_i to w_j $(i = 1, \ldots, m - 1, j = 1, \ldots, vscore(c))$ with capacity one if and only if c_i is potentially or definitely vetoed aside from c by w_j.
- From v_j to t, there is an edge with capacity 2 $(j = 1, \ldots, n - vscore(c))$. These edges certify that all voters vetoing two non-distinguished candidates may give only two vetoes to non-distinguished candidates.
- From w_j to N, there is an edge with capacity 1 $(j = 1, \ldots, vscore(c))$.
- There is an edge from N to t with capacity $vscore(c) - \ell$. N certifies that only $vscore(c) - \ell$ of these votes (not bribed) initially vetoing c actually assign vetoes to non-distinguished candidates. The other ℓ voters initially vetoing c can give arbitrarily vetoes to non-distinguished candidates. These are modeled by B.
- From c_i to B, there is an edge with capacity ℓ $(i = 1, \ldots, m - 1)$.
- From B to t, there is an edge with capacity 2ℓ. These latter two groups of edges ensure that each c_i receives at most ℓ vetoes by bribed voters, and bribed voters can give 2ℓ vetoes to non-distinguished candidates in total. These two capacity constraints are necessary and sufficient conditions for 2-Veto concerning how 2ℓ vetoes can be admissibly assigned to the c_i, in particular, these conditions ensure that no candidate is vetoed twice by the same voter.

From the reasoning above, it follows that the possible bribery problem has a solution for c if and only if there is a maximum flow of size $(m-1)(vscore(c)-\ell)$. If such a flow exists, all feasibility constraints are satisfied and there is a way to bribe such that each non-distinguished candidate receives enough vetoes - at least as many as c. This maximal flow surely does not fill all open veto positions in general, but c stays the winner if the other open veto positions are arbitrarily filled with further vetoes for some c_i. Conversely, suppose such a flow does not exist (or has a smaller value). This implies that some c_i does not get enough vetoes and thus beats c.

As computing a maximal flow is in P and all constraints are bounded above by a polynomial, the overall problem is in P. $\square$

For the two remaining partial information models TOS and 1TOS we could establish an easiness result for 1TOS, the case of TOS remains open.

Theorem 5. 2-VETO-1TOS-POSSIBLE BRIBERY *is in* P.

For 3-Veto, we again obtain seven P results. We will omit the proofs due to space restrictions. Similarly to 2-Veto, the TOS case remains an open problem.

Theorem 6. 3-VETO-X-POSSIBLE BRIBERY *is in* P *for each model* $X \in \{GAPS, FP, 1GAP, BTO, TTO, CEV, 1TOS\}$.

The proof for all models is based on a similar construction as in the proof of Theorem 2 with the difference, that here we reduce to the generalized b-edge *cover* problem, i.e., a bribery is successful if and only if there is a *minimum* matching with at most a certain number of edges.

3.2 Destructive Possible Bribery

In this section we turn to the destructive case where we ask if the briber can prevent a despised candidate from being a winner in at least one complete profile. The following theorem provides not only results for the classes of k-Approval and k-Veto, but for the class of all scoring rules.

Theorem 7. DESTRUCTIVE-$\mathcal{E}$-X-POSSIBLE-BRIBERY *is in* P *for every scoring rule* $\mathcal{E}$ *and each model* $X \in PIM$.

Proof. Our proof is only for $X \in \{PC, FP\}$. All remaining models are special cases of at least one of them and thus inherit the polynomial time upper bound.

Basically it suffices to check for each non-distinguished candidate c_i ($1 \leq i \leq m - 1$) if c_i beats c whenever c_i is fixed as good as possible in each vote and c as bad as possible. This way we will construct a profile completion as good as possible for c_i and as bad as possible for c.

Let's start with FP. If c's or c_i's position is fixed in a vote, this position is unique and unchangeable. If c is not assigned to any position in a vote v, we fix c onto the worst possible open position. Analogously, we fix c_i onto the best possible open position. Thus, we can uniquely determine the score of c and c_i in each vote.

Now we can apply the algorithm under full information which is basically the same as the algorithm in [24].

For PC, almost the same algorithm can be applied. The only difference arises from the preprocessing. If $c_i \succ c$ for a vote v or at least $\neg(c \succ c_i)$, we fix c_i on position $\alpha_v + 1$ and c on position $m - \beta_v$. α_v here denotes the number of candidates $d \in C \setminus \{c_i, c\}$ with $d \succ_v c_i$ and counts the number of definitely lost pairwise comparisons of c_i in v, whereas β_v sums up all pairwise comparisons of c which are definitely won. We suppose that c_i wins all pairwise comparisons in jeopardy and c loses all such comparisons.

If $c \succ_v c_i$ for a vote v, we again fix c as low as possible (on position $m - \beta_v$ as before) and c_i as high as possible below c, i.e., we assume that c_i loses the same pairwise comparisons as c and all other definitely lost pairwise comparisons, but wins all remaining duels in jeopardy. Following this, we keep the potential score of c_i as high as possible. After applying this subroutine, we have unique scores for c and c_i and can apply the algorithm for FP.   ☐

4   Conclusion

We have studied the complexity of the constructive and destructive possible bribery problem for k-Approval and k-Veto. In contrast to the necessary bribery version investigated in [5], we mostly achieved P results. We could also prove that destructive possible bribery is easy for the family of scoring rules.

References

1. Baumeister, D., Faliszewski, P., Lang, J., Rothe, J.: Campaigns for lazy voters: truncated ballots. In: Proceedings of the 11th International Joint Conference on Autonomous Agents and Multiagent Systems, pp. 577–584. IFAAMAS, June 2012
2. Baumeister, D., Rothe, J.: Taking the final step to a full dichotomy of the possible winner problem in pure scoring rules. Inf. Process. Lett. **112**(5), 186–190 (2012)
3. Betzler, N., Bredereck, R., Niedermeier, R.: Partial kernelization for rank aggregation: theory and experiments. In: Proceedings of the 3rd International Workshop on Computational Social Choice, pp. 31–42 (2010)
4. Betzler, N., Dorn, B.: Towards a dichotomy for the possible winner problem in elections based on scoring rules. J. Comput. Syst. Sci. **76**(8), 812–836 (2010)
5. Briskorn, D., Erdélyi, G., Reger, C.: Bribery in k-approval and k-veto under partial information (extended abstract). In: Proceedings of The 15th International Joint Conference on Autonomous Agents and Multiagent Systems, pp. 1299–1301. IFAAMAS, May 2016
6. Chevaleyre, Y., Lang, J., Maudet, N., Monnot, J.: Possible winners when new candidates are added: the case of scoring rules. In: Proceedings of the 24th AAAI Conference on Artificial Intelligence, pp. 762–767. AAAI Press, July 2010
7. Conitzer, V., Walsh, T., Xia, L.: Dominating manipulations in voting with partial information. In: Proceedings of the 25th AAAI Conference on Artificial Intelligence, pp. 638–643. AAAI Press, August 2011
8. Dwork, C., Kumar, R., Naor, M., Sivakumar, D.: Rank aggregation methods for the web. In: Proceedings of the 10th International World Wide Web Conference, pp. 613–622. ACM Press, May 2001
9. Ephrati, E., Rosenschein, J.: The Clarke Tax as a consensus mechanism among automated agents. In: Proceedings of the 9th National Conference on Artificial Intelligence, pp. 173–178. AAAI Press (1991)
10. Ephrati, E., Rosenschein, J.: Multi-agent planning as a dynamic search for social consensus. In: Proceedings of the 13th International Joint Conference on Artificial Intelligence, pp. 423–429. Morgan Kaufmann (1993)
11. Ephrati, E., Rosenschein, J.: A heuristic technique for multi-agent planning. Ann. Math. Artif. Intell. **20**(1–4), 13–67 (1997)
12. Fagin, R., Kumar, R., Sivakumar, D.: Efficient similarity search and classification via rank aggregation. In: Proceedings of the 2003 ACM SIGMOD International Conference on Management of Data, pp. 301–312. ACM Press (2003)
13. Faliszewski, P., Hemaspaandra, E., Hemaspaandra, L.A.: The complexity of bribery in elections. In: Proceedings of the 20th AAAI Conference on Artificial Intelligence, pp. 641–646. AAAI Press (2006)
14. Faliszewski, P., Hemaspaandra, E., Hemaspaandra, L.A.: How hard is bribery in elections? J. Artif. Intell. Res. **35**, 485–532 (2009)

15. Faliszewski, P., Hemaspaandra, E., Hemaspaandra, L.A., Rothe, J.: Llull and Copeland voting computationally resist bribery and constructive control. J. Artif. Intell. Res. **35**, 275–341 (2009)
16. Faliszewski, P., Reisch, Y., Rothe, J., Schend, L.: Complexity of manipulation, bribery, and campaign management in Bucklin and fallback voting. Auton. Agents Multi-Agent Syst. **29**(6), 1091–1124 (2015)
17. Ghosh, S., Mundhe, M., Hernandez, K., Sen, S.: Voting for movies: the anatomy of recommender systems. In: Proceedings of the 3rd Annual Conference on Autonomous Agents, pp. 434–435. ACM Press, May 1999
18. Grötschel, M., Lovász, L., Schrijver, A.: Geometric Algorithms and Combinatorial Optimization. Springer, Heidelberg (1988)
19. Konczak, K., Lang, J.: Voting procedures with incomplete preferences. In: Proceedings of IJCAI-05 Multidisciplinary Workshop on Advances in Preference Handling, pp. 124–129 (2005)
20. Lin, A.: The complexity of manipulating k-approval elections. In: Proceedings of the 3rd International Conference on Agents and Artificial Intelligence, pp. 212–218 (2011)
21. Parkes, D.C., Xia, L.: A complexity-of-strategic-behavior comparison between schulze's rule and ranked pairs. In: Proceedings of the 26th AAAI Conference on Artificial Intelligence. AAAI Press (2012)
22. Reger, C.: Voter control in k-approval and k-veto under partial information. In: Proceedings of the 14th International on Artificial Intelligence and Mathematics, January 2016
23. Reisch, Y., Rothe, J., Schend, L.: The margin of victory in Schulze, cup, and copeland elections: complexity of the regular and exact variants. In: Proceedings of the 7th European Starting AI Research Symposium, pp. 250–259. IOS Press (2014)
24. Xia, L.: Computing the margin of victory for various voting rules. In: Proceedings of the 13th ACM Conference on Electronic Commerce, pp. 982–999. ACM Press, June 2012
25. Xia, L.: Designing social choice mechanisms using machine learning. In: Proceedings of the 12th International Conference on Autonomous Agents and Multiagent Systems, pp. 471–474. IFAAMAS, May 2013
26. Xia, L., Conitzer, V.: Determining possible and necessary winners given partial orders. J. Artif. Intell. Res. **41**, 25–67 (2011)

An Adjusted Recommendation List Size Approach for Users' Multiple Item Preferences

Serhat Peker[(✉)] and Altan Kocyigit

Department of Information Systems, Middle East Technical University, 06800 Ankara, Turkey
{speker,kocyigit}@metu.edu.tr

Abstract. This paper describes the design and implementation of a novel approach to dynamically adjust the recommendation list size for multiple preferences of a user. By considering users' earlier preferences, machine learning techniques are employed to estimate the optimal recommendation list size according to current conditions of users. The proposed approach has been evaluated on real-life data from grocery shopping domain by conducting a series of experiments. The results show that the proposed approach achieves better overall recommendation quality than the standard approach and it outperforms the benchmark method in efficiency by shortening the recommendation list while maintaining the effectiveness.

Keywords: Top-N recommender systems · Recommendation list size · Recommendation length · Recommendation quality · Recommendation efficiency

1 Introduction

Nowadays, recommender systems are commonly used by commercial companies for the purpose of helping their customers on decision making process. These systems usually use two main recommendation strategies: "find all good items" and "recommend top-N items" [1]. The recommender systems using "find all good items" approach offer all the recommendable items that can suit the user's tastes, whereas in the "recommend top-N items" approach, only the top ranked N items are recommended to the user. The latter one is the most common solution for product recommendations and many commercial companies take the take advantage of this technique in their recommender systems [2, 3].

In the top-N recommender systems that employ "showing top k matching items" as the recommendation strategy, of the length recommendation list usually ranges from 5 to 20 [4]. It is possible to increase the proportion of items that are correctly identified (recall) by providing more items in the recommendation list. likely to However, users are not be overwhelmed by recommendation lists containing a large number of items, and it is also important to fit recommendation lists in small display devices such as mobile phones [4]. Moreover, increasing the number of recommended items, N, improves recall, but it is likely to deteriorate precision [1, 5, 6] which is the proportion of recommended items that result in matches. Together with recall, precision shows the

© Springer International Publishing Switzerland 2016
C. Dichev and G. Agre (Eds.): AIMSA 2016, LNAI 9883, pp. 310–319, 2016.
DOI: 10.1007/978-3-319-44748-3_30

quality of the recommendation, and high precision is more preferable in recommender systems using "recommend top-N items" approach [6].

Although, top-N recommender systems typically return a fixed number of items in each recommended list, individuals may have multiple preferences at a time or in a specific time interval, and the number of such preferences may vary depending on cases and context in which the individual is. For example, consider a man being time pressed during weekday daytime in the grocery shopping. He prefers to buy a couple of items (e.g., 3 unique items) on weekday evenings for daily needs such as bread, eggs, milk, etc., whereas he has too many items (e.g., 15 unique items) in his shopping basket on a weekend afternoon.

To evaluate the quality of existing recommender systems producing a fixed number of items for users with multiple preferences at a time, let us consider the above example again. Suppose that this man uses a recommendation agent in his smart phone for the grocery shopping and this application generates a fixed length of recommendation list containing 10 items for his next visit. For the first case, agent may probably perform with high recall and low precision, since it produces a long recommendation list for a number of purchased items, but agent also overwhelms the user with many irrelevant items. For the second case, on the other hand, recall is most probably lower than the one in first case and precision may higher, because the number of recommended items is closer to the number of preferred items. However, recommendation agent also misses some relevant items in this case, since it recommends fewer items than the user prefers.

As explained in the examples, for multiple preferences at a time, recommender systems returning a fixed number of items may cause some issues which are undesirable from the users' point of view, and it is obvious that the length of recommendation list has a significant impact on the recommendation quality. In this respect, this paper aims to dynamically adjust the recommendation list size of a user with multiple preferences by employing machine learning techniques. The proposed approach dynamically determines the optimal recommendation list size based on the previous preferences of the user. The applicability of the approach is experimentally evaluated by using real-life data obtained from the grocery shopping domain and the results show the effectiveness of our approach.

The remainder of this paper is organized as follows. Section 2 describes proposed approach. In Sect. 3, evaluation methodology is presented together with the experimental results. Finally, Sect. 4 concludes the study and points out directions for future work.

2   Proposed Approach

The major steps of our approach for adjusted recommendation list size are depicted in Fig. 1. First, a predictive model is constructed based on previous preferences of users. Then, identified features' values pertaining to the user and the current recommendation are used as input for the model to estimate an adjusted recommendation list size that will be finally used as a reference in the construction of recommendation list.

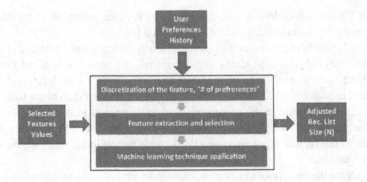

Fig. 1. Schematic overview of our approach

In our predictive model, "number of preferences", the numeric target value, is firstly discretized into a set of intervals in order to use machine learning classifiers. The reason behind this step is to construct a categorical variable in order to transform the problem into a classification one, and to employ powerful classification techniques. In case of a continuous target attribute, unsupervised discretization techniques are used to divide the variable into discrete intervals. Famous representatives of unsupervised techniques are equal-width and equal-frequency discretization [7, 8]. In this study, we use an equal-frequency discretization method, because the equal-width method does not perform well when the variable observations are not distributed evenly, and thereby may cause information loss after the discretization process [9]. Equal-frequency as an unsupervised method requires the user to specify the number of discrete intervals and fewer number of intervals is preferable in order to avoid the fragmentation problem occurring in classifiers [10]. Because of that, we set the number of intervals to three (as low, medium and high) in this study.

Not surprisingly, the same number of preferences may mean different things to different people. For instance, in grocery shopping, buying 10 items could be taken as too many to a user with an average number of purchased items of 5, whereas too few to a consumer with an average number of purchased items of 20. Hence, it is important to apply the discretization technique for the number of preferences at the individual level. Therefore, the proposed approach employs equal-frequency technique at the individual level to produce personalized cut points for each user.

After the "number of preferences" is transformed to categorical variable via equal-frequency discretization, features related to users' preferences are identified to build a model that produces better performance. At the final stage of the model construction, a machine learning technique is trained with users' previous preferences with selected features. In this study, we use two classifiers and a regression technique in order to compare their performance for recommendation quality. The classifiers used are decision tree (J48) and KNN algorithms. These classifiers are chosen because they are in the top 10 list of classification techniques [11]. For KNN method, we searched the optimal k value and identified it as 75. We also use multiple regression by coding the values of discretized "number of preferences" variable (low, medium and high) as 1, 2, and 3.

In this way, the feature is ordinal and the continuous nature of it is preserved which allow the researchers to employ regression analysis [12].

After training the model, identified features' values of the user and the current recommendation are used as input for the model. Then, it outputs the estimated number of preferences of the user for the recommendation list size. Note that, machine learning classifiers produce the output as a categorical value such as low, medium or high. On the other hand, regression model generates the output as numerical nominal value as 1, 2, or 3. Since, our approach employs the equal-frequency discretization method for "number of preferences", these produced values correspond to bins. Because of that, our approach replaces each bin value by its mean. Note that, if this value is decimal, it is rounded to nearest integer. Therefore, the proposed approach returns this final result as the possible recommendation list size.

3   Evaluation

In this study, we performed an offline evaluation technique to measure the effectiveness of our approach. We chose this evaluation method, since it does not require interaction with real users, and the performance of different recommender algorithms and approaches can be easily evaluated by using existing datasets at low cost [13]. Grocery shopping was chosen as the application domain, since customers in the grocery industry prefer multiple products in a single transaction. A comprehensive set of experiments were conducted on a dataset obtained from a Turkish retail grocery store. In the following subsections, we first describe the dataset and data pre-processing task. Next, we explain the experimental settings and evaluation measures. Then, the recommendation method that used in the experiments and a benchmark method for the comparison are introduced. Finally, we present the results we obtained.

3.1   Dataset

The dataset we used in the experiments contains purchase transactions gathered over 104 weeks (January 1, 2013 – December 31, 2014). In the dataset, many of the customers have visited the store irregularly (only a few times during the period) and transactions of these customers most probably mislead the results of the study. Thus, we pre-processed the dataset by excluding customers who have visited the store less than 25 times. After this elimination, the dataset is left with 46 customers with 534616 purchase records. In the dataset, products are categorized according to a four-level hierarchy. The lowest level includes information about the product brand, amount and etc. For ketchup, for example the third level category is "ketchup", and one of the examples of fourth level category for this product is "Tat Ketchup, 600 Grams". "Tat" is the brand of product, whereas "600 Grams" is the amount of it. Since the fourth level is highly specific, we considered third-level categories in the experiments. There are unique 335 third-level categories in the dataset. Customers within our sample bought 47.63 distinct third-level categories on average with the standard deviation of 26.07 and average basket size for our sample is 11.43 with the standard deviation of 6.61.

3.2 Data Pre-processing

This process mainly includes outlier elimination and feature extraction steps. One way to define an outlier is using interquartile range (IQR). If a value is more than (Q3 + 3*IQR) or less than (Q1 - 3*IQR), then it is considered as an extreme outlier. Note that, Q1 and Q3 represent the lower and upper quartiles, respectively. Outlier detection is applied to the "number of preferences" (number of unique products purchased). However, this variable may have different distribution for different customers. Because of that, we applied the outlier detection for each customer separately. Therefore, for each customer, we remove the instances having an outlier value for the corresponding variable.

Among existing contextual dimensions, temporal information is precious and easy-to-collect feature for increasing the performance of recommendations [14]. As a result, we considered timestamp of transactions as one of the features. Since timestamp is a composite variable in our dataset, we extracted two distinct features from this variable, which are day of the week and time of the day. We also categorized the values of these variables and Table 1 shows both categorical and actual values of each feature.

Table 1. Time features

Feature	Categorical values	Range of actual values
Day	Weekend	Monday to Friday
	Weekday	Saturday, Sunday
Time	Morning	08:00 to 11:59
	Afternoon	12:00 to 17:59
	Evening	18:00 to 20:59
	Night	21:00 to 22:59

In the experiments, categorical values were used for the features of time and day. Moreover, simple n-visit moving average feature which is the average number of purchased unique products in last n visit is calculated. We selected n as 5. The number of purchased unique products in last visit is also computed and formed as another feature.

3.3 Experimental Settings and Design

To generate the training and test sets, we sorted each customer's shopping trips according to their timestamps. For each customer's purchase history, we use the first 80 % of the visits as training and the latter 20 % as test data for all set of experiments. The training data is utilized to train the predictive model of our approach. In the testing set, for each visit of the customer a ranked list of recommended products is generated and the recommended products are compared with the actual ones purchased by the customer in the test set to compute the corresponding evaluation metric.

3.4 Evaluation Metrics

Recall and precision, which are first originated in the field of information retrieval, have been widely used in the performance evaluation of top-N recommender systems [1, 15]. These metrics are defined as follows:

$$Recall@N = \frac{|rec@N \cap purc|}{|purc|} \tag{1}$$

$$Precision@N = \frac{|rec@N \cap purc|}{|rec@N|} \tag{2}$$

where rec@N donates the top-N recommended products for the test instance and purc is the actual product set that the customer has purchased in the same test instance. However, as known well, there is a tradeoff between these two measures. For instance, increasing the number N tends to increase recall but is likely to reduce precision. Since both are critical measures in the quality assessment, F-measure [16] which is the harmonic mean of precision and recall was used in the performance evaluation. It is computed as:

$$F\text{-}measure = 2 \times \frac{recall \times precision}{recall + precision} \tag{3}$$

We first computed F-measure for each customer separately by averaging all computed values in the customer's test set, and then average these customers' personal values to get overall F-measure value for a given recommendation list size, N.

3.5 Recommendation Method

In the experimental evaluation, most-frequent item recommendation approach [5] was used to generate top-N recommendation lists for the customers' visits in the test set. The reasons behind this choice are that it is simple and well-known method and it was also employed in similar studies on predicting grocery shopping lists [5, 17, 18]. This approach sorts the products in customer's purchase history according to their frequency count and simply returns the N most frequently products as the current shopping list. This approach may not be so efficient in the recommendation performance, but it is not our aim in this study. Most-frequent item recommendation method is only considered as a reasonable baseline to predict top-N recommendation lists and to evaluate the effectiveness of our proposed approach.

3.6 Benchmark Method

To compare the performance of our proposed approach on the recommendation efficiency, we selected last visit's N as a benchmark. This method identifies the number of products purchased in the previous visit of the customer as the recommendation list size for the next visit of that customer. Additionally, we also conducted experiments by varying the recommendation list size, N from 5 to 15 (in increments of 1) for comparing our approach on adjusted recommendation list size with the standard one using different fixed length of recommendation list sizes.

3.7   Results

Figure 2 shows the performance of typical recommendation technique with fixed size recommendation list for consecutive experiments which were conducted by using different recommendation list sizes (N) varying from 5 to 15.

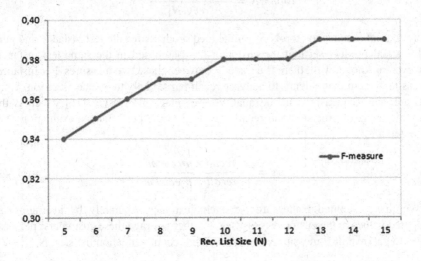

Fig. 2.   F measure w.r.t recommendation list size

As shown in Fig. 2, F-measure increases slightly at the beginning, but stay constant as the recommendation list size increases. The experimental results of our proposed approach with different ML methods and benchmark methods are listed in Table 2. In the table, there are three different methods that adjust recommendation list size using different machine learning techniques and a benchmark method that adjusts the recommendation list size by using last visit's item count. In addition to these, only two relevant methods using fixed recommendation list size were added to the table for the purpose of comparison.

Table 2.   Comparison of methods for recommendation list size

Method	Avg Rec. List Size	F-Measure
Adjusted with J48	8.96	0.4
Adjusted with KNN	8.94	0.39
Adjusted with Multi. Reg.	11.98	0.44
Last Visit N	11.18	0.4
Fixed N	9	0.37
Fixed N	12	0.38

In Table 2, "Avg Rec. List Size" column indicates the average number of recommended items to per customer. The results listed in Table 2 reveals the following findings:

- Our approach with J48 classifier achieves 8 % improvement in terms of F-measure, compared to the standard method using fixed recommendation list size (N = 9). Similarly, our approach with KNN classifier also improves F-measure by 5 %, compared to related standard method with the same size of recommendation list.
- Our approach with multiple regression achieves 16 % improvement in terms of F-measure, compared to the standard method with the same size of recommendation list (N = 12).
- Our approach with J48 classifier achieves same F-measure as the method of Last Visit N by shortening the average recommendation list size by nearly 20 % (i.e., with average recommendation list size of 8.96, down from 11.18 of Last Visit N approach).
- When we compare our approach based on KNN classifier with the method of Last Visit N, our approach provides 20 % reduction in the recommendation list size with only a small amount of accuracy loss (2.5 %)
- By increasing recommendation list size by 7 % (from 11.18 of Last Visit N approach to 11.98), our approach based on multiple regression achieves 10 % improvement in the recommendation performance, compared to the method of Last Visit N.

The above findings indicate that for the same size of recommendation list, our approach provides a better quality of recommendation than the standard one does. It also outperforms the benchmark, Last Visit N method in efficiency by providing reduction in the recommendation list size while preserving the accuracy.

4 Conclusion

This research proposes an approach to dynamically adjust the recommendation list size of a user with multiple preferences in order to improve the recommendation quality. By taking advantage of users' previous preferences and machine learning techniques, the proposed approach adjusts the number of items that will be preferred by the user according to his changing conditions. We evaluated our approach by conducting extensive experiments on a real-life dataset in the grocery shopping domain. According to the experimental results, our approach with all three selected machine learning techniques outperforms the traditional and widely used standard approach in effectiveness and it also provides better performance than the benchmark, Last Visit N method in efficiency by shortening the recommendation list while maintaining the effectiveness.

The proposed approach has the advantages of recommending a reasonable number of items without overwhelming the users and increasing recommendation quality. In this manner, recommender systems can utilize our approach to determine the optimal number of items to be recommended for especially the cases where the user has multiple preferences at a time. Therefore, such an efficient recommender system creates a win-win opportunity for both companies and customers by raising both customer satisfaction and company's revenue.

In this study, we presented the effectiveness of our approach for grocery shopping and the features related to this domain were identified as input for the proposed approach. However, our approach can potentially be applied to other domains as well, and the selected feature set may be different depending on the application domain and the chosen data set. Further, we used a set of machine learning methods in this study and other different machine learning techniques can also be employed in the proposed approach. In this context, a possible direction in the future work might be to apply the proposed approach in other application domains and using other machine learning techniques. In addition to this, in order to measure the user satisfaction, evaluating the proposed approach by conducting a live user study would be another future research issue.

Acknowledgments. This work is partially supported by the Scientific and Technological Research Council of Turkey (TUBITAK).

References

1. Herlocker, J.L., Konstan, J.A., Terveen, L.G., Riedl, J.T.: Evaluating collaborative filtering recommender systems. ACM Trans. Inf. Syst. (TOIS) **22**, 5–53 (2004)
2. Ricci, F., Rokach, L., Shapira, B.: Introduction to Recommender Systems Handbook. Springer, New York (2011)
3. Schafer, J.B., Konstan, J., Riedl, J.: Recommender systems in e-commerce. In: Proceedings of the 1st ACM Conference on Electronic Commerce, pp. 158–166. ACM (1999)
4. Pu, P., Faltings, B., Chen, L., Zhang, J., Viappiani, P.: Usability guidelines for product recommenders based on example critiquing research. In: Ricci, F., Rokach, L., Shapira, B., Kantor, P.B. (eds.) Recommender Systems Handbook, pp. 511–545. Springer, New York (2011)
5. Sarwar, B., Karypis, G., Konstan, J., Riedl, J.: Analysis of recommendation algorithms for e-commerce. In: Proceedings of the 2nd ACM Conference on Electronic Commerce, pp. 158–167. ACM (2000)
6. Gunawardana, A., Shani, G.: A survey of accuracy evaluation metrics of recommendation tasks. J. Mach. Learn. Res. **10**, 2935–2962 (2009)
7. Dougherty, J., Kohavi, R., Sahami, M.: Supervised and unsupervised discretization of continuous features. In: Machine Learning: Proceedings of the Twelfth International Conference, pp. 194–202 (1995)
8. Catlett, J.: On changing continuous attributes into ordered discrete attributes. In: Kodratoff, Y. (ed.) EWSL 1991. LNCS(LNAI), vol. 482, pp. 164–178. Springer, Heidelberg (1991)
9. Kotsiantis, S., Kanellopoulos, D.: Discretization techniques: a recent survey. GESTS Int. Trans. Comput. Sci. Eng. **32**, 47–58 (2006)
10. Quinlan, J.R.: C4.5: Programs for Machine Learning. Elsevier, Amsterdam (2014)
11. Wu, X., Kumar, V., Quinlan, J.R., Ghosh, J., Yang, Q., Motoda, H., McLachlan, G.J., Ng, A., Liu, B., Philip, S.Y.: Top 10 algorithms in data mining. Knowl. Inf. Syst. **14**, 1–37 (2008)
12. Rucker, D.D., McShane, B.B., Preacher, K.J.: A researcher's guide to regression, discretization, and median splits of continuous variables. J. Consum. Psychol. **25**, 666–678 (2015)
13. Shani, G., Gunawardana, A.: Evaluating recommendation systems. In: Ricci, F., Rokach, L., Shapira, B., Kantor, P.B. (eds.) Recommender Systems Handbook, pp. 257–297. Springer, New York (2011)

14. Campos, P.G., Díez, F., Cantador, I.: Time-aware recommender systems: a comprehensive survey and analysis of existing evaluation protocols. User Model. User-Adapt. Interact. **24**, 67–119 (2014)
15. Cremonesi, P., Koren, Y., Turrin, R.: Performance of recommender algorithms on top-n recommendation tasks. In: Proceedings of the Fourth ACM Conference on Recommender Systems, pp. 39–46. ACM (2010)
16. Yang, Y.M., Liu, X.: A re-examination of text categorization methods. In: Proceedings of 22nd International Conference on Research and Development in Information Retrieval, SIGIR 1999, pp. 42–49 (1999)
17. Cumby, C., Fano, A., Ghani, R., Krema, M.: Predicting customer shopping lists from point-of-sale purchase data. In: Proceedings of the Tenth ACM SIGKDD International Conference on Knowledge Discovery and Data Mining, pp. 402–409. ACM (2004)
18. Cumby, C., Fano, A., Ghani, R., Krema, M.: Building intelligent shopping assistants using individual consumer models. In: Proceedings of the 10th International Conference on Intelligent User Interfaces, pp. 323–325. ACM (2005)

mRHR: A Modified Reciprocal Hit Rank Metric for Ranking Evaluation of Multiple Preferences in Top-N Recommender Systems

Serhat Peker[(✉)] and Altan Kocyigit

Department of Information Systems, Middle East Technical University, 06800 Ankara, Turkey
{speker,kocyigit}@metu.edu.tr

Abstract. Average reciprocal hit rank (ARHR) is a commonly used metric for ranking evaluation of top-n recommender systems. However, it suffers from an important shortcoming that it cannot be applied when the user has multiple preferences at a time. In order to overcome this problem, a modified version of ARHR metric is introduced and applied to grocery shopping domain by conducting a series of experiments on real-life data. The results show that the proposed measure is feasible for ranking evaluation of Top-N recommender systems in the cases where the users have multiple preferences at a time or a specific time interval.

Keywords: Top-N recommender systems · Ranking evaluation · ARHR · Reciprocal rank · Hit rank

1 Introduction

Recommender systems have become an important tool to help users in making the best choice among a huge set of alternatives by providing useful and relevant offers [1, 2]. Today, recommender systems are extremely common and applied in a broad range of applications including recommending movies, music, TV programs, books, and similar many products and services. In recent years, recommender systems research has embraced the effectiveness of Top-N recommendation technique, which produces a ranked list of recommended items where the top ranked items are predicted to be the most preferred.

In the context of top-N recommendations, recommendation quality is an important issue. For this purpose, there are two widely used metrics, which are Hit Rate (HR) and the Average Reciprocal Hit-Rank (ARHR) [3]. Both HR and ARHR are regularly employed for measuring the quality of such recommender systems. HR returns the percentage of users whose top-N recommendation list includes at least one item in the test set. This metric is used when each user's test set includes only one test instance. Some of the previous studies [4–6] refer to this metric as recall@N and some of these studies [4, 5] also modified it to be applicable for the cases when there is more than one test instance in users' test sets. This modified metric returns the percentage of test instances that appears in top-N recommendation list. Please note that, in both HR and

C. Dichev and G. Agre (Eds.): AIMSA 2016, LNAI 9883, pp. 320–329, 2016.
DOI: 10.1007/978-3-319-44748-3_31

recall@N calculation, if only one item in the test instance appears in top-N recommendation list, it is qualified as a hit and other items are not considered at all.

On the other hand, ARHR is a ranking measure that considers the item position in Top-N list. It measures how close the predicted items to the top of the recommendation list. This metric is especially important for the cases in which there are too many alternatives to recommend and the cases in which it is required to present a limited number of items for users. In a grocery shopping, for example, there is a tremendous diversity of products and forming a reasonable recommended item set is extremely difficult. If, for example, the display size of mobile devices is limited, the users may not want to browse through the entire list. It is needed to provide a small set of items that fit in such a small display. These two different examples illustrate two basic challenges in Top-N recommender systems.

In addition to above cases, users usually investigate items in Top-N list starting at the top and most probably they do not go down to the end of the list, when the list is not short [7]. Because of that, users do not like long recommendation lists containing too many items. In order to overcome all of the aforementioned problems, the solution is to shorten the recommendation list by identifying the most appealing items for the user. In this manner, it is important to predict highly relevant items in the top positions of recommendation list. Therefore, ARHR measures closeness of correctly predicted item to the top of the recommendation list, and thereby perfectly suits for evaluating the quality of Top-N recommender systems.

Despite the usefulness of ARHR for the quality assessment of Top-N recommender systems, it cannot be applied for the cases where the user has multiple preferences at a time or a specific time interval. In many of the cases, however, users have multiple preferences at a time or a specific time interval, such as a list of purchased products for a single transaction of grocery shopping, a list of websites being visited throughout the day or a list of TV programs being watched throughout the evening. In such cases, there may be multiple correct matches (hits) and present ARHR metric cannot evaluate these cases because of its shortcomings.

The primary goal of this paper is to propose a new metric called mRHR to evaluate the quality of Top-N recommender systems for the users with multiple preferences at a time or a specific time interval. For this purpose, the standard ARHR metric is modified accordingly. The use of the proposed metric is demonstrated in an example and its performance is also investigated by conducting an exhaustive set of experiments on the real-life data from the grocery shopping domain. Experimental and statistical test results are reported to show that our new measure is consistent and appropriate for evaluating quality of Top-N recommender system.

The remainder of this paper is organized as follows. In Sect. 2, the original ARHR metric is described. Section 3 introduces our proposed metric and its usage is illustrated on an example in Sect. 4. Section 5 describes comprehensive experimental evaluation. Finally, the paper is concluded with a summary in Sect. 6.

2 ARHR

ARHR is based on reciprocal rank (RR) [8] in Information Retrieval (IR) research and it is a popular ranking metric to measure the quality by finding out how far from the top of the list the first relevant document is. ARHR is defined as [3]:

$$ARHR = \frac{1}{n} \sum_{i=1}^{h} \left(\frac{1}{p_i} \right) \tag{1}$$

where h is the number of hits which is the number of items in the test set that were also present in the top-N recommended items returned for each user, p_i is the position of the item in the ranked recommendation list for the i-th hit, and n is the number of users. According to the Eq. 1, RR, which is equal to $1/p_i$, is calculated for each hit and at the end all computed RR values are averaged.

ARHR rewards each hit based on its position in the recommendation list, and hits that occur earlier in the ranked list are assigned higher weights than that of the ones that occur later in the list. Thus, ARHR measures how close the correctly predicted items to the top of the recommendation list. The value returned by ARHR is between 0 and 1. Higher values of ARHR are more desirable as they indicate that the algorithm is able to predict items in the earlier positions of Top-N lists. Conversely lower values of ARHR indicate that the algorithm predicts items in the later positions of Top-N lists.

Many previous studies [9–12] used ARHR successfully in the evaluation of top-N recommendations. These studies used only one instance for each user in the test set and the users prefer a single item in the test instance. This means that ARHR was used when the user prefers only an item at a time. When the user prefers multiple items at a time, it is not possible to compute and use ARHR metric directly. Therefore, a new metric is required to evaluate the user's multiple preferences at a time and the next section presents mRHR, a modified version of ARHR which can be used for this purpose.

3 Proposed Metric

In order to cover multiple preferences at a time or a specific time interval, we proposed a novel metric called *mRHR* which overcomes the shortcomings of the ARHR mentioned in the previous section.

Let the user prefer more than one item in a time interval or at a time, *mRHR* is defined as follows:

$$mRHR = \frac{1}{\# \, of \, preferences} \sum_{i=1}^{N} \left(\frac{hit_i}{rank_i} \right) \tag{2}$$

where hit_i donates if the recommended item is preferred by the user. Like in original ARHR, if the recommended item is preferred by the user, then it gets true (1), otherwise it gets false (0). N is the length of ordered recommendation list and $rank_i$ is the ranking position of the preferred item in the recommendation list. We employ a mapping function for $rank_i$ variable as follows:

$$rank_i = \begin{cases} rank_{i-1}, hit_{i-1} = 1 \\ rank_{i-1} + 1, hit_{i-1} = 0 \end{cases} \tag{3}$$

As shown in the function, $rank_i$ value of an item is determined by whether or not the earlier item is preferred by the user. That is, if the hit value pertaining to preceding item in the recommendation list is true (1), then $rank_i$ is equal to $rank_{i-1}$, otherwise it equals $rank_{i-1}$ plus one. By definition, $rank_i$ for the first recommended item ($rank_1$) is always set to 1 and the above function is computed for the following items.

Our proposed metric, mRHR requires a ranked list of the recommended items as an input and starts the process from the top of the list. For the top (first) recommended item, i is set to 1. Moreover, in the original ARHR, the Eq. 2 is divided by number of users, since it is only used for users' test sets including single instance and the overall score is computed directly. mRHR can be used when the test sets contain more than one test instance. Thus, it is computed for each test instance of the user. In our proposed metric, the denominator is replaced by the number of preferences of the user at a time, rather than number of users in order to evaluate the multiple items preferred at a time. Then, the overall mRHR of the user is computed by averaging over all computed mRHR values of the user and finally, the overall mRHR for all users is computed by averaging these personal mRHR values of the users.

4 Illustrative Example

The following example illustrates the whole computation process of mRHR. Suppose that Alice uses a recommendation agent in her smart phone for the grocery shopping and this application generates a ranked list of recommended products. Suppose further that the application delivers the following recommendation list for the next grocery shopping of Alice: 1. milk, 2. pasta, 3. egg, 4. sausage, 5. ketchup. The numbers indicate the rank of the recommended product in the recommendation list and 1 means the first item in the list. In her next visit, she purchases the following products: egg, cheese, pasta, ketchup. Then, the mRHR metric to evaluate the recommendation performance is computed step by step as shown in Table 1.

Table 1. mRHR computation for multiple preferred items by an example

Step	Rec. rank	Recommended product	hit_i	$rank_i$	$\frac{hit_i}{rank_i}$
1	1	Milk	0	1	0
2	2	Pasta	1	2	1/2
3	3	Egg	1	2	1/2
4	4	Sausage	0	2	0
5	5	Ketchup	1	3	1/3

Please note that, the computation process starts with the product at the top of the recommended list and goes on until the end of the list. There are three relevant products in the recommendation list and these products are marked as 1 on the hit_i column of the Table 1. $rank_i$ is calculated by using Eq. 3. For example, $rank_2$ and $rank_3$ values for the second and third recommended products respectively are calculated as follows:

Set $rank_1 = 1$
Since $hit_1 = 0$, then $rank_2 = rank_1 + 1 = 1 + 1 = 2$
Since $hit_2 = 1$, then $rank_3 = rank_2 = 2$
Finally, by using $hit_i / rank_i$ values, mRHR is computed as:
mRHR $= (1/2 + 1/2 + 1/3)/4 = 0.33$

The denominator in the above calculation is four, because the user prefers to purchase four products in her visit. Thus, as shown in the example, our proposed metric, mRHR is capable of evaluating the cases where the user has multiple preferences at a time or a specific time interval.

We used a mapping function for $rank_i$ variable in the computation of mRHR. The following example demonstrates the reasoning behind this strategy. Consider a typical recommender application and its user. Assume that a recommender agent generates a ranked recommendation list containing four items for the next preference of the user. Then, the user prefers four items in that transaction and suppose that Table 2 shows the hits for five different cases. "Rec. Rank" in the table indicates the order of a recommended item. For instance, the second case in Table 2 indicates that the first three items in the recommendation list were preferred by the user, whereas fifth case indicates that the last three items in the recommendation list were preferred by the user.

Table 2. The effect of mapping function for $rank_i$ variable on the mRHR computation

Rec. rank	hit_i				
	(1)	(2)	(3)	(4)	(5)
1	1	1	1	1	0
2	1	1	1	0	1
3	1	1	0	1	1
4	1	0	1	1	1
mRHR without mapping	0.52	0.45	0.43	0.4	0.27
mRHR with mapping	1	0.75	0.63	0.5	0.38

The last two rows in Table 2 show mRHR values calculated by using two different techniques. The first one computes mRHR without using mapping function for $rank_i$ variable in the Eq. 3, whereas the second one uses this function to compute mRHR. For example, mRHR values using both options for the first case is computed as follows:

Without using mapping function for $rank_i$: mRHR $= (1/1 + 1/2 + 1/3 + 1/4)/4 = 0.52$
With using mapping function for $rank_i$: mRHR $= (1/1 + 1/1 + 1/1 + 1/1)/4 = 1$

Although all four recommended items are preferred by the user, mRHR value without using mapping function for *rank$_i$* is computed as 0.52. Actually, this case should be evaluated as perfect matching. When we look at the mRHR value with using mapping function for *rank$_i$*, it is 1, which shows the actual evaluation of the case. The similar situation also exists in other cases in Table 2. mRHR values without using proposed technique for cases 2–5 is very low, although three of the four items preferred by the user is also in the recommendation list. On the other hand, mRHR values with using the related function are more consistent and reasonable. Therefore, this example demonstrates the effect of using mapping function and clarifies why we employed such a mapping function for *rank$_i$* in the computation of our proposed metric, *mRHR*.

5   Evaluation

In order to evaluate the performance of newly proposed metrics, most studies explored that whether they correlate well with other standard metrics used for same purpose [8, 13, 14]. In this manner, we chose recall which is the most widely used metric for evaluating top-N recommender systems. We investigated the correlation between recall and our metric in the evaluation of predicting top-N recommendation lists by applying Pearson's correlation test. Grocery shopping was chosen as application domain, since customers in the grocery industry prefers multiple products in a single transaction. Two distinct datasets from two Turkish retail grocery stores were used in the experiments.

In the following experiments, most-frequent item recommendation method [15] was chosen to be the baseline for evaluating and comparing the performance of our proposed metric and recall. The reasons behind this choice were that it is simple and well-known method and it was also employed in the studies of predicting grocery shopping lists [15–17]. This approach is not so efficient in the recommendation performance, but it is not our aim in this study. Most-frequent item recommendation method is only considered as a reasonable baseline to predict top-N recommendation lists and to evaluate and compare the performance of corresponding metrics.

In this section, we first describe the datasets used in the experiments. Then, we explain the experimental settings and procedures. Finally, we present the results of the experiments carried out and correlation analysis to explore how well mRHR and recall metrics are correlated.

5.1   Datasets

We used two datasets obtained from two different Turkish retail grocery stores. The first company operates in Central Anatolia region, the other one operating in Mediterranean region. Throughout the report, we call the datasets with the region names that they operate in. Mediterranean dataset contains purchase transactions gathered over 100 weeks (January 10, 2012 – August 31, 2014), whereas Anatolia dataset covers the period of 104 weeks (January 1, 2013 – December 31, 2014). In the datasets, many of the customers have visited the stores irregularly (only a few times during the period) and transactions of these customers most probably mislead the results of the study. Thus, we pre-processed the dataset to exclude customers who have visited the store less than 25 times. After this process, the Anatolia and Mediterranean datasets are left with 46 and 2121 customers with 120467 and 534616 purchase records, respectively. Therefore, we have two different legitimate samples to predict top-N recommendation lists.

In both datasets, products are categorized according to a four-level hierarchy. The lowest level includes information about the product brand, amount and other attributes. For ketchup, for example the third level category is "ketchup", and one of the examples of fourth level category for this product is "Tat Ketchup, 600 Grams". "Tat" is the brand of product, whereas "600 Grams" is the amount of it. Since the fourth level is overly unique, we only considered third-level categories, instead of exact products in the experiments.

5.2   Experimental Settings and Design

In both datasets, we sorted each customer's shopping trips according to their timestamps. For each customer's purchase history, we use the first 80 % of the visits as training and the latter 20 % as test data. In the testing set, for each visit of the customer a size-N ranked list of recommended products is generated and the recommended products are compared with the actual ones purchased by the customer in the test set to compute the corresponding evaluation metrics. We define recall as follows:

$$Recall@N = \frac{|rec@N \cap purc|}{|purc|} \tag{4}$$

where $rec@N$ donates the top-N recommended products for the test instance and $purc$ is the actual product set that the customer has purchased in the same test instance.

We first computed results of each metric for each customer separately by averaging all computed values in the customer's test set, and then average these customers' personal values to get overall metric values for a given recommendation list size, N for both datasets.

In the experiments, most-frequent item recommendation method was used to generate top-N recommendation lists for the customers' visits in the test set. This approach sorts the products in customer's purchase history according to their frequency count and simply returns the N most frequently products as his current shopping list. We repeated our experiments 10 times for each dataset by varying recommendation list size, N from 1 to 10 (in increments of 1) in order to investigate the performance of related metrics for different values of N.

5.3   Experimental Results

On the customers' personal metric values, Pearson's correlation analysis was applied to examine the relationship between mRHR and recall, and Table 3 shows the correlation coefficient results for different values of N (i.e., 2, 4, 6, 8 and 10) for both datasets.

Table 3. Correlation statistics between mRHR and recall metrics

Dataset	N				
	2	4	6	8	10
Anatolia	.967	.893	.867	.826	.776
Mediterranean	.971	.897	.831	.776	.729

According to Table 3, as N increases, the correlation coefficients decrease for both datasets. The main reason behind this is that the difference between the values of two metrics increase, as N increases. Despite this, the output shows that there is a strong, positive correlation between mRHR and recall for both datasets. Note that the significance value is less than .001 for all of the correlation coefficients in the table and this indicates that all of these correlations are statistically significant.

Figure 1 also demonstrates the overall performance results for different values of N for both datasets. It can be observed from the plots that mRHR and recall are positively correlated. This figure also verifies our previous statement, which is that the difference between the values of two metrics raises, as N increases. This is because mRHR is a rank measure, and so it decays faster than recall with the growth of N.

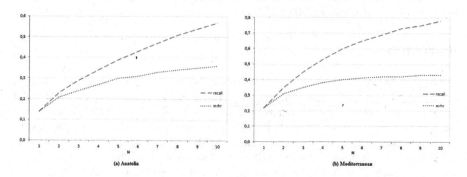

(a) Anatolia   (b) Mediterranean

Fig. 1. Overall performance results for different values of N

As in many prior studies [3, 5, 6, 10–12], with the growth of the recommendation list size N, ARHR and recall or hit rate (HR) generally tend to increase, and many studies [3, 9–12] in the literature also show that recall or HR takes mostly higher scores than ARHR for the same N value. In this manner, it is obvious that similar patterns exist in the plots of both datasets in Fig. 1, in which our new metric mRHR takes the place of ARHR in the previous studies. Therefore, all of these results indicate that mRHR can take the place of ARHR for evaluating performances of top-n recommender systems in the cases where the users have multiple preferences at a time.

6 Conclusion

In this paper, we proposed mRHR, a modified version of ARHR metric that can be used for the purpose of ranking evaluation of multiple preferences in top-n recommender systems. The standard ARHR metric is modified in order to make it applicable to multiple preferences of the user in a time interval or at a time. We further demonstrated the applicability of the proposed metric on a grocery shopping domain using real-life data. A series of experiments were conducted to investigate the performance of the proposed metric and the relationship between the performances of the proposed metric and recall was also explored via performing Pearson's correlation analysis.

Experimental results confirmed that our proposed metric, mRHR is significantly correlated with recall and the performance measured by mRHR is consistent with the performance measured by ARHR used in the prior studies in the literature. These findings indicate that our proposed metric overcomes the main shortcoming of standard ARHR metric and supports the ranking evaluation of Top-N recommender systems in the cases where the users have multiple preferences at a time.

Grocery shopping is chosen as experimental context, but the proposed metric can be applied to other domains as well. A possible direction in the future work might be to evaluate the proposed metric in other application domains and using other recommendation methods.

Acknowledgements. This work is partially supported by the Scientific and Technological Research Council of Turkey (TUBITAK).

References

1. Resnick, P., Iacovou, N., Suchak, M., Bergstrom, P., Riedl, J.: GroupLens: an open architecture for collaborative filtering of netnews. In: Proceedings of the 1994 ACM Conference on Computer Supported Cooperative Work, pp. 175–186. ACM (1994)
2. Hill, W., Stead, L., Rosenstein, M., Furnas, G.: Recommending and evaluating choices in a virtual community of use. In: Proceedings of the SIGCHI Conference on Human Factors in Computing Systems, pp. 194–201. ACM Press/Addison-Wesley Publishing Co. (1995)
3. Deshpande, M., Karypis, G.: Item-based top-N recommendation algorithms. ACM Trans. Inf. Syst. **22**, 143–177 (2004)
4. Cremonesi, P., Koren, Y., Turrin, R.: Performance of recommender algorithms on top-n recommendation tasks. In: Proceedings of the Fourth ACM Conference on Recommender Systems, pp. 39–46. ACM (2010)
5. Ostuni, V.C., Di Noia, T., Di Sciascio, E., Mirizzi, R.: Top-N recommendations from implicit feedback leveraging linked open data. In: Proceedings of the 7th ACM Conference on Recommender Systems, pp. 85–92. ACM (2013)
6. Li, W., Matejka, J., Grossman, T., Konstan, J.A., Fitzmaurice, G.: Design and evaluation of a command recommendation system for software applications. ACM Trans. Comput. Hum. Interact. (TOCHI) **18**, 6 (2011)
7. Bobadilla, J., Ortega, F., Hernando, A., Gutiérrez, A.: Recommender systems survey. Knowl. Based Syst. **46**, 109–132 (2013)
8. Moffat, A., Zobel, J.: Rank-biased precision for measurement of retrieval effectiveness. ACM Trans. Inf. Syst. **27**, 27 (2009)
9. Zheng, N., Li, Q.D.: A recommender system based on tag and time information for social tagging systems. Expert Syst. Appl. **38**, 4575–4587 (2011)
10. Ning, X., Karypis, G.: Slim: sparse linear methods for top-n recommender systems. In: 2011 IEEE 11th International Conference on Data Mining (ICDM), pp. 497–506. IEEE (2011)

11. Cheng, Y., Yin, L.A., Yu, Y.: LorSLIM: low rank sparse linear methods for top-n recommendations. In: 2014 IEEE International Conference on Data Mining (ICDM), pp. 90–99. IEEE (2014)
12. Kang, Z., Cheng, Q.: Top-N recommendation with novel rank approximation. arXiv preprint (2016). arXiv:1602.07783
13. Chapelle, O., Metlzer, D., Zhang, Y., Grinspan, P.: Expected reciprocal rank for graded relevance. In: Proceedings of the 18th ACM Conference on Information and Knowledge Management, pp. 621–630. ACM (2009)
14. Kekalainen, J.: Binary and graded relevance in IR evaluations - comparison of the effects on ranking of IR systems. Inf. Process. Manage. **41**, 1019–1033 (2005)
15. Sarwar, B., Karypis, G., Konstan, J., Riedl, J.: Analysis of recommendation algorithms for e-commerce. In: Proceedings of the 2nd ACM Conference on Electronic Commerce, pp. 158–167. ACM (2000)
16. Cumby, C., Fano, A., Ghani, R., Krema, M.: Predicting customer shopping lists from point-of-sale purchase data. In: Proceedings of the Tenth ACM SIGKDD International Conference on Knowledge Discovery and Data Mining, pp. 402–409. ACM (2004)
17. Cumby, C., Fano, A., Ghani, R., Krema, M.: Building intelligent shopping assistants using individual consumer models. In: Proceedings of the 10th International Conference on Intelligent User Interfaces, pp. 323–325. ACM (2005)

A Cooperative Control System
for Virtual Train Crossing

Bofei Chen[✉] and Franck Gechter

University of Technologie of Belfort Montbliard, UBFC,
F-90010 Belfort Cedex, France
{bofei.chen,franck.gechter}@utbm.fr
http://www.multiagent.fr/

Abstract. The use of individual vehicles is increasing in inner cities involving a large number of unwanted side effects. However, the rise of intelligent vehicle technologies allows finding a solution to some of these problem bringing new vehicle usage. For traffic jam management, aside to widespread traffic light control, one of the promising solutions is the possibility to form virtual trains with vehicles so as to increase road capacities. This solution, however relevant it is, suffer from limitations when the road network has many interconnections where interactions between trains are hard to define. This requires thus a control process which can deal with train, vehicles, and hardware control issues. The goal of this paper is to propose a multi-level cooperative control system. This proposal relies on a dynamic adaptation of train parameters in runtime and allows crossroads and roundabout sharing without stopping any vehicle. The proposal is tested in simulation coping with the roundabout scenario.

Keywords: Multi-level decision process · Virtual train · Platoon control · Roundabout

1 Introduction

The use of individual vehicles is becoming more and more important in inner cities, leading to many side problems such as traffic jam air pollution and an increase of accidents. Advanced intelligent vehicle technologies have been studied so as to overcome these problems. Some of the solutions provided by these works are already available to the general public market such as city safety systems, lane assists,... Many research works are still focusing on finding suitable and acceptable solutions to these problems and particularly dealing with traffic jam management which is the central issue of this paper. The proposed solutions can tackle the traffic jam problem whether on system level focusing on traffic light management or on individual vehicle providing better control and perception systems aimed at reducing time response and/or at increasing the road capacity [9,10]. On the system side, the traffic light cycle optimization attracts many researches' attention [7,11]. For instance, in [8], a multi-agent traffic light

© Springer International Publishing Switzerland 2016
C. Dichev and G. Agre (Eds.): AIMSA 2016, LNAI 9883, pp. 330–339, 2016.
DOI: 10.1007/978-3-319-44748-3_32

control system based on a multi-objective sequential decision-making framework and a traffic light controller are developed. On the vehicle side, one of the most promising solutions is to group vehicles into platoons (virtual train) making possible a huge reduction of the longitudinal distance between vehicles and thus allowing an increase of roads capacity. Basically, two main trends can be found literature. On one side, global approaches are based on a common reference frame, generally tied to the vehicles playground, shared by all vehicles of the train. Then, each vehicle behaves according to this shared reference which can be either the trajectory of the first vehicle of the train or a reference trajectory built offline [4,6]. On the other side, local approaches are based on vehicle local perception abilities. Some methods, based on classical control algorithms [3,12] or physical-inspired and inter-vehicular interaction link [2,13], are developed. Despite numerous research works on this subject, which are focusing on individual vehicle control, few of them consider the platoon control solution on the system point of view. However, this system point of view is particularly important when several platoons have to share the road network and meet at critical nodes such as crossroads and runabouts. This introduces new issues such as the interactions between trains of vehicles and requires the development of strategies so as to make the sharing of the road infrastructure efficient, reliable and safe. Solving these problems will then allow considering virtual train solution as a good candidate for solving traffic flow issues at transportation system level [1]. Several projects studied this solution among which one can cite FUI/FCE CRISTAL, FP7-CATS[1] or ANR-VTT-2010-SafePlatoon[2]. The key element of the virtual-train solution is the ability to control each vehicle to obtain a coherent global behavior at train level.In this paper, we propose a cooperative control system which relies on multi-level decision processes aimed at dealing with the interaction of platoons at road network nodes. This decision system allows both to maintain the coherence and the safety condition of each involved train of vehicles and to adapt each train components behavior so as to make train shared the road, and especially the roundabout, efficiently (i.e. without stopping any vehicle). This multi-level decision system is divided into three different levels. The global train state is managed at the top-level based on the train level perceptions. The middle-level process makes the decision concerning each individual vehicle according to data provided by the top-level and to the interaction between vehicles. Finally, the low-level process makes the link between the middle-level command and hardware level of vehicles. When encountering, trains exchange information such as one part of their perceptions. The proposal is then tested in the roundabout scenario Fig. 2(a) (i.e. several trains of vehicles meet at a roundabout and have to share it without implying too many disturbances to the nominal train behavior). This paper is structured as follows: the multi-level decision system is detailed in Sect. 2. Then, experimental platform, algorithm, and results are presented in Sect. 3. Finally, the conclusion and future work are given in Sect. 4.

---

[1] http://www.parc-innovation-strasbourg.eu/index.php/CATS-Project/welcome-on-cats-webpage.html.

[2] http://web.utbm.fr/safeplatoon/.

2 Multi-level Decision System

The structure of the multi-level decision system is shown in Fig. 1(a). Each vehicle mission is associated with a trip along the road network defined by a departure position, a final position, and several way-points. Vehicles that share the same mission are assigned to the same group and compose one platoon (also called virtual-train). The top-level decision process produces the virtual train command (i.e. roles and priorities of its components). The middle-level is devoted to platoon control algorithm application. Then, the low-level decision applies the vehicle command to the hardware controller. This section explains the multi-level decision process in detail.

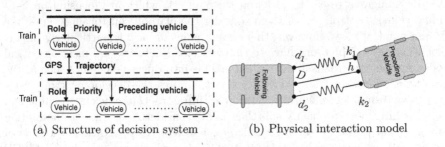

(a) Structure of decision system   (b) Physical interaction model

Fig. 1. System structure and interaction model

2.1 Top-Level Decision Process

The goal of the top level decision process is to make sure all vehicles assigned with proper priority and role. Base on that, the preceding vehicle for every vehicle is planed correctly. Information is exchanged between trains: the path of each vehicle and the number of vehicles in each train. Figure 2(a) depicts the situation of deciding priorities, and preceding vehicles.

Inputs: Trains' information (i.e. GPS, planned path) is exchanged between them when arriving at a node. The planned paths predict the next motion of all vehicles and is used to order the sequence in trains.

- *GPS*: Trains exchange GPS information including positions, directions and speed. This allows to keep an efficient synchronization and are the key elements in the computation of the priorities.
- *Trajectory*: Each vehicle owns their planned path made up by a series of way-points. Way-points are corresponding to the nodes of the road network. According to the trajectory, the decision system arranges the order in each train and the priority when necessary.

Outputs:

– *Role and priority*: The role includes leader and follower. The head of each train, adjusting speed according to the mission, holds the leader role. Followers perform the platooning control and follow their preceding vehicle which is given by the top-level decision process. Roles may change when trains meet at crossroads and roundabouts. Priorities are decided according to assume-finish time, required for one vehicle to finish the encountering mission, in this paper corresponding to crossing the intersection, depending on distances to the intersection point d, vehicles velocities v and vehicle lengths l. For all vehicles, the information matrix I is computed according to the calculation of assume-finish time $f(d, l, v)$ as follows:

$$
I = \left\{ \begin{array}{c} V_{i1} \\ V_{i2} \\ V_{i3} \\ \vdots \\ V_{m1} \\ V_{m2} \\ V_{m3} \\ \vdots \end{array} \right\} = \left\{ \begin{array}{ccc} d_{i1} & l_{i1} & v_{i1} \\ d_{i2} & l_{i2} & v_{i2} \\ d_{i3} & l_{i3} & v_{i3} \\ \vdots & \vdots & \vdots \\ d_{m1} & l_{m1} & v_{m1} \\ d_{m2} & l_{m2} & v_{m2} \\ d_{m3} & l_{m3} & v_{m3} \\ \vdots & \vdots & \vdots \end{array} \right\}, t = f(d, l, v) = \frac{d + l}{v} \tag{1}
$$

where V_{ij}, d_{ij}, l_{ij} and v_{ij} are respectively the representative vector, the distance to the node, the length and the speed of the jth vehicle of ith train. Then, the matrix T, composed by all vehicle assume-finish times, is:

$$
T = \{t_{i1}, t_{i2}, , t_{i3}, \dots, t_{m1}, t_{m2}, t_{m3}, \dots\} \tag{2}
$$

Then, times are sorted in ascending order:

$$
T_{order} = \{\dots, t_{ij}^{k-1}, \dots, t_{mn}^{k}, \dots\} \tag{3}
$$

Corresponding to the time matrix, the Vehicle Order matrix VO can be built as follows:

$$
VO = \{\dots, V_{ij}^{k-1}, \dots, V_{mn}^{k}, \dots\} \tag{4}
$$

where $j, n = 1, 2, 3, \dots$ V_{ij}^{k-1} is the $(k-1)^{th}$ vehicle in priorities array as the jth vehicle in platoon i associated to the vehicle V_{ij} and the time t_{ij}^{k-1}. Coordinately, V_{mn}^{k} is the kth vehicle in priorities array as the nth vehicle in platoon m associated to the vehicle V_{mn} and the time t_{mn}^{k}. In this paper, velocities and vehicle lengths are set to be the same for each vehicle; hence, only distances are compared. As shown in Fig. 2(a), virtual-train A is turned to the same side of B around the intersection point O. Priorities are shown previously according to relative positions. If assume-finish times for two vehicles are the same, the rule: "right vehicle first" is adopted.

- *Preceding vehicle:* For every following vehicle, the uniqueness of the preceding vehicle should be pre-set. Preceding vehicles, related to priorities, may change during trains meetings depending on the path of every individual. The preceding vehicle information includes position, direction, and speed. It could be real or virtual one. In the nominal situation (i.e. without any conflict in priorities), each following vehicle follows one real vehicle keeping the security distance. The virtual vehicle concept has been introduced so as to be able to adapt inter-vehicle longitudinal distance when trains are crossing without interfering on platoon control algorithm, shown in Fig. 2(b). Consequently, the adaptation of distance between vehicles is made by introducing a virtual vehicle at a suitable position and speed (i.e. when two successive vehicles in the VO array are not belonging to the same train). Virtual vehicles have the same properties and behavior as real ones. In the vehicle priorities array V_{ij}^k, for $n_{th}(n > 1)$ vehicle V_{ij}^n, the closest last one in the same virtual-train is $V_{i(j-1)}^m$. The number of virtual vehicles inserted before V_{ij}^n is:

$$N = n - m - 1$$

By contrast, if N is equal to zero, the preceding vehicle of V_{ij}^n is a real one, actually is $V_{i(j-1)}^m$.

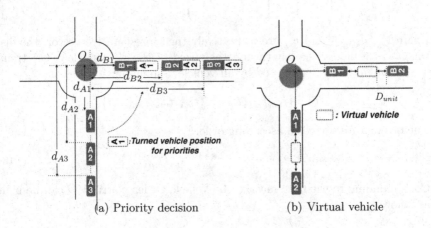

(a) Priority decision   (b) Virtual vehicle

Fig. 2. Priorities decision and virtual vehicle

2.2   Middle-Level Decision Process

The middle-level decision process deals with the result from top-level and provides commands corresponding to a speed and a steering angle commands. The speed is computed according to the role of the vehicle. The leader speed is determined on its priority, distance to intersection and other leaders situation. The follower speed is computed by platoon control algorithm based on one physical model [13].

Physical Model: The interaction model, presented in [13], is used in this paper for two main reasons. First, maintaining stable the desired vehicles distance. Second, guaranteeing a good trajectory matching. The virtual link, between two closed vehicles, is described by a Physics-inspired interaction model composed of two springs and a damper, shown in Fig. 1(b). The model parameters are learned so as to ensure safety and stability, see [13] for more details.

Speed of Leader: Speeds of leaders are calculated according to priorities set by the top-level. When different trains encounter a new mission is triggered. New role and priority are produced for every vehicle. Leaders adjust their speeds as soon as possible for preparing, then hold constant speeds to finish the mission. When trains separate, the mission is finished and leaders recall suitable speeds.

- *Speed before encountering:* As shown in Fig. 2(a), the intersection information could be inferred according to positions and directions of the two trains. Consequently, we must calculate the distance between virtual-train head and intersection point. $B1$ has the priority, hence $B1$ keep a normal speed set to be a constant. The key point is to adjust $A1$'s speed to follow $B1$. According to the length of $B1$, l_{B1}, and to the security distance, D_{unit}, the speed of $A1$ should be:

$$v_{A1} = \frac{d_{A1} + D_{unit}}{d_{B1} + l_{B1}} v_{B1} \tag{5}$$

- *Encountering speed:* When virtual trains are encountering, a maneuverable method is to keep all vehicles at the same speed. On one side, the same speed keeps steady constant distance between vehicles in the same virtual train. On the other side, identical speed avoids disturbance of another vehicle in the transportation system. However, it is also necessary to set a maximum speed so as to ensure safety.

Steering Angle: Every single vehicle owns their planned path. The steering angle is tied to the trajectory to follow, especially when trains encounter each other. For path following control, the detail is given in [2].

2.3   Low-Level Decision Process

The low-level decision process is based on vehicle kinematic, dynamic model and hardware constraint. First, the process should shorten vehicle response time, e.g. increasing the acceleration. Second, the vehicle motor system is steadied under the low-level decision process. In our proposal, the low-level decision process includes a standard Proportional Integral controler to reduce the response time. Besides, the use of the algorithm, presented in [2], limits the step-movement scale for precise and smooth tracking.

3 Experiment and Results

3.1 Simulation Platform

Simulations are now a mandatory step in intelligent vehicle development. It allows to reduce the time and financial cost and limits the risk of false manipulation. Furthermore, in the topics dealt by the paper, testing crossing algorithms for several platoons is nearly impossible for a laboratory due to a large number of required vehicles and their associated staff for ensuring security. Consequently, we decided to test our proposal using a simulation tool developed by the System and Transportation Laboratory of UTBM already used in several academic project[3] [5]. This can simulate vehicles and sensors within their physical properties.

3.2 Simulation Results

The goal of the simulations, presented in this paper, is to compare the multi-level decision algorithm with a normal behavior, in which each platoon is behaving as a whole (i.e. it passes entirely before or after another platoon so as to maintain the structure of each). In this normal behavior, the first train in the roundabout has got the priority and the second is waiting for the release of the roundabout. For the multi-level decision algorithm, different speeds have been tested. During tests, two virtual train features (i.e. distance and speed) have been measured. The passing time and normal distance in different speed are compared as well. Figure 3 shows the way where two trains are sharing the roundabout using multi-level decision process. In this part, the two methods are compared. The recording of multi-level decision algorithm is shown in Fig. 4. Meanwhile, the normal behavior is shown in Fig. 5. The train, running from right to left, is called train B, the other is called train A. Figure 3(a) shows the situation where vehicles are preparing for the road crossing. In this situation, distances between vehicles are adjusted according to virtual vehicles given by top-level decision process. In Fig. 3(b), the mission is being executed. After all the vehicles passed roundabout, distances are changed to platoon distance as shown in Fig. 3(c). The distances changes of two trains are also shown in Fig. 4.

- *Speeds*: In both of methods, the normal speed is set to 40 km/h and the max speed to 50 km/h. The train B has the priority. For the multi-level decision algorithm, the vehicle B1, leader of train B, accelerates to 40 km/h and keeps this normal speed. In the train B, vehicles B2 and B3 adjust their speeds in order to hold the necessary distance, with a max speed set to 50 km/h as shown in Fig. 4(d). Meanwhile, vehicle A1, the leader of train A, moves using a new speed so as to pass the roundabout after B1. Vehicle A2 and A3 are also adjusting their speed as B2 and B3 (cf. Fig. 4(d)). With normal behavior, train B passes roundabout first. B1 is accelerating to normal speed, B2 and B3 are following B1 as shown in Fig. 5(d). As opposed to the previous solution,

---

[3] http://www.vivus-simulator.org/Main_Page.

train A is waiting for the release of the roundabout and then enters into it (cf. Fig. 5(c)).

- *Distances:* Normally, vehicles keep a safety distance. For the multi-level decision algorithm, one virtual vehicle is inserted, when necessary, between two vehicles when trains encounter. Each distances was then doubled as shown in Fig. 4(a) and (b). Nevertheless, for trains under traffic light control, distances were kept in safety distance as shown in Fig. 5(a) and (b).
- *Times:* Passing times in different situation were also counted. The time under normal behavior is about 23 s as opposed to 10 s for the multi level solution.

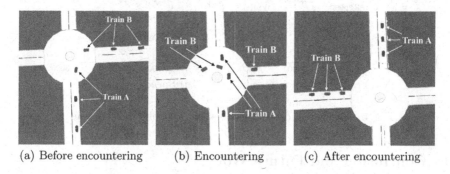

(a) Before encountering   (b) Encountering   (c) After encountering

Fig. 3. The dynamic of encountering

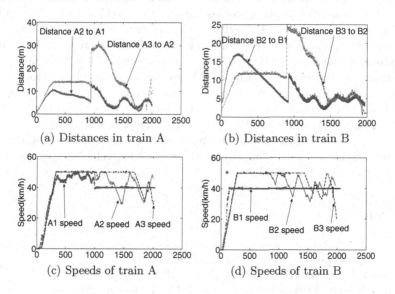

(a) Distances in train A   (b) Distances in train B

(c) Speeds of train A   (d) Speeds of train B

Fig. 4. Speeds and distances under multi-level decision algorithm

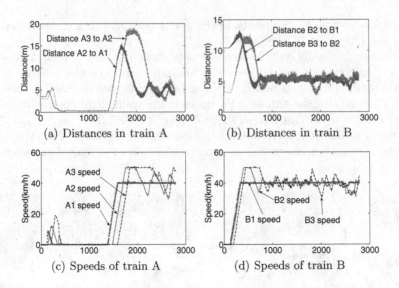

(a) Distances in train A   (b) Distances in train B

(c) Speeds of train A   (d) Speeds of train B

Fig. 5. Speeds and distances under the control of normal rules

4 Conclusion and Future Work

In this paper, a platoon cooperative control problem was addressed. A multi-level decision strategy based on platoon control was proposed. The top-level decision process gives the command to virtual trains including vehicles roles and priorities. The middle-level decision process adjusts every vehicle to fit the top-level decision and platoon control algorithm requirements. The low-level deals with the hardware application of the commands. Experiments were carried out in the roundabout scenario under two algorithms: one was normal behavior algorithm, the second was the cooperation method based on the strategy developed in this paper. In addition, safety distances in different speeds were studied. The results proved that the cooperation control saved more time than the normal method. For the future work, on the one hand, the proposal will be tested with harder conditions, e.g. various numbers of trains and vehicles. On another hand, dynamic train reconfiguration will be implemented, i.e. allowing vehicles to change trains when they cross in the roundabout.

References

1. Chen, B., Gechter, F., Koukam, A.: Multi-level decision system for the crossroad scenario. Procedia Comput. Sci. **51**, 453–462 (2015)
2. Dafflon, B., Chen, B., Gechter, F., Gruer, P.: A self-adaptive agent-based path following control lateral regulation and obstacles avoidance. In: International Conference on High Performance Computing and Simulation (HPCS), pp. 452–459 (2014)

3. Pascal (INRIA) Daviet and Michel (INRIA) Parent. Longitudinal and lateral servoing of vehicles in a platoon. In: Proceedings of Conference on Intelligent Vehicles, pp. 41–46. IEEE (1996)
4. Fritz, H.: Longitudinal and lateral control of heavy duty trucks for automated vehicle following in mixed traffic: experimental results from the CHAUFFEUR project. In: Proceedings of the 1999 IEEE International Conference on Control Applications (Cat. No. 99CH36328), vol. 2, pp. 1348–1352. IEEE (1999)
5. Gechter, F., Contet, J.M., Galland, S., Lamotte, O.: VIVUS: virtual intelligent vehicle urban simulator: application to vehicle platoon evaluation. Simul. Model. Pract. Theor. 24(Complete), 103–114 (2012)
6. Guillet, A., Lenain, R., Thuilot, B., Martinet, P.: Adaptable robot formation control: adaptive and predictive formation control of autonomous vehicles. IEEE Robot. Autom. Mag. 21(1), 28–39 (2014)
7. Hu, C.,Wang, Y.: A novel intelligent traffic light control scheme. In: 2010 Ninth International Conference on Grid and Cloud Computing, pp. 372–376. IEEE, November 2010
8. Khamis, M.A., Gomaa, W., El-Shishiny, H.: Multi-objective traffic light control system based on Bayesian probability interpretation. In: 2012 15th International IEEE Conference on Intelligent Transportation Systems, pp. 995–1000. IEEE, September 2012
9. Levinson, J., Askeland, J., Becker, J., Dolson, J., Held, D., Kammel, S., Kolter, J.Z., Langer, D., Pink, O., Pratt, V., Sokolsky, M., Stanek, G., Stavens, D., Teichman, A., Werling, M., Thrun, S.: Towards fully autonomous driving: systems and algorithms. In: IEEE Intelligent Vehicles Symposium (IV), pp. 163–168 (2011)
10. Luettel, T., Himmelsbach, M., Wuensche, H.-J.: Autonomous ground vehicles - concepts and a path to the future. Proc. IEEE 100(Special CentennialIssue), 1831–1839 (2012)
11. Perronnet, F., Abbas-Turki, A., Buisson, J., El Moudni, A., Zeo, R., Ahmane, M.: Cooperative intersection management: real implementation and feasibility study of a sequence based protocol for urban applications. In: 2012 15th International IEEE Conference on Intelligent Transportation Systems, pp. 42–47. IEEE, September 2012
12. Sheikholeslam, S., Desoer, C.A.: Longitudinal control of a platoon of vehicles with no communication of lead vehicle information: a system level study. IEEE Trans. Veh. Technol. 42(4), 546–554 (1993)
13. El Zaher, M.: Approche réactive pour la conduite en convoi desvéhicules autonomes: Modélisation et vérification. Ph.D. thesis, Université de Technologie deBelfort-Montbéliard (2013)

Posters

Artificial Intelligence in Data Science

Lillian Cassel[1(✉)], Darina Dicheva[2], Christo Dichev[2], Don Goelman[1],
and Michael Posner[1]

[1] Department of Computing Sciences,
Villanova University, Villanova, PA, USA
{lillian.cassel,don.goelman,
michael.posner}@villanova.edu
[2] Department of Computer Science, Winston Salem State University,
Winston-Salem, NC, USA
{dichevad,dichevc}@wssu.edu

Abstract. Data Science programs are emerging in many areas and are related to many disciplines. This includes sciences, social sciences, business, journalism, history, and any other area dealing with massive amounts of data. People may understand that the quantity of data now available has changed the nature of research and has begun to impact the way students must prepare to be part of their discipline. However, they may not understand that artificial intelligence is a key component of the new reality. Massive amounts of data require more than computational power from computers. The size of the data collections also requires machine intelligence to organize and cluster data.

Keywords: Data science · Machine learning · Big data

1 Introduction

The emergence of Data Science as an academic subject dates only to the 1980s. Even then, the few programs in existence focused on graduate study in Knowledge Discovery in Databases (KDD). By 2013, of 61 academic programs related to KDD in the US and Canada, only four served undergraduates [1].

Unlike Data Science, Artificial Intelligence (AI) has a longer history dating back to 1956 when its name was coined. However, both disciplines have a strong, mutually enriching relationship stirring mutual advancements. The revival of AI in the last decade is partially due to the emergence of Data Science and the advancements in Data Science are partially due to AI. With the rapid growth of the Internet, social media, mobile devices, and low-cost sensors, the volume of data is increasing dramatically. The growth of the Web at scale has resulted in some tremendous knowledge bases such as Freebase and Wikipedia. With text mining, incorporating different AI techniques, these data can be combined and made to interoperate. Unlike the early years of mostly algorithmic and logic based approach, AI has evolved to explicitly embrace Web-scale data. Colossal datasets of entities, relationships and facts are being extracted from the Web, combining machine learning with statistical techniques. Examples of extractions from the open Web such as KnowItAll or Open IE are growing in parallel with datasets

C. Dichev and G. Agre (Eds.): AIMSA 2016, LNAI 9883, pp. 343–346, 2016.
DOI: 10.1007/978-3-319-44748-3_33

derived from Web-scale pages or Google books. Thus the recent data science progress is defining AI in a new light which in turn contributes to the evolution of Data Science. It is via the combination of the approaches developed in these two fields that we are seeing the most impressive AI and Data Science results.

As we have discussed previously, the rapid development of large data collections now creates a much broader need for students to gain some familiarity with the nature of data that is collected and available, and an understanding of what can be done with that data [3]. Given the common use of machine learning to analyze data and to make predictions and classifications based on the rules that are learned, data science, and its dependence on artificial intelligence, now approaches a status of critical general knowledge. For example, Microsoft's Kate Crawford describes a problem of bias in artificial intelligence, with specific reference to machine learning [2].

To address the need for broader understanding of data science, and the artificial intelligence that is key to the topic, the authors proposed to develop course materials for a general education course or for use as modules in other courses. The students may come from many disciplines, and will not be burdened with heavy prerequisite structures that demand they already understand computing and statistics. The goal of the project is to provide course materials that can be used in several ways:

- The materials combine to form a stand-alone course that provides a meaningful introduction to the most important concepts needed for managing large quantities of data. The course includes fundamental concepts from computer science and from statistics. Harnessing machine intelligence to support human effort forms a significant component of the course.
- The materials also can be used in smaller units to augment another course. For example, a course that includes data handling laboratories might choose to use a module on machine learning to do some clustering of the data before some other processing or interpretation happens.

2 Project Status

The project began in 2014 with funding from the United States National Science Foundation. The team includes faculty in Computer Science at Villanova University and Winston Salem State University, and a member of the Statistics faculty at Villanova. The director of the Villanova Institute for Teaching and Learning (VITAL) serves as project evaluator. The student populations at the two universities are quite different and allow exploration of how best to develop materials that serve a good variety of students. A member of the team taught an initial version of the course in Spring 2016 at Winston Salem.

The project evaluator worked with the instructor to obtain initial evaluation data. The class consisted of students from several different majors, with varying levels of mathematical and computing background. The diversity of the student backgrounds provides valuable insight into the module development, as the plan is to have modules that will serve students from very weak technical backgrounds through stronger backgrounds with gaps in the specific tools and techniques associated with data science. Two general observations informed the course design: [3]

- Data science is emerging as an academic discipline, defined not just as a combination of units from other disciplines, but as a distinct body of knowledge
- Students from many, or perhaps all, disciplines benefit from the ability to use appropriate tools to collect, organize, visualize, and analyze data.

The pilot course "Introduction to Data Science" offered in the spring semester at WSSU was different in both instruction and content compared to typical introductory courses in statistics or programming. It had no prerequisites, and students were not expected to have any programming experience. The undergraduates enrolled in this class included students majoring in Computer Science, Business Administration, Psychology, Nursing, Social Work and Political Sciences, ranging from freshmen to seniors. The primary challenge resulted from the need to find the right balance for students with diverse motivations, programming abilities and statistical backgrounds. The adopted strategy was to make the course different from typical introductory computing courses that emphasize programming concepts by focusing on developing skills for using computational environments for data science tasks. The course was based on Python programming language with emphasis on data preparation and presentation. The targeted objectives with this choice were to demonstrate the practical value of data science by using real data sets and using Python's data science library with focus on visualizing and communicating data science concepts. Using this approach allows not only to demonstrate in a realistic setting how to recognize relationships among data items or how to detect hidden patterns in complex data sets but also to develop skills for applying Python's data science capabilities for deeper data exploration. The instructional strategy was based on pseudo flipped classroom model aimed at developing practical skills through well sequenced exercises.

In the fall semester 2016, the Winston Salem course will run a second time, building on lessons from the first instance. At the same time, a version of the course at Villanova will be co-taught by a computer scientist and a statistician. The enrolled students are all incoming freshmen who are undeclared Arts majors. This course will build on the Winston Salem experience, but will put extra emphasis on the flipped classroom teaching approach.

The Villanova course will involve students in at least three multi-week projects. In these projects, the process of knowledge discovery in data (KDD) will be repeated, with increasing levels of complexity. The repetition of the process is intended to emphasize that knowledge discovery is a process and to explore the techniques and challenges inherent in the steps. The preliminary plan, as of summer 2016, is to include a project based on political data (given the election year in the U.S.), a project related to environmental data, and a project built on access to Twitter data. All projects will involve out-of-class and in-class components designed to keep the students actively involved in every step.

3 The Poster

The poster summarizes the experiences in the first course offering, describes the planned modules and their status, and includes examples of how the topics are presented in a flipped classroom environment.

The poster contains a complete sample module, including pre-class material for students, in-class activities, post-class reinforcement activities, as well as assessment materials. Also available at the poster session are detailed syllabi for the first and second Winston-Salem courses and for the first Villanova course.

Acknowledgments. This material is based upon work supported by the NSF Grant 1432438: IUSE Collaborative Research: Data Computing for All: Developing an Introductory Data Science Course in Flipped Format (09/01/2014-08/31/2017).

References

1. Anderson, P., Bowring, J., McCauley, R., Pothering, G., Starr, C.: An undergraduate degree in data science: curriculum and a decade of implementation experience. In: Proceedings of the 45th ACM Technical Symposium on Computer Science Education, SIGCSE 2016. ACM (2016)
2. Crawford, K.: Artificial Intelligence's White Guy Problem. New York Times (2016). http://www.nytimes.com/2016/06/26/opinion/sunday/artificial-intelligences-white-guy-problem.html?login=email&emc=edit_tu_20160628&nl=bits&nlid=19620297&ref=technology&te=1
3. Dichev, C., Dicheva, D., Cassel, L., Goelman, D., Posner, M.A.: Preparing all students for the data-driven world. In: Proceedings of the Symposium on Computing at Minority Institutions, ADMI 2016 (2016)

Exploring the Use of Resources in the Educational Site Ucha.SE

Ivelina Nikolova[1]([⊠]), Darina Dicheva[2], Gennady Agre[1], Zhivko Angelov[3], Galia Angelova[1], Christo Dichev[2], and Darin Madzharov[4]

[1] Institute of Information and Communication Technologies,
Bulgarian Academy of Sciences, Sofia, Bulgaria
{iva,galia}@lml.bas.bg, agre@iinf.bas.bg
[2] Winston-Salem State University, Winston-Salem, NC 27110, USA
{dichevad,dichevc}@wssu.edu
[3] ADISS Ltd., Sofia, Bulgaria
angelov@adiss-bg.com
[4] UCHA.SE, Sofia, Bulgaria
darin@ucha.se

Keywords: Educational data mining · e-Learning · User modelling

1 Motivation

Educational Data Mining allows the discovery of new knowledge based on learners usage data in order to help validate and/or evaluate educational systems, to potentially improve some aspects of the quality of education and to lay the groundwork for a more effective learning process [4,5]. Its potential explains the significant interest to it in the educational community and the developers of online learning environments [1]. For online educational websites/learning environments, EDM can provide valuable information on user behaviour on the site pages, such as how often are students using the site, what they are interested in. These findings can be effectively used to improve customer experience and increase the engagement and the rate at which people learn.

The present paper discusses some results from an on-going pilot project aiming at analysing and improving the quality of the educational services and respectively the revenue generation for the educational site UCHA.SE. UCHA.SE is an online learning environment, aimed at supporting the K-12 National Bulgarian Curricula as well as Introductory level English, German, French, and Spanish, and Introduction to Programming, through offering interactive instructional materials videos and practice exercises for all subjects. It implements also some gamification elements [2] to engage and motivate the users, including points, levels, badges, progression and status, as well as social features such as commenting, voting and liking. Currently UCHA.SE offers more than 4,000 videos in 17 subjects.

We aim at assessing the quality and the user interest of the different subject categories in UCHA.SE where a category is meant as a combination of a subject

© Springer International Publishing Switzerland 2016
C. Dichev and G. Agre (Eds.): AIMSA 2016, LNAI 9883, pp. 347–351, 2016.
DOI: 10.1007/978-3-319-44748-3_34

and a grade such as Chemistry 7th grade, Mathematics 10th grade etc. The data used for evaluating the quality of educational resources in UCHA.SE was extracted from over 3 mln user accesses to 3,797 resources within 65 categories. The analysis is based on the video materials only[1] and is a follow up of [3] where user behavior was studied.

2   Approach

Each category in the set we analyse e.g. "Chemistry 7th grade", "Mathematics 10th grade" is described by a set of features constructed on the base of the site's system logs and students interactions stored directly in the system database. These features include: (i) total number of video materials in the category; (ii) average number of accesses to the resource category for a 3-months period; (iii) percentage of all **students** who have accessed the resource category; (iv) number of teachers comments for this resource category. Based on these characteristics we perform statistical analysis on the resource availability in the educational site UCHA.SE and the user interest in all 65 resource categories in the portal.

3   Discussion

We analyse the distribution of resource categories by subjects, as well as by school grades. Thus we identified that the most popular subject in means of grades is Mathematics with resources for each of the 11 grades, this is followed by Bulgarian Language, History, Geography, and Chemistry. The distribution of subjects by school grades is much more uniform: for the majority of grades there are resources in 5 or 6 subjects. The leaders are 7th and 8th grade with 8 and 7 subjects. At the opposite end are 1st and 11th grade with 4 and 2 subjects respectively. When we look at the relation between the number of resources offered in UCHA.SE and the average number of accesses for 3 months (Fig. 1), we can see that the following categories have really high access rate with respect to the number of resources in that category: German Language for Grade 1, Mathematics for 7 grade, History for 7th grade, English for Beginners 2nd grade. Which indicates that there is a demand for learning resources from these categories. In the same time Chemistry for university acceptance exams, History for 11th grade and Math for 4th grade have comparatively high number resources with regard to the number of user accesses.

Considering the distribution of the number of average accesses per resource for 3 months period by grade, the most accesses are to 7th grade (Mathematics followed by History), 6th grade (Mathematics and History), and 5th grade (History and Geography). As for 1st and 2nd grades popular are only German for 1st grade and English for Beginners for 2nd grade.

---

[1] We present here only relative data in consent with the non violation of sensitive data policy of UCHA.SE.

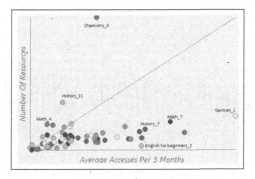

Fig. 1. Relation between the number of resources and the average number of accesses/3 months.

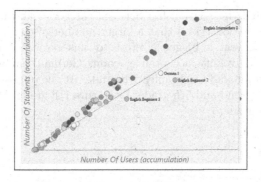

Fig. 2. Relation between the total number of users and the number of students among them.

We explored also the relationship between the total number of users and the number of student-users. These numbers indicate that learning materials for the language courses are attractive and used not only by students but by many non-student users. The two correlate well, but we can still see a few outliners where the ratio number of students to number of all users diverts from that trend (Fig. 2). These are: Intermediate English (62 % students); English Grade 2 (50 % students); German Language Grade 1 (60 % students), English beginners Grade 3 (55 % students); The Man and the Nature Grade 4 (66 % students).

As shown on Fig. 3, the largest number of student users accessed learning materials in Chemistry, English, Biology, Physics and category Fun. On the chart 100 on y-axis is the average number of student users among all categories. E.g. there are 3 times more students accessing Chemistry 7 than the average number of students and Mathematics 10 is about the average. This shows that there are groups of users which may be approached in a more specific way e.g. student/non-student/combined. The course designers and developers at UCHA.SE may redirect their efforts to creating learning resources of their particular interests.

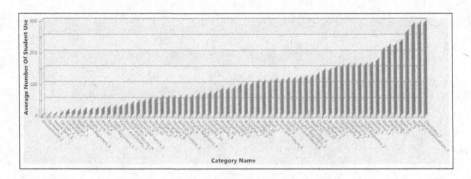

Fig. 3. Relation between the categories and number of student users accessing them.

We also explored the relationship between categories and number of teachers comments. We hypothesized that a larger number of comments on a learning resource will reveal a bigger interest to this resource. The leader here is Chemistry for university acceptance exam, German Language 1st grade, Chemistry 10th grade and Geography 6th grade. At the opposite end, we have: Bulgarian Language 2nd and 5th grade; Literature 7th grade; Mathematics 9th, 10th and for 11th grade.

4   Conclusion

In summary, this is a statistical exploration of the use of resources in UCHA.SE. It aims at revealing gaps between the demand and supply that suggest possible improvement of the content and help identifying groups of users, which could be approached in a specific way. We point out the categories which are outliers and need more attention. In order to get a more comprehensive evaluation of the resources, further exploration of the user logs which includes also semantic analysis of user comments and more detailed per resource assessment will be performed.

Acknowledgements. This research was partly supported by grant DO1-192/2014 (national co-financing of AComIn project) of the Bulgarian Ministry of Education and Science.

References

1. Baker, R., Inventado, P.: Educational data mining and learning analytics. In: Larusson, J.A., White, B. (eds.) Learning Analytics, pp. 61–75. Springer, New York (2014)
2. Dichev, C., Dicheva, D., Angelova, G., Agre, G.: From gamification to gameful design and gameful experience in learning. Cybern. Inf. Technol. **14**(4), 80–100 (2014). doi:10.1515/cait-2014-0007. Sofia. ISSN 1311-9702

3. Nikolova, I., Dicheva, D., Agre, G., Angelov, Z., Angelova, G., Dichev, D., Madzharov, D.: Emerging applications of educational data mining in Bulgaria: the case of UCHA.SE. In: Margenov, S., Angelova, G., Agre, G. (eds.) Innovative Approaches and Solutions in Advanced Intelligent Systems. Studies in Computational Intelligence, vol. 648, pp. 113–131. Springer, Heidelberg (2015)
4. Romero, C., Ventura, S.: Educational data mining: a review of the state-of-the-art. IEEE Trans. Syst. Man Cybern. Part C: Appl. Rev. **40**(6), 601–618 (2012)
5. Siemens, G., Baker, R.: Learning analytics and educational data mining: towards communication and collaboration. In: Shum, S.B., Gasevic, D., Ferguson, R. (eds.) Proceedings of the 2nd International Conference on Learning Analytics and Knowledge (LAK 2012), pp. 252–254. ACM, New York (2012). doi:http://dx.doi.org/10.1145/2330601.2330661

Expressing Sentiments in Game Reviews

Ana Secui[1], Maria-Dorinela Sirbu[1], Mihai Dascalu[1(✉)],
Scott Crossley[2], Stefan Ruseti[1], and Stefan Trausan-Matu[1]

[1] Computer Science Department, University Politehnica of Bucharest,
Bucharest, Romania
{ana.secui,maria.sirbu}@cti.pub.ro, {mihai.dascalu,
stefan.ruseti,stefan.trausan}@cs.pub.ro
[2] Applied Linguistics and ESL, Georgia State University, Atlanta, USA
scrossley@gsu.edu

Abstract. Opinion mining and sentiment analysis are important research areas of Natural Language Processing (NLP) tools and have become viable alternatives for automatically extracting the affective information found in texts. Our aim is to build an NLP model to analyze gamers' sentiments and opinions expressed in a corpus of 9750 game reviews. A Principal Component Analysis using sentiment analysis features explained 51.2 % of the variance of the reviews and provides an integrated view of the major sentiment and topic related dimensions expressed in game reviews. A Discriminant Function Analysis based on the emerging components classified game reviews into positive, neutral and negative ratings with a 55 % accuracy.

Keywords: Natural Language Processing · Sentiment analysis · Opinion mining · Lexical analysis

1 Introduction

The domain of opinion mining and sentiment analysis refers to extracting information about feelings, ideas and emotions by analyzing textual productions using Natural Language Processing (NLP) techniques [1, 2]. There is now a lot of interest in this direction due to the tremendous volume of messages, reviews and discussion forum posts on social networks like Facebook or Twitter, and on various web portals (for example, Amazon.com or Youtube.com). Having an application that can identify and extract opinions from this huge amount of data, and provide an estimation of users' preferences about a manufactured good is of great interest to companies. The same interest is also encountered in politics in terms of candidate elections or for a government that wants to introduce new regulations, in order to have a glimpse on people opinions about the organization and their acts.

Natural Language Processing techniques may be applied for extracting sentiments and opinions in two ways: based on lexicons and through machine learning techniques. Both approaches have drawbacks. The first category is based on the polarity of sets of specific words (for example, the word "like" expresses a positive sentiment), but fails when modifiers are used, which may be at a distance in text (for example, negations:

© Springer International Publishing Switzerland 2016
C. Dichev and G. Agre (Eds.): AIMSA 2016, LNAI 9883, pp. 352–355, 2016.
DOI: 10.1007/978-3-319-44748-3_35

"I don't agree that from my previous post you can infer that I like the new phone launched on the market"). For the second category, supervised machine learning approaches (for example, Naïve Bayes, Maximum Entropy and Support Vector Machines) are used to classify texts into positive and negative opinions [3]. Their disadvantage is the need of human annotation for large volumes of training examples [4]. In terms of structure, the paper continues with details on the performed experiment. The third section presents the obtained results and the last section is centered on conclusions.

2 Details of the Experiment

This study was performed in the context of the RAGE H2020 EC project (http://rageproject.eu/), which focuses on serious games for e-learning. Our corpus consists of 9750 game reviews from 44 games, all written in English language and extracted from Amazon.com using crawl4j. The reviews were ranked on a Likert scale from 1 to 5 and were considered relevant if they contained more than 50 content words. The reviews were used to develop component scores from which to determine differences in positive, neutral, and negative game reviews.

Various vectors or word lists covering both general meaning and particular linguistic traits were combined with a Principal Component Analysis in order to determine the latent variables that define the specificities of gamer reviews following the techniques reported in Crossley et al. [5]. To develop our component scores, the following word categories were selected from the *General Inquirer* (GI, http://www.wjh.harvard.edu/~inquirer/homecat.htm) [6]: words referring to role, words indicating overstatement, words reflecting a sociological perspective, words expressing arousal, and general references to humans.

From the *Laswell* dictionary [7], we extracted words talking about skills, respect, power, wealth and gain. From *SenticNet* [8, 9], we selected words used to describe feelings based on four dimensions: attention, sensitivity, aptitude and pleasantness. *GALC* (Geneva Affect Label Coder, [10]) was used for selecting specific word categories about emotions, which were split into: boredom, anger, depression, amusement, admiration, positive, and negative. Word features related to arousal, dominance and affective variables were selected from ANEW (Affective Norms for English Words, [11]).

In addition, word lists that incorporated affective, perceptual, and cognitive processes, as well as personal concerns and relativity were extracted from the *Linguistic Inquiry and Word Count* (LIWC, [12]). The Hu-Liu polarity lists containing 2.000 positive and 4.500 negative words [13], and the Stanford Core NLP [14] sentiment analysis model based on recursive deep networks were also integrated. Only lemmas of content words were considered and multiple indices were computed in order to express the linguistic coverage of each word list for a given review.

3 Results

Eight affective components were identified using a Principal Component Analysis (PCA) which explained 51.22 % of the variance in the selected game reviews. The derived components were related to:

- *Negative Emotions:* the most powerful component, contains words with negative loadings including user frustration or game mechanics that are not working well;
- *Relations and Power:* includes words about interpersonal relationships (including relations between game characters or with other human players in multiplayer sessions), descriptions of actions, gameplay and achievements;
- *Positive Emotions:* reflects positive loadings and emotions, i.e., General Positive Words, GI Positive or GI Virtue;
- *Activities and Skills:* refers to actions within the game, as well as activities and their characteristics, i.e., GI Expressivity and LIWC Leisure activity;
- *Motivation:* reflects the overall impression induced by the game with regards to trust, surprise, attention;
- *Human and Roles:* depicts human functions (e.g., leader or authority) from reviews debating about characters with specific roles (e.g., commanders) or multiplayer modes (e.g., "Call of Duty");
- *Communication:* includes words that present ways and types of communication from GI Human and GI Role lists;
- *Ambiguous and Passive Language:* contains words with no active meaning (e.g., "admire", "passive", or "fell").

These components were used in a multi-variate analysis of variance (MANOVA) and six components yielded significant differences between the three classes of game reviews: *positive emotions* ($F = 1004.72$, $p < .01$, $\eta^2 = .171$), *negative emotions* ($F = 272.10$, $p < .01$, $\eta^2 = .053$), *relations and power* ($F = 39.22$, $p < .01$, $\eta^2 = .008$), *activities and skills* ($F = 32.13$, $p < .01$, $\eta^2 = .007$), *human and roles* ($F = 9.01$, $p < .01$, $\eta^2 = .002$), *ambiguous and passive language* ($F = 3.67$, $p - .025$, $\eta^2 = .001$). A stepwise discriminant function analysis using these six variables retained the first 5 variables and correctly allocated 5,371 of the 9,750 game reviews in the total set, $\chi2$ ($df = 4$, $n = 9,750$) $= 51.750$, $p < .001$, for an accuracy of 55.1 % (the chance level for this analysis is 33.3 %).

4 Conclusions

Opinion mining and sentiment analysis are of great interest nowadays in many domains of economy, commerce and society. Natural Language Processing techniques can be used to provide useful insights; however, there are limitations. Up to date, only a few studies exist that focus on gaming, despite its huge popularity among people of all ages.

The research described in this paper presents a linguistic analysis centered on extracting language traits used by gamers when expressing opinions about game quality. The PCA analysis explained more than 50 % of the variance in language across all game reviews, while the DFA classification highlighted promising insights into game quality.

Acknowledgement. The work presented in this paper was partially funded by the EC H2020 project RAGE (Realising and Applied Gaming Eco-System) http://www.rageproject.eu/ Grant agreement No 644187.

References

1. Liu, B.: Sentiment Analysis and Opinion Mining. Morgan & Claypool Publishers, San Rafael (2012)
2. Jurafsky, D., Martin, J.H.: An Introduction to Natural Language Processing. Computational Linguistics, and Speech Recognition. Pearson Prentice Hall, London (2009)
3. Pak, A., Paroubek, P.: Twitter as a corpus for sentiment analysis and opinion mining. In: LREC 2010, Valletta, Malta (2010)
4. Cheng, O.K.M., Lau, R.Y.K.: Probabilistic language modelling for context-sensitive opinion mining. Sci. J. Inf. Eng. **5**(5), 150–154 (2015)
5. Crossley, S., Kyle, K., McNamara, D.S.: Sentiment Analysis and Social Cognition Engine (SEANCE): An Automatic Tool for Sentiment, Social Cognition, and Social Order Analysis. Behavior Research Methods (in press)
6. Stone, P., Dunphy, D.C., Smith, M.S., Ogilvie, D.M.: The General Inquirer: A Computer Approach to Content Analysis. The MIT Press, Cambridge (1966)
7. Lasswell, H.D., Namenwirth, J.Z.: The Lasswell Value Dictionary. Yale University Press, New Haven (1969)
8. Cambria, E., Grassi, M., Poria, S., Hussain, A.: Sentic computing for social media analysis, representation, and retrieval. In: Ramzan, N., Zwol, R., Lee, J.S., Clüver, K., Hua, X.S. (eds.) Social Media Retrieval, pp. 191–215. Springer, New York (2013)
9. Cambria, E., Schuller, B., Xia, Y.Q., Havasi, C.: New avenues in opinion mining and sentiment analysis. IEEE Intell. Syst. **28**(2), 15–21 (2013)
10. Scherer, K.R.: What are emotions? And how can they be measured? Soc. sci. Inf. **44**(4), 695–729 (2005)
11. Bradley, M.M., Lang, P.J.: Affective Norms for English Words (ANEW): Stimuli, Instruction Manual and Affective Ratings. The Center for Research in Psychophysiology, University of Florida, Gainesville (1999)
12. Pennebaker, J.W., Booth, R.J., Francis, M.E.: Linguistic inquiry and word count: LIWC [Computer software] (2007)
13. Hu, M., Liu, B.: Mining and summarizing customer reviews. In: ACM SIGKDD International Conference on Knowledge Discovery and Data Mining (KDD-2004). ACM, Seattle (2004)
14. Socher, R., Perelygin, A., Wu, J.Y., Chuang, J., Manning, C.D., Ng, A.Y., Potts, C.P.: Recursive deep models for semantic compositionality over a sentiment treebank. In: Conference on Empirical Methods in Natural Language Processing (EMNLP 2013). ACL, Seattle (2013)

The Select and Test (ST) Algorithm and Drill-Locate-Drill (DLD) Algorithm for Medical Diagnostic Reasoning

D.A. Irosh P. Fernando[1,2,3] and Frans A. Henskens[1,2,4(✉)]

[1] School of Electrical Engineering and Computer Science, University of Newcastle, Callaghan, NSW, Australia
[2] Distributed Computing Research Group, University of Newcastle, Callaghan, NSW, Australia
irosh.fernando@uon.edu.au, frans.henskens@newcastle.edu.au
[3] School of Medicine and Public Health, University of Newcastle, Callaghan, NSW, Australia
[4] Health Behaviour Research Group, University of Newcastle, Callaghan, NSW, Australia

Abstract. Two algorithms for medical diagnostic reasoning along with their knowledgebase design and a method known as orthogonal vector projection for determining differential diagnoses are described. Whilst further research is necessary to achieve an effective full automation of medical diagnostic reasoning, these two algorithms provide the necessary initial theoretical foundations.

Keywords: Seléct and Test (ST) algorithm · Drill-Locate-Drill (DLD) algorithm · Orthogonal vector projection method · Medical expert systems

1 Introduction

Medical diagnostic reasoning can be conceptualised as a process consisting of two stages. The first stage involves a search for clinical information driven by diagnostic hypothesis, whereas the second stage involves testing these diagnostic hypotheses using the clinical information gathered after the completion of clinical information search [1]. Achieving reliable clinical reasoning including an exhaustive search for all relevant clinical information can be a challenging task even for expert clinicians given humans' limited cognitive capacity compared to the complexity of testing diagnostic hypotheses and vastness of the search space. Clinical information searches not only need to ensure that all the likely diagnoses are considered, but also that even the less likely diagnoses are considered when those diagnoses are associated with critical or life-threatening outcomes. Also, various factors including fatigue are known to have an adverse effect on cognitive capacity, and as a result, even expert clinicians are not immune from committing various diagnostic errors [2]. Whilst it is therefore important to achieve automation, previous attempts to automate medical diagnostic reasoning have been unsuccessful [3], and lack of an adequate theoretical foundation with an efficient way to represent knowledge can be considered as one of the main reasons for such failures. For example, rule-based and Bayesian based approaches become less effective as the size of the knowledgebase become larger because of the unmanageable number of diagnostic rules and joint probability distributions that are required; also the rule-based approaches are not able to handle missing values [4].

© Springer International Publishing Switzerland 2016
C. Dichev and G. Agre (Eds.): AIMSA 2016, LNAI 9883, pp. 356–359, 2016.
DOI: 10.1007/978-3-319-44748-3_36

2   Diagnostic Inferences and the Algorithms

The knowledgebase required for medical diagnostic reasoning can simply be conceptualised as a graph consisting of the nodes of clinical features $x_1, x_2, \ldots, x_m$ and the nodes of diagnoses $d_1, d_2, \ldots, d_n$, in which there are diagnostic relations between any pair (x_i, d_j) as described in the algorithms, where $i = 1, \ldots, m$ and $j = 1, \ldots, n$. The process of mapping what a patient describes as clinical features using the patient's own terminology into defined elements of clinical features in the set $\{x_1, x_2, \ldots, x_m\}$ is a challenging process known as abstraction.

The first stage of diagnostic reasoning involves a search of the knowledge graph driven by diagnostic hypothesis using abduction and deduction as the main inferences. The abduction can be described as the process of generating diagnostic hypothesis (i.e. differential diagnoses) based on clinical features (e.g. clinical symptoms, signs and investigation results). It can be modelled using posterior probability $P(d_j|x_i)$, which is the probability of having diagnosis d_j given symptom x_i, and a threshold value $t_d \in [0, 1]$. That is, given any x_i all d_j with $P(d_j|x_i) \geq t_d$ can be considered as differential diagnoses. Also, the model needs to take into account the fact that even though $P(d_j|x_i) < t_d$ it may still need to consider some d_j in differential diagnosis because of potential serious implications of missing diagnosis d_j. For example, when a child presents with a fever, which can mostly be due to a self-limiting viral infection, more serious causes such as bacterial meningitis need to be excluded. This is modelled using a criticality function: $\{d_1, d_2, \ldots, d_n\} \to R$ which assigns a real value to each d_j, and another threshold value $t_c \in R$; all diagnoses with a criticality above a chosen threshold value $c(d_j) \geq t_c$ must also be considered as differential diagnoses.

The deduction can simply be considered as the opposite of abduction, and attempts to derive the expected clinical features of a given a diagnosis d_j. Even though this can simply be modelled as the posterior probability, $P(x_i|d_j)$, it may be inadequate in some situations. This is because even though the probability for having x_i may be low given d_j, the presence of x_i in a patient may be highly confirmative of diagnosis d_j compared to some commonly found clinical features. This is modelled using $w_{ij} = P(x_i|d_j).W(x_i|d_j)$ where $W(x_i|d_j) \in R$ assigns a weight for x_i according to its diagnostic importance in relation to d_j

It is important to note that eliciting clinical features also requires quantification (e.g. how high body temperature is), which can be achieved via a real function $q:\{x_1, x_2, \ldots, x_m\} \to R$ with $q(x_i)$ representing the severity of clinical feature x_i. Whilst each diagnosis can also be quantified in a similar way in relation to its severity, there exists a functional relationship between the severity of each diagnosis and $q(x_i)$ [5]. For example, as $q(x_i)$ increases the related diagnosis (i.e. illness) may become more severe. At the end of the search for clinical information, the elicited and quantified clinical features can be presented as a vector $X = < q(x_1), q(x_2), \ldots, q(x_m) >$ with $q(x_i) = 0$ if x_i was not found in the patient or is unknown. Then, based on the orthogonal vector projection method [6], deriving the likelihood (i.e. how likely, not in statistical terms) of each diagnosis d_j can be modelled as a comparison of X with

$D_j = <q(x_1), q(x_2), \ldots, q(x_m)>$ which consists of the highest quantities of each clinical feature expected in the most severe form of diagnosis d_j. Whilst the reader may refer elsewhere for details of the orthogonal vector projection method [6], deriving the likelihood of diagnosis d_j involves projecting X on to D_j, and then calculating $\left|X_{proj}\right| / \left|D_j\right|$, which is the ratio of length of the projected X to length of D_j. This gives a clinically intuitive measure, which outperformed cosine similarity and Euclidean distance methods [6] since the angle between the two vectors is a measure of diagnostic similarity whilst the length of X is a measure of overall severity of symptoms.

The ST algorithm models standard diagnostic consult with an expert clinician, which often starts with the patient expressing the reasons for consult (i.e. health complaints or presenting clinical features) [7]. It uses iteratively: (1) abstraction to establish the presence of clinical features; (2) abduction to derive likely diagnoses; (3) deduction to derive the symptoms expected in each likely diagnoses. In contrast, the DLD algorithm [8] starts with eliciting screening symptoms followed by the abduction and deduction without iterations. This results in improved efficiency from $O(n^3)$ of the ST algorithm to $O(n^2)$ nonetheless at the cost of a compromised search that can potentially miss diagnoses (unless they are on a search path starting from screening symptoms). After completion of the search for clinical features both algorithms use the orthogonal vector projection method to derive the likelihood of each diagnosis.

Both the ST and DLD algorithms were implemented in java and evaluated in clinical psychiatry using patient data. The knowledgebase consisted of 44 psychiatric diagnoses $\times$ 70 clinical features and screening symptoms. In order to reduce the size of the knowledgebase, based on the conceptualisation that clinical features can be represented has a hierarchical structure [9], the 70 clinical features used in the knowledgebase included clusters of related clinical features. Details of both implementations including the knowledgebase and evaluations have been described separately elsewhere [7, 8]. Both algorithms produced comparable results in relation to diagnostic sensitivity and specificity.

3   Conclusion

This paper has described two algorithms, which provides a theoretical foundation for automating medical diagnostic reasoning based on logical inferences including abduction, deduction, and induction. Both algorithms conceptualise knowledge representation as a bipartite graph consisting of clinical features and diagnoses. Such simple representation enables use of an orthogonal vector projection method in diagnostic reasoning, which is more efficient method compared to rule-based and probabilistic approaches. Whilst the two algorithms can be used individually or in combination depending on the type the diagnostic consult required (e.g. the DLD algorithm can be used for initial assessments or triaging patients), their further evaluation in other specialties of clinical medicine is required. We acknowledge that complete automation of the abstraction step in diagnostic reasoning is a challenging task requiring a complex human computer interface consisting of natural language and multimodal sensory processing, and future research should focus on achieving this.

References

1. Ramoni, M., Stefanelli, M., Magnani, L., Barosi, G.: An epistemological framework for medical knowledge-based systems. IEEE Trans. Syst. Man Cybern. **22**, 1361–1375 (1992)
2. Nendaz, M., Perrier, A.: Diagnostic errors and flaws in clinical reasoning: mechanisms and prevention in practice. Swiss Med Wkly **142**, w13706 (2012)
3. Wolfram, D.A.: An appraisal of INTERNIST-I. Artif. Intell. Med. **7**, 93–116 (1995)
4. Onisko, A., Lucas, P., Druzdzel, M.J.: Comparison of rule-based and bayesian network approaches in medical diagnostic systems. In: Quaglini, S., Barahona, P., Andreassen, S. (eds.) AIME 2001. LNCS (LNAI), vol. 2101, pp. 283–292. Springer, Heidelberg (2001)
5. Fernando, I., Henskens, F., Cohen, M.: An approximate reasoning model for medical diagnosis. In: Lee, R. (ed.) SNPD 2013. SCI, vol. 492, pp. 11–24. Springer, Heidelberg (2013)
6. Fernando, D.A.I., Henskens, F.A.: A modified case-based reasoning approach for triaging psychiatric patients using a similarity measure derived from orthogonal vector projection. In: Chalup, S.K., Blair, A.D., Randall, M. (eds.) ACALCI 2015. LNCS, vol. 8955, pp. 360–372. Springer, Heidelberg (2015)
7. Fernando, I., Henskens, F.: Select and test algorithm for inference in medical diag-nostic reasoning: implementation and evaluation in clinical psychiatry. In: 15th IEEE/ACIS International Conference on Computer and Information Science Okayama, Japan 2016
8. Fernando, D.A.I.P., Henskens, F.A.: The Drill-Locate-Drill (DLD) algorithm for automated medical diagnostic reasoning: implementation and evaluation in psychiatry. In: Lee, R. (ed.) Computer and Information Science. Studies in Computational Intelligence, vol. 656, pp. 1–14. Springer, Switzerland (2016)
9. Fernando, I., Cohen, M., Henskens, F.: A systematic approach to clinical reasoning in psychiatry, vol. 21, pp. 224–230. Australasian Psychiatry, 1 June 2013

How to Detect and Analyze Atherosclerotic Plaques in B-MODE Ultrasound Images: A Pilot Study of Reproducibility of Computer Analysis

Jiri Blahuta$^{(\boxtimes)}$, Tomas Soukup, and Petr Cermak

The Institute of Computer Science, Silesian University in Opava,
Bezruc Sq. 13, 74601 Opava, Czech Republic
jiri.blahuta@fpf.slu.cz
http://www.slu.cz/fpf/en/institutes/the-institute-of-computer-science

Abstract. This pilot study is focused on recognition and digital analysis of atherosclerotic plaques in ultrasound B-images. The plaques are displayed as differently echogenic regions depending on plaque composition. The first goal is to find significant features to plaque analysis in B-images. We developed software to finding hyperechogenicity of substantia nigra to Parkinson's Disease evaluation in B-images. We try to discover how to use this software also for atherosclerotic plaques analysis. The software has a function of intelligent brightness detection. We use a set of 23 images, each of them was analyzed five times. The primary goal is to verify the reproducibility of this software to atherosclerotic plaques analysis in medical practice.

Keywords: Ultrasound · Atherosclerotic plaques · B-MODE · B-images · Stroke ultrasound · Plaques B-images

1 Introduction

The goal of the study is to find a way how to analyze atherosclerotic plaques and their risk depending on composition, shape and size. This pilot study is focused on using B-MODE [1] to find distinguishable features of plaques in B-images using our developed software.

1.1 A Set of Images Used for This Study

Totally of 23 images of atherosclerotic plaques [2] in transversal section have been used. All images have the same initial settings.

1.2 Homogeneous or Heterogeneous Atherosclerotic Plaques

Homogeneous and heterogeneous plaques are distinguished depending on composition. The aim of this study is to investigate how to distinguish heterogeneous and homogeneous plaques in B-MODE, see Fig. 1.

© Springer International Publishing Switzerland 2016
C. Dichev and G. Agre (Eds.): AIMSA 2016, LNAI 9883, pp. 360–363, 2016.
DOI: 10.1007/978-3-319-44748-3_37

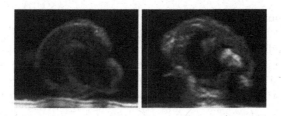

Fig. 1. Homogeneous (left) and heterogeneous (right) atherosclerotic plaque in B-image.

2   Developed Application B-MODE Assist System

We developed a software tool B-MODE Assist System to analysis of echogenicity in substantia nigra to Parkinson's Disease diagnosis [3–5,9]. The reproducibility of the algorithm has been published in the past. The core algorithm can be expressed as follows.

1. Automatic or manual selection of window from image native axis
2. Select a predefined ROI or draw free-hand ROI (for atherosclerotic plaques)
3. Binary thresholding for all thresholds $T \in \langle 0; 255 \rangle$
4. For each threshold T is computed the area in mm^2
5. Graphical representation of computed values

Let H is brightness level of a pixel and T is the threshold, then is computed

$$\text{if } H > T \text{ then } output = 1 \tag{1}$$

Figure 2 shows the predefined ROI in substantia nigra and graphical representation of computed values.

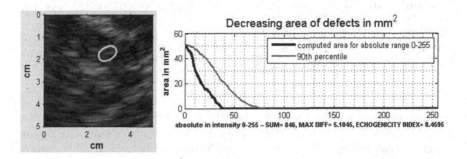

Fig. 2. Predefined ROI for ipsilateral substantia nigra and graphical representation.

In the software is implemented a subsystem [7] to check window size of 20 × 20 mm from native axis. The subsystem also checks gray levels in the window as follows:

1. check if the number of loaded images ≥ 50
2. more than 70 % of pixels of the image must meet $H \geq 10$
   (a) if the condition is met then is set as initial
   (b) otherwise is set 65 % (5 % less) and repeat the step 2
3. initial value of non-black pixels has been set on checked %
4. activation of this rule into the application until is not set a new value

It is required against incorrect window size and the window outside the atherosclerotic plaque.

3   Statistical Analysis of Examined Data

Statistical analysis [8,9] is needed to reproducibility [10] assessment. We observe changes for the same plaque and between different plaques. Each image was measured five times by 4 non-experienced observers and 1 experienced observer in sonography. Variation and coefficient of variation are used to reproducibility assessment, see Table 1.

Table 1. Variance, count of zeros and coefficient of variation of analyzed data

variance	zeros	var coef	variance	zeros	var coef	variance	zeros	var coef
p2 SHE	**p2 SHE**	**p2 SHE**	**p9 SHE**	**p9 SHE**	**p9 SHE**	**p16 LHE**	**p16 LHE**	**p16 LHE**
296649,280	77	1,63	2248191,81	30	1,44	704533,19	70	1,63
481849,860	13	1,35	2454192,91	39	1,47	986188,08	67	1,57
423055,710	7	1,29	2223271,38	39	1,47	917606,09	67	1,60
439065,710	7	1,30	2207635,87	39	1,47	879972,52	67	1,58
498293,220	7	1,34	2298301,93	40	1,48	912075,56	67	1,58
p3 SHE	**p3 SHE**	**p3 SHE**	**p10 SHE**	**p10 SHE**	**p10 SHE**	**p17 HO**	**p17 HO**	**p17 HO**
1101680,390	73	1,36	1626783,08	21	1,21	1428527,69	84	1,75
1041866,370	59	1,33	1915156,73	18	1,22	1612813,42	82	1,74
1165585,670	63	1,34	2021941,60	18	1,23	1323303,02	87	1,88
1068539,020	38	1,32	1772949,79	18	1,21	1344977,95	87	1,89
1162451,270	76	1,36	1755481,51	18	1,21	1398099,77	87	1,81
p4 SHE	**p4 SHE**	**p4 SHE**	**p11 SHE**	**p11 SHE**	**p11 SHE**	**p18 HO**	**p18 HO**	**p18 HO**
1281303,870	9	1,29	1296769,81	17	1,09	1476636,06	32	1,55
1425791,120	9	1,31	1502142,60	17	1,13	1591963,06	32	1,51
1336806,030	9	1,30	1579343,14	17	1,13	1734303,82	32	1,52
1568830,650	9	1,31	1480590,39	17	1,12	1741655,31	32	1,51
1316964,610	9	1,30	1558315,94	17	1,12	1356243,51	32	1,51

An experienced sonographer classified the images into the groups according to visual assessment. Homogeneous (HO), lightly heteregenenous (LHE) and strongly heterogeneous (SHE) plaques, see Table 1 (only part of all computed values).

4   Conclusions and Future Work

The paper is focused on reproducibility of atherosclerotic plaques analysis in cross-section view on B-images. For this study were analyzed 23 images.

To reproducibility was observed values for the same plaque and changes between homogeneous and heterogeneous plaques. Each plaque was analyzed five times by 5 independent observers (4 non-experienced and 1 experienced sonographer). The principal fact is that measurements of the same plaque show small differences to reproducibility appraisal. No reliable features are known to distinguish heterogeneous and homogeneous plaques in B-images. The study shows the method is generally reproducible but is needed to find features to distinguish heterogeneous and homogeneous atherosclerotic plaques in B-images in the future work.

Acknowledgments. This work was supported by The Ministry of Education, Youth and Sports from the National Programme of Sustainability (NPU II) project IT4Innovations excellence in science - LQ1602.

References

1. Edelman. S.K.: Understanding Ultrasound Physics, 4th edn. E.S.P. Ultrasound (2012)
2. Griffin, M., Kyriakou, E., Nikolaidou, A.: Normalization of ultrasonic images of atherosclerotic plaques and reproducibility of gray-scale media using dedicated software. Int. Angiol. **26**(4), 372–378 (2007)
3. Blahuta, J., Soukup, T., Cermak, P., Vecerek, M., Jakel, M., Novak, D.: ROC and reproducibility analysis of designed algorithm for potential diagnosis of Parkinson's disease in ultrasound images. In: Mathematical Models and Methods in Modern Science, 14th WSEAS International Conference on Mathematical Methods, Computational Techniques and Intelligent Systems (MAMECTIS 2012) (2011)
4. Blahuta, J., Soukup, T., Cermak, P., Rozsypal, J., Vecerek, M.: Ultrasound medical image recognition with artificial intelligence for Parkinson's disease classification. In: Proceedings of the 35th International Convention, MIPRO 2012 (2012)
5. Blahuta, J., Soukup, T., Cermak, P., Novak, D., Vecerek, M.: Semi-automatic ultrasound medical image recognition for diseases classification in neurology. In: Kountchev, R., Iantovics, B. (eds.) MedDecSup 2012. Studies in Computational Intelligence, vol. 473, pp. 125–133. Springer, Switzerland (2013)
6. Blahuta, J., Soukup, T., Jelinkova, M., Bartova, P., Cermak, P., Herzig, R., Skoloudik, D.: A new program for highly reproducible automatic evaluation of the substantia nigra from transcranial sonographic images. Biomed. Pap. **158**(4), 621–627 (2014)
7. Blahuta, J., Cermak, P., Soukup, T., Vecerek, M.: A reproducible application to B-MODE transcranial ultrasound based on echogenicity evaluation analysis in defined area of interest. In: Soft Computing and Pattern Recognition, 6th International Conference on Soft Computing and Pattern Recognition (2014)
8. Skoloudik, D., Jelinkova, M., Bartova, P., Soukup, T., Blahuta, J., Cermak, P., Langova, K., Herzig, R.: Transcranial sonography of the substantia nigra: digital image analysis. Am. J. Neuroradiol. **35**(9), 2273–2278 (2014)
9. Skoloudik, D., Fadrna, T., Bartova, P., Langova, K., Ressner, P., Zapletalova, O., Hlustik, P., Herzig, R., Kanovsky, P.: Reproducibility of sonographic measurement of the substantia nigra. Ultrasound Med. Biol. **9**, 1347–1352 (2007)
10. Riffenburgh, R.H.: Statistics in Medicine, 3rd edn. Academic Press, Cambridge (2012)

Multifactor Modelling with Regularization

Ventsislav Nikolov[✉]

EuroRisk Systems Ltd., Varna, Bulgaria
v.g.nikolov@gmail.com

Keywords: Multifactor · Polynomial formula · Basis functions · Genetic algorithm · Least squares regression · Regularization

1 Introduction

Suppose we are given a finite number of discrete time series x_i called factors. They can represent arbitrary physical, social, financial or other indicators. All factors are with equal length and their values correspond to measurements performed in equal time intervals. One of the series is chosen to be a target factor and some of the others are chosen to be explanatory factors. The aim is to create a formula by which a series can be generated, using the explanatory factors for the given historical period, that should be as close as possible to the given target series, using a chosen criterion [4]. For simplicity such a criterion can be the Euclidean distance between the target and generated factor for all data points. Such a created formula can be used for different purposes in the financial instruments modelling, sensitivity analysis, etc. In the case of predictable explanatory factors and unpredictable target factor analysis can be performed about the influence of the explanatory factors changes to the target factor.

The formula can be created in different forms but simplifying the solution the following polynomial form is used:

$$y = \beta_1 f_1(x_1) + \beta_2 f_2(x_2) + \ldots + \beta_m f_m(x_m) + \beta_{m+1} \tag{1}$$

where f_1, f_2, ...f_m are arbitrary basis functions, and β_1, β_2, ... β_m are regression coefficients, β_{m+1} is a free term without explanatory factor.

2 Formula Generation

First of all the target factor is selected according to the specific purposes. After that the explanatory factors are selected amongst the all available series. In our solution a few alternative approaches can be used as selection of the most correlated factors to the target factor or minimal correlated each other or so on. When both the target and explanatory factors are selected the automatic modelling stage is performed by repeating the stages of applying basis functions to explanatory factors and after that calculation of the regression coefficients.

C. Dichev and G. Agre (Eds.): AIMSA 2016, LNAI 9883, pp. 364–367, 2016.
DOI: 10.1007/978-3-319-44748-3_38

Taking into account that for all selected factors all basis functions can be applied, there are k^m combinations, where k is the number of the basis functions and m is the number of the explanatory factors. Usually in the practice the factors are a few hundred and the functions are a few dozen. Thus the brute force searching of the best basis functions combination is practically impossible. That is why for that purpose we chose to apply heuristic approach by usage of a genetic algorithm. It is realized as a software library written in Java.

2.1   Finding the Best Combination of the Basis Functions

Initial Population. The genetic algorithm is used to determine the combination of the basis functions to the explanatory factors. And a function can be used for more than one factor. Thus an individual in terms of the genetic algorithms is a sequence of integer values representing the indices of the basis functions and the goodness of fit is the distance between the generated and the given target factor [2]. In the realized system a random integer sequence generator was created to generate the initial population of the sequences. Applying the functions to the explanatory factors and calculating the regression coefficients produces a set of target factors which are compared to the given target in order to select the best individuals.

Selection. Given a set of the generated individuals the best of them should be selected according to their goodness of fit. We have implemented two alternative approaches: roulette wheel and truncation selection [3]. The first one is preferred as default because it allows every individual to continue even with less chance.

Recombination and Mutation. The recombination is performed by splitting the selected L individuals in a given point and randomly combining their parts. In our implementation the splitting point is randomly generated at every step within the interval from 25 % to 75 % of the individuals length rounded to the nearest integer.

Coefficients Determination. The calculation of the regression coefficients is done for every combination of basis functions. In our case the ordinary least squares error is used according to which the coefficients are obtained in matrix form calculating the following matrix equation [1]:

$$B = (A^T A)^{-1} A^T Y \qquad (2)$$

where B is the matrix of the regression coefficients, A is the matrix of factors with applied functions and Y is the target factor.

Having B calculated the generated target factor is:

$$\hat{Y} = A \times B \qquad (3)$$

and the distance between the generated and given target is:

$$d = \|Y - \hat{Y}\| \qquad (4)$$

Coefficients Reduction. The formula terms with small coefficients can be removed because they do not significantly influence the formula results. Removing or not the small coefficients is an optional setting in our system and if it is chosen the second regression coefficients calculation must be performed at every step after the reduction.

Calibration. Using the generated formula for future calculations and modelling must be periodically reconsidered and the formula must be calibrated because its accuracy decreases. This can be done either by using the same explanatory factors or by other factors.

Regularization. The multifactor formula provides good results in the cases when there are explanatory factors similar to the target factor. Otherwise often the future calculations are not very accurate because of the overfitting. In order to avoid overfitting a regularization parameter is used in the following form:

$$B = (A^{T}A + \lambda I)^{-1}A^{T}Y \qquad (5)$$

where I is the identity matrix and λ is the regularization parameter.

This is L2-regularization or ridge regularization [5]. In the formula searching stage the set of the data points is separated in training and validation subsets. The formula functions and coefficients are determined using the training set but the error is

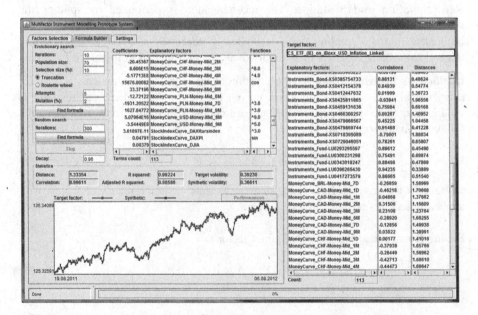

Fig. 1. The multifactor modelling prototype

calculated using the validation set. In order to separate these two sets the factors values are shuffled together and the last, for example, 20 % or 30 % of the length are used as validation set. When the training and validation sets are determined and the basis functions are fixed to explanatory factors an appropriate value of λ should be found. Our investigation shows that there is a single global minimum of the validation error which allows searching it with adaptive step starting from a random point.

3 Conclusions and Future Work

The built software prototype system can be seen on Fig. 1. The experimental results show that the best results are obtained when the number of the explanatory factors is near to, but not exceeding, the number of the historical dates.

The system also confirms that the greater the regularization parameter is the greater the penalization is which produces better results in the future calculations with generated formula in cases when the target factor is different to some extent than anyone of the explanatory factors. But this is not a general rule and taking into account that often in practice there are indicators with similar behavior sometimes the regularization parameter should not be used.

References

1. Hamilton, J.: Time Series Analysis. Princeton University Press, Princeton (1994)
2. Koza, J.: Genetic Programming. MIT Press, Cambridge (1992)
3. Mitchell, M.: An Introduction to Genetic Algorithms. MIT Press, Cambridge (1999)
4. Rosen, K.: Discrete Mathematics and Its Applications, 4th edn. AT&T (1998)
5. Rosenberg, A.: Machine Learning Lectures, CUNY Graduate Center (2009). (http://eniac.cs. qc.cuny.edu/andrew/gcml/lecture5.pdf)

Erratum to: A Novel Method for Extracting Feature Opinion Pairs for Turkish

Hazal Türkmen, Ekin Ekinci, and Sevinç İlhan Omurca$^{(\boxtimes)}$

Computer Engineering Department, Kocaeli University, İzmit, Kocaeli, Turkey
hazalturkmen91@gmail.com,
{ekin.ekinci, silhan}@kocaeli.edu.tr

Erratum to:
Chapter "A Novel Method for Extracting Feature Opinion
Pairs for Turkish" in: C. Dichev and G. Agre (Eds.):
Artificial Intelligence: Methodology, Systems,
and Applications, LNAI 9883,
https://doi.org/10.1007/978-3-319-44748-3_16

In an earlier version of this paper, the acknowledgement was missing. This has now been corrected.

The updated online version of this chapter can be found at
https://doi.org/10.1007/978-3-319-44748-3_16

C. Dichev and G. Agre (Eds.): AIMSA 2016, LNAI 9883, p. E1, 2016.
https://doi.org/10.1007/978-3-319-44748-3_39

Author Index

Printed in the United States
By Bookmasters